Professional

Microsoft® Smartphone Programming

Professional Microsoft® Smartphone Programming

Baijian Yang
Pei Zheng
Lionel M. Ni

Wiley Publishing, Inc.

Professional Microsoft® Smartphone Programming

Published by
Wiley Publishing, Inc.
10475 Crosspoint Boulevard
Indianapolis, IN 46256
www.wiley.com

Published simultaneously in Canada

ISBN: 978-0-471-76293-5

Manufactured in the United States of America

10 9 8 7 6 5 4 3 2 1

1B/QX/RS/QW/IN

Library of Congress Cataloging-in-Publication Data:

Yang, Baijian, 1972–
Microsoft smartphone programming / Baijian Yang, Pei Zheng, Lionel Ni.
p. cm.
Includes index.
ISBN-13: 978-0-471-76293-5 (paper/website)
ISBN-10: 0-471-76293-8 (paper/website)
1. Cellular telephones—Computer programs. 2. Pocket computers—Computer programs. 3. Microsoft .NET Framework. I. Zheng, Pei, 1972– II. Ni, Lionel M. III. Title.
TK6570.M6Y37 2006
621.3845—dc22

2006033469

For general information on our other products and services please contact our Customer Care Department within the United States at (800) 762-2974, outside the United States at (317) 572-3993 or fax (317) 572-4002.

Wiley also publishes its books in a variety of electronic formats. Some content that appears in print may not be available in electronic books.

About the Authors

Baijian Yang is an assistant professor in the Computer Technology program at Ball State University. He became a Microsoft Certified Systems Engineer (MCSE) in 1998 and was one of the core software designers/developers for etang.com. He received his Ph.D. in Computer Science from Michigan State University in 2002. He is now engaged in research and development in the area of wireless networks and distributed systems.

Pei Zheng received his Ph.D. in Computer Science from Michigan State University in 2003. He joined Microsoft as a software engineer in 2005. Before that he was an assistant professor of Computer Science at Arcadia University, and a member of the technical staff at Bell Laboratories, Lucent Technologies. His research interests include distributed systems, network simulation and emulation, and mobile computing.

Lionel M. Ni is Chair Professor, Head of the Computer Science and Engineering Department, and Director of the Digital Life Research Center at the Hong Kong University of Science and Technology. Dr. Ni earned his Ph.D. in electrical and computer engineering from Purdue University in 1980. He has been involved in many projects related to wireless technologies, 2.5G/3G cellular phones, and embedded systems. He is co-author of the book *Interconnection Networks: An Engineering Approach* (Morgan Kaufmann, 2002), and *Smart Phone and Next Generation Mobile Computing* (Morgan Kaufmann, 2006).

Credits

Senior Acquisitions Editor
Jim Minatel

Development Editor
John Sleeva

Production Editor
William A. Barton

Copy Editor
Luann Rouff

Editorial Manager
Mary Beth Wakefield

Production Manager
Tim Tate

Vice President and Executive Group Publisher
Richard Swadley

Vice President and Executive Publisher
Joseph B. Wikert

Graphics and Production Specialists
Carrie A. Foster
Brooke Graczyk
Denny Hager
Barbara Moore
Rashell Smith
Alicia B. South

Quality Control Technicians
Laura Albert
Brian H. Walls

Project Coordinator
Erin Smith

Proofreading
Techbooks

Indexing
Infodex Indexing Services, Inc.

To my wife, Chen Wen

—Baijian Yang

To my wife, Ning Liu, for her understanding, professional support, and encouragement

—Pei Zheng

To my dear children, Elaine and Wayland

—Lionel M. Ni

Acknowledgments

Writing this book required a great deal of effort that went beyond our initial expectations. We would first like to thank our family members, without whose support and encouragement we simply could not finish this book.

We also would like to thank the folks at Wiley. Senior Acquisitions Editor Jim Minatel helped us shape our proposal and provided us with professional help at every stage of the project. Thank you very much, Jim, for believing in us and pushing to keep the project on the right track. Development Editor John Sleeva provided us with tons of valuable comments and suggestions, including insightful technical feedback as well as general writing and formatting guidance. Thank you very much, John, for your extensive professional support and your dedication to keeping the project on a tight schedule. Thanks, also, to Production Editor Bill Barton and Copy Editor Luann Rouff for thoroughly checking and fixing the details of the book, including grammar, format, and layout. Our appreciation also extends to other members of the Wiley team for your hardworking and consistent contributions to the book.

Our special thanks go to Ya-Qin Zhang, the Vice President of Microsoft Corporation. It was his vision and passion in mobile computing that initiated our book project. Besides his inspirations to our work, Ya-Qin has also offered many helps to the book and directed our technical questions to the fellows at Microsoft. We would also like to thank Yong Rui at Microsoft Research for reviewing our work and providing constructive professional opinions.

Contents

Foreword

Computing has been continuously advancing for half a century. In the early stages, mainframes and mini-computers drove the revolution, where they realized the computing transition from analog to digital. Then, in the 1980s and 90s, the personal computer (PC) became the dominant force, where its open framework enabled the widespread integration of desktop computing into people's work and play. The third computing wave came in the mid 1990s and continues to evolve today. A key feature of this wave is the integration of computing, communication, and storage technologies. Cellular phones are at the center of this wave. It is fair to say that a cellular phone is the most ubiquitous personal gadget ever devised: For the first time, a single device, less than the size of a wallet, captures the whole spectrum of one's daily activities.

This book is about smartphones. To be precise, *smartphone* is an overloaded word. From a customer's point of view, a smartphone is a "smart" phone—an electronic handheld device that integrates the functionality of a mobile phone, personal digital assistant (PDA), or other information appliance. A key feature of a smartphone is that additional native applications can be installed on it. For the content of this book, Smartphone is the software platform running on the physical smartphones. Here the notion of "software platform" refers to an integrated computing environment that consists of an operating system, the .NET runtime environment, a set of applications, and related application development tools. Microsoft began its foray into the mobile software platform about a decade ago, and has recently picked up the pace significantly. The Microsoft .NET Compact Framework and Microsoft Smartphone platform have demonstrated strong potential in commanding a significant share of the mobile OS market. In 2005, Windows Mobile held the number one worldwide volume share of the PDA market, had 40 device makers and 68 mobile operators in 48 countries, 640,000 developers worldwide, and more than 18,000 applications. Microsoft's strategy with Windows Mobile is to make it a powerful and open platform; emphasize the integration between devices, PCs, servers, and the Web; and build a rich ecosystem that inspires innovation.

The book is unique in several ways. First, although a few books address the Microsoft .NET Compact Framework or Windows Mobile, they discuss these two topics in isolation. This is the first book dedicated to Smartphone software development with sufficient programming details. Second, this book targets a wide audience. On the one hand, it covers the basics of Windows Mobile and the .NET Compact Framework, so it is a good textbook for students in school. On the other hand, it has an entire part on advanced topics, which is valuable to both veteran practitioners and experienced developers. Last but not least, it offers a good mix of both the authors' experience and expertise. Professor Yang and Dr. Zheng are two of the most passionate young researchers in the field, with a lot of hands-on experience. Professor Ni is a veteran in wireless technologies, 2.5G/3G cellular phones, and embedded systems. This combination of energy, hands-on experience, and long-term vision ensure that the book is of highest quality.

I continue to be pleasantly surprised by how powerful smartphones become. I call a smartphone a *C3 device,* because it seamlessly combines the functions of communication (voice call, e-mail, and IM), computing (entertainment and location-based services), and control (a universal remote control and micropayment). As wireless technologies—e.g., 3G cellular systems, wireless LANs, Bluetooth, WiMAX, and Ultra-Wideband—continue to mature, I am confident that in the near future we will enjoy the power of mobile computing anywhere, anytime, and on any devices.

Ya-Qin Zhang, Ph.D.
Corporate Vice President
Microsoft Corporation
Beijing

Introduction

The smartphone segment of the worldwide mobile wireless industry is growing rapidly, largely due to the strong demand for converged mobile devices from enterprises and consumers. ABI Research predicts that smartphone sales worldwide will reach 150 million by 2008. Many enterprises are considering deploying mobile applications, and many consumers want a converged device for both communication and computing. Both of these markets have created enormous opportunities for mobile application design and development, and the migration of desktop applications to mobile devices, with enabling and powerful programming tools on a variety of mobile software platforms.

Microsoft began its quest for mobile markets over a decade ago but was not able to draw much of the attention until recently, partly due to its powerful developing tools for its mobile operating system. The Microsoft .NET Compact Framework and Microsoft Smartphone platform have demonstrated a strong potential to realize a significant share of the mobile operating system market. In 2005, sales of Windows Mobile–powered devices grew by 40 percent. As a result, both the end-user community and the developer community of Microsoft Smartphone have grown significantly. The upcoming Windows Mobile 6.0, which is estimated to be shipped in 2007, will surely further boost Microsoft Smartphone software development.

Although you can find some books that address the Microsoft .NET Compact Framework or Windows Mobile, they tend to focus on Pocket PC devices or a general discussion of .NET Compact Framework programming. Moreover, none of them is dedicated to Microsoft Smartphone software development with sufficient programming details. The MSDN website and some online resources do provide in-depth articles about mobile programming, but the documents are not systematically organized, making it difficult for developers to efficiently use them. It was our intention to provide the first comprehensive book dedicated to Smartphone programming.

The major goals of this book are as follows:

- To help you understand the software design guidelines for Smartphone devices
- To demonstrate how to develop, debug, and deploy Smartphone applications with Microsoft Visual Studio 2005 in C#
- To discuss security and performance issues in Smartphone programming
- To provide you with essential programming skills that you can apply when the next version of Smartphone, the .NET Compact Framework, and Visual Studio are released

Who This Book Is For

As a detailed reference to Microsoft Smartphone programming with the Microsoft .NET Compact Framework, the core audience for this book includes software architects and developers working in the area of mobile application development with intermediate programming skills in C/C++ or C#, and professionals seeking a thorough overview of the Microsoft Smartphone software development platform.

Students (assuming some C/C++ or C# programming experience) who would like to gain some experience with Microsoft Windows Mobile programming will also find this book valuable. Experienced developers familiar with the .NET framework and C# can skim or skip the first three introductory chapters and jump to Chapter 4.

What This Book Covers

This book is a comprehensive guide to Microsoft Smartphone programming with the Microsoft .NET Compact Framework. It provides in-depth coverage of key architectural concepts, application design guidelines, and programming techniques for Microsoft Smartphone software developers, and includes extensive hands-on examples and code listings. Visual Studio 2005 and the .NET Compact Framework 2.0 are used as the underlying programming environment (although a number of chapters touch on issues in the .NET Compact Framework 1.0, Smartphone 2002, and Smartphone 2003.

How This Book Is Structured

The topics covered in the book can be divided into three categories:

Part I, "Smartphone and .NET," presents an overview of the Microsoft Smartphone platform from a software developer's perspective. It also covers the .NET Compact Framework, the Smartphone programming environment, and a quick get-started guide to Microsoft Smartphone programming.

Part II, "Smartphone Application Development," discusses Microsoft Smartphone–related application design and programming in the domains of the .NET Compact Framework (managed code). Each topic starts out with a brief overview of key concepts and tasks covered in the chapter, followed by a detailed discussion of the programming framework and classes available in Windows Mobile and the Smartphone SDK.

Part III, "Advanced Topics," covers application development topics such as security, globalization and localization, graphics, and performance considerations.

What You Need to Use This Book

Because this book focuses heavily on Smartphone programming with Visual Studio 2005, it is expected that readers have this tool installed on a desktop computer running Windows XP or Windows Vista. In addition, it would be better if a Windows Smartphone device were available for developing and testing. However, if that is not available, readers can simply use the Smartphone Emulator that comes with Visual Studio 2005.

Conventions

To help you get the most from the text and keep track of what's happening, we've used a number of conventions throughout the book.

Boxes like this one hold important, not-to-be forgotten information that is directly relevant to the surrounding text.

Tips, hints, tricks, and asides to the current discussion are offset and placed in italics like this.

As for styles in the text:

- We *highlight* new terms and important words when we introduce them.
- We show keyboard strokes like this: Ctrl+A.
- We show filenames, URLs, and code-related terms within the text like so: `persistence.properties`
- We present code in two different ways:

```
In code examples we highlight new and important code with a gray background.
```

```
The gray highlighting is not used for code that's less important in the present
context, or has been shown before.
```

Source Code

As you work through the examples in this book, you may choose either to type in all the code manually or to use the source code files that accompany the book. All of the source code used in this book is available for download at `www.wrox.com`. Once at the site, simply locate the book's title (either by using the Search box or by using one of the title lists) and click the Download Code link on the book's details page to obtain all the source code for the book.

Because many books have similar titles, you may find it easiest to search by ISBN; this book's ISBN is 978-0-471-76293-5.

Once you download the code, just decompress it with your favorite compression tool. Alternately, you can go to the main Wrox code download page at `www.wrox.com/dynamic/books/download.aspx` to see the code available for this book and all other Wrox books.

Errata

We make every effort to ensure that there are no errors in the text or in the code. However, no one is perfect, and mistakes do occur. If you find an error in one of our books, such as a spelling mistake or faulty piece of code, we would be very grateful for your feedback. By sending in errata you may save another reader hours of frustration, and at the same time you will be helping us provide even higher quality information.

To find the errata page for this book, go to `www.wrox.com` and locate the title using the Search box or one of the title lists. Then, on the book's details page, click the Book Errata link. On this page you can view all errata that has been submitted for this book and posted by Wrox editors. A complete book list, including links to each book's errata, is also available at `www.wrox.com/misc-pages/booklist.shtml`.

If you don't spot "your" error on the Book Errata page, go to `www.wrox.com/contact/techsupport.shtml` and complete the form there to send us the error you have found. We'll check the information and, if appropriate, post a message to the book's errata page and fix the problem in subsequent editions of the book.

p2p.wrox.com

For author and peer discussion, join the P2P forums at `p2p.wrox.com`. The forums are a Web-based system for you to post messages relating to Wrox books and related technologies and interact with other readers and technology users. The forums offer a subscription feature to e-mail you topics of interest of your choosing when new posts are made to the forums. Wrox authors, editors, other industry experts, and your fellow readers are present on these forums.

At `http://p2p.wrox.com` you will find a number of different forums that will help you not only as you read this book, but also as you develop your own applications. To join the forums, just follow these steps:

1. Go to `p2p.wrox.com` and click the Register link.
2. Read the terms of use and click Agree.
3. Fill in the required information to join as well as any optional information you wish to provide and click Submit.
4. You will receive an e-mail with information describing how to verify your account and complete the joining process.

You can read messages in the forums without joining P2P but in order to post your own messages, you must join.

Once you join, you can post new messages and respond to messages other users post. You can read messages at any time on the Web. If you would like to have new messages from a particular forum e-mailed to you, click the Subscribe to this Forum icon by the forum name in the forum listing.

For more information about how to use the Wrox P2P, be sure to read the P2P FAQs for answers to questions about how the forum software works as well as many common questions specific to P2P and Wrox books. To read the FAQs, click the FAQ link on any P2P page.

Part I
Smartphone and .NET

1

Introduction to Microsoft Smartphone

Mobile computing is everywhere today. The number of cell phones, PDAs, and other handheld mobile devices has exceeded the number of computers in the world. The Yankee Group estimates that there are approximately 1.8 billion mobile devices in use worldwide, used for a variety of tasks, including the still predominant voice communication, text messaging, web surfing, e-mail, gaming, and so on. As wireless technologies such as 3G cellular systems, wireless LANs, Bluetooth, WiMAX, and Ultra-Wideband continue to mature, empowering those mobile devices and the network infrastructure, we soon will be able to enjoy the power of mobile computing anywhere, anytime, and on any devices.

What will ultimately cause this vision to materialize are new mobile applications and services built on the pervasive computing infrastructure, along with a set of software platforms and programming tools. Motivated by this vision, we decided to write a book on one of the most powerful software platforms for mobile computing: the Microsoft Smartphone. This chapter begins with a brief introduction to smartphones and the challenges of developing smartphone applications. Then we introduce Microsoft Windows Mobile and some related technologies. Readers will likely find this section informative, as it clarifies some terms and concepts pertinent to Microsoft Smartphone.

What Is a Smartphone?

The worldwide mobile wireless industry is quickly moving from traditional, voice-based cellular phone services to combined voice and data services, as a result of increasing demand for mobile data access and the deployment of high-speed wireless data services utilizing a variety of wireless technologies. For example, 2.5G/3G wireless services are being rolled out and used by a rapidly growing number of subscribers, and the number of WiFi hotspots and residential wireless LANs continues to grow substantially. The trend is clear: Cell phones, PDAs, and portable consumer electronic devices will likely merge into a single, handheld device as a universal personal communicator and computing platform (generally called a *smartphone*). Indeed, the market has seen a dramatic increase in smartphone sales when compared to the fairly slow growth of PDA sales worldwide.

Generally, a smartphone is a powerful, multi-function cell phone that incorporates a number of PDA functionality, such as a personal scheduler, calendar, and address book, as well as the ability to access Internet services and applications using either a keypad or a stylus. In addition to making a call from a smartphone, users can surf the web, check e-mail, create documents, play online games, update schedules, or access an enterprise network via a virtual private network (VPN). Wireless Internet access is enabled by means of cellular wireless networks — such as GSM/GPRS, CDMA, CDMA2000, or WCDMA, among others.

Bill Gates, Chairman and then Chief Software Architect of Microsoft, introduced his vision for the smartphone at the 2004 Mobile Developer Conference:

> *The pocket devices, phone and PDA, really the trend is to have the best of both together. The phone is no longer just a voice-only device; more and more it has that rich, color screen. A PDA is no longer a disconnected device; more and more it's got the ability to make calls and connect up to wireless data networks. In many cases that will be both the wide area data networks, 2 ½ G or 3G networks, but also increasingly you'll have WiFi connectivity built into the device as well. So it will be able to connect up to whichever network is available, whichever one provides the best bandwidth and economics there.*

An increasing number of high-end cell phones and smartphones are equipped with powerful mobile processors (such as ARM processors), 64–128MB memory, 256–512MB flash storage, and even 2–4GB hard drives. Examples include the Motorola Q, SPV C600, and O2 XDA II. Some smartphones are PDA-based with handwritten recognition or a tiny keyboard, and phone functions as add-on features, such as Palm Xplore and Palm Treo. In fact, cell phone manufacturers and PDA manufacturers have different views regarding the future of smartphone devices. Unsurprisingly, each camp believes their device will prevail, with add-on functionality of devices from the other side. As wireless technologies and the mobile market continue to evolve, it is still too early to tell which approach will finally win. Nevertheless, one thing is certain: They both need reliable, high-performance, low-power consumption operating systems and software to leverage the wireless services.

Microsoft Smartphone refers to Microsoft's platform for next-generation cell phones — basically a software architecture with Windows CE as the operating system, plus a rich set of applications such as Pocket Internet Explorer and Pocket Outlook and powerful software development tools such as .NET Compact Framework and Visual Studio 2005. (We use *Smartphone* to refer to Microsoft Smartphone throughout this book and *smartphone* to refer to general multifunction cell phones.)

This book focuses on software development issues and practices on smartphones running Microsoft Windows Mobile software. There are, of course, other software development solutions. For example, Palm Inc., also provides a software development kit (SDK) for Palm OS smartphones, and you can find an SDK and supporting tools for Symbian OS, another popular cell phone operating system.

Smartphone Applications and Services

With the vision of mobile convergence supporting communication and computing on a single set of hardware components, mobile wireless network operators, cell phone manufactures, and independent software vendors are working together to create new applications and services with the hope of taking a lead position in the next wave of mobile computing. These services and applications essentially leverage the increasingly high computing capability supplied by the cell phone and the flexible, high-speed wireless connectivity to offer an efficient, reliable, and rich experience to the end user. This section summarizes the potential services and applications in this domain.

Mobile Commerce

This category includes mobile banking, location-based business information service and shopping assistance, mobile advertising, and mobile payment, among other services. Japan and Korea already offer widespread mobile payment applications that enable consumers to make purchases at a convenience store by waving the cell phone past a reading device. Numerous startup companies in the United States are developing applications that enable credit card payments to be verified, parking fees to be paid at the meter, and social networking. Industries involved in this category include banks, credit card companies, retail stores, stock trading agencies, and online businesses.

Mobile Enterprise

Services and applications in this category are concerned with mobile worker assistance such as real-time job scheduling, route planning, package delivery updates, mobile collaboration and communication, and mobile business transaction. Moreover, enterprise resource planning (ERP) applications and supply chain management (SCM) systems can be extended to support mobile access and onsite processing. In addition to mobile enterprise, law enforcement, educational, and healthcare organizations may also utilize these services and solutions to improve productivity and reduce costs.

Other Mobile Software Platforms

Symbian OS is developed by Symbian, a company supported by several cell phone manufacturers, including Nokia, Ericsson, Sony Ericsson, and Samsung. Originally based on the EPOC operating system, Symbian OS defines several UI reference models for different types of devices. Symbian OS uses EPOC C++, a pure object-oriented language, as the supporting programming language for both system services implementations and APIs. It also allows Java applications for mobile devices (Java 2 Micro Edition, J2ME, applications) to run on top of a small Java runtime environment. The Symbian Developer website (`www.symbian.com/developer`) provides numerous technical documents for Symbian OS, SDKs, and sample code, as well as information on Symbian OS development and the Symbian developer community.

Palm OS, developed by Palm Inc., is a preemptive, multitasking operating system for Palm PDAs and cell phones. Palm OS supports both the ARM and Motorola 68000 architectures. Developers can choose a programming language from C, C++, Visual Basic, or Java, although C is most widely used for Palm OS software development. Interested readers can visit the Palm OS developer site (`www.palmsource.com/developers`) for more technical details. Palm OS application development is facilitated by the Palm OS 68K and Protein SDKs and some commercial developer suites. A developer suite is an integrated software tool that enables developers to create both ARM-native and Palm OS Protein-powered applications for Palm OS Cobalt and 68K applications.

Mobile Data Service and Entertainment

This category includes real-time, location-based navigation assistance coupled with traffic data access, mobile gaming, rich media services, and so on. Mapping and GPS-based navigation services are increasingly being integrated into general-purpose smartphone platforms. Mobile television services have been available in the United States and some Asian countries; online music download services (such as Apple's iTunes service) are available on some high-end smartphones; mass media companies, music and movie companies, online gaming service providers, and of course the consumer, will be involved in this category of services and applications.

Needless to say, the aforementioned summary is by no means exhaustive; however, it is indicative of the broad range of new services and applications with tremendous potential for businesses. Indeed, the enormous opportunity of next-generation mobile computing has created myriad services and applications that will likely continue to mature and succeed in the foreseeable future. What all these services and applications share is a reliance on software running on smartphones to reach the end user. To this end, software designers and developers have to be aware of the challenges and obstacles involved in smartphone-based application development.

Challenges of Smartphone Application Development

The application design paradigm for smartphones not only differs largely from that of desktop applications, it also has some inherent requirements that separate it from application development on common mobile devices. First, the hardware constraints of a smartphone, such as processor speed, persistent storage capacity, battery, and wireless connection, significantly affect the application design principles. Second, the input method of a smartphone, either a telephone keypad with additional navigation keys or a soft keyboard, forces application developers to pay more attention to the GUI of a smartphone application than that of desktop applications. Third, the cost of wireless data service remains a major factor for mass adoption of Smartphone technology, as it is still quite high compared with the cost of landline Internet services. Moreover, the application must be easily ported to various hardware platforms using different processor architectures and peripherals. Following are the salient factors to consider while developing applications for smartphones:

- **Efficient storage and adaptive networking.** Despite the advancements in mobile embedded and wireless technologies, smartphones usually have neither a large memory capacity nor a reliable, high bandwidth wireless connection. It is crucial, therefore, to make better use of precious local storage for high performance. The networking functionality of a smartphone must be able to adapt to the comparatively low bandwidth and high drop rate of wireless links. Finally, remote data access must be able to function in disconnected mode. For a more detailed discussion of data access on a smartphone, see Chapters 5 and 6.
- **Simple and user-friendly GUI.** Because entering characters on a smartphone is not as easy as typing on a PC keyboard (although some teenagers will disagree), a program's GUI has to deal with most of the input by using graphical components. Due to a smartphone's relatively small screen, the layout of these components on a window, as well as the grouping of components into different windows, has to been designed carefully. For example, you generally should eliminate horizontal scrolling for a smartphone application. GUI design is discussed in more detail in Chapter 4.

- **Lightweight computation and power management.** A smartphone is not designed for CPU-intensive applications; instead, most CPU-intensive computation should be done on the server side, which is typically much more powerful than a mobile processor, of an application whenever possible, enabling the client side on a smartphone to run faster. This creates a scenario for peer-to-peer mobile applications, whereby mobile nodes essentially serve each other without relying on a central server. Web services is a good example of offloading computation to the server. Chapter 9, "XML and Web Services," provides more details about accessing web services on a smartphone. Power management is also crucial to increasing battery life. Smartphone operating systems must be able to put the device into power-save state whenever possible (subject to a user's configuration) to save battery power.
- **High security.** In a sense, a smartphone can be viewed as another form of personal identification, with sensitive data stored and transmitted over the air. There have been a number of cell phone–related security hacks utilizing various vulnerabilities of cell applications. As the smartphone evolves to function not only as a telephone but also as a credit card or ID card, the compromise of a cell phone will result in serious problems. Low-overhead authentication and authorization must be enforced for critical smartphone applications. Chapter 12, "Device and Application Security," and Chapter 13, "Data and Communication Security," cover smartphone security issues.

Introducing Microsoft Windows Mobile

To facilitate the design and development of smartphone applications, hardware manufactures and software vendors have teamed up to provide powerful programming tools for smartphone developers. One of the solutions is Microsoft's Windows Mobile, a unified platform specifically designed to enable developers to leverage Windows desktop programming experiences for mobile application design and development targeting both smartphones and PDAs. Because of the innovative .NET Compact Framework and bundled programming tools, Microsoft Windows Mobile for Smartphone has gained a sizable amount of popularity among software developers worldwide.

Windows Mobile is Microsoft's software platform for Pocket PCs (PDAs running Microsoft's software platforms) and Smartphones (smartphones running Microsoft's software platform). Here the notion of "software platform" refers to an integrated computing environment that consists of an operating system, the .NET runtime environment, a set of applications, and related application development tools. The concepts of "Pocket PC" and "Smartphone" are also commonly referred to as software platforms targeting PDAs and smartphones, respectively.

Windows Mobile 5.0

The latest version of Microsoft Windows Mobile is Windows Mobile 5.0. Windows Mobile consists of a tailored Windows CE operating system; the Microsoft .NET Compact Framework (the runtime and the class libraries); a set of tools and APIs for native code development; a device software emulator; and for developers, an IDE (Integrated Development Environment) component for Visual Studio 2005.

Windows Mobile 5.0 is based on the Windows CE .NET 5.0 operating system. As part of the Smartphone offering, Microsoft also provides the Smartphone 2005 SDK for ISVs (independent software vendors). Developers can use the Smartphone SDK in conjunction with Visual Studio 2005, the flagship programming environment tool from Microsoft, for .NET Compact Framework–based managed code application development. The Smartphone SDK also allows C or C++ unmanaged application programming utilizing Win32 APIs using Microsoft embedded Visual C++ (eVC). The next version of Windows Mobile is called Windows Mobile 6.0 and code named Photon.

Windows Mobile 5.0 unifies application development for Smartphones and Pocket PCs by providing the same Win32 APIs and a set of development tools. Unlike the previous version of Windows Mobile, Windows Mobile 2003, which does not allow developers to use SQL Server CE in Smartphone applications, now both Pocket PC and Smartphone platforms can take advantage of the same set of common data services provided by SQL Server 2005 Mobile Edition, the next release of SQL Server CE. In addition, both platforms share the same security model and the same common application installer. The unification of these two platforms reflects Microsoft's mobile platform strategy in response to the imminent convergence of these two types of mobile devices.

In addition, Windows Mobile 5.0 provides a new set of standardized WinCE APIs, including multimedia APIs supporting Windows Media 10 Mobile, Direct 3D Mobile APIs, new GPS APIs, DirectDraw APIs, and camera APIs. These APIs essentially extend the functionality of a Windows Mobile device and offer an improved user experience. On the managed application environment side, Windows Mobile 5.0 comes with .NET Compact Framework 2.0 and provides a set of .NET managed APIs for messaging and telephony, which are not available in .NET 1.0 and 1.1. Chapter 2 discusses the .NET Compact Framework in detail.

Application development on Windows Mobile 5.0 can further take advantage of the latest release of Microsoft's flagship IDE, Visual Studio 2005. The software emulator in Visual Studio 2005 has been rewritten to eliminate the gap between emulation and physical device deployment. In addition, both managed and unmanaged code can be developed within the same Visual Studio 2005 environment. No eVC is needed anymore. Additionally, Visual Studio 2005 includes a number of improvements in the form designer and deployment tools, as well as common programming and debugging support. This book uses Visual Studio 2005 as the application development platform, so you will see screenshots taken from it and code developed using a set of smart device tools integrated into Visual Studio 2005.

Microsoft Smartphone from a User's Perspective

The market for cell phone operating systems is highly segmented. Many cell phones run operating systems other than Microsoft Smartphone; indeed, Microsoft is relatively new to the cell phone platform market. The stronghold in this market is Symbian OS, which is funded and supported by cell phone makers such as Nokia, Sony Ericsson, Siemens, and others. Palm is historically strong in the PDA market. For people with a background in those types of platforms, it is necessary to present a quick tour into Microsoft Smartphone. Those of you who have used a Smartphone device for some time can feel free to skip this section.

To a typical user, a Smartphone is a much more powerful cell phone. It provides many more types of applications, and the user interface is more sophisticated than a traditional cell phone. Like most other cell phones, a Microsoft Smartphone actually refers to a combination of the handset and its running applications. The aesthetic design of a smartphone handset may vary significantly, but the principle user interface — how a user interacts with the device — is almost the same across different devices from different manufacturers. Let's take a look at the Smartphone emulator, a software tool that helps developers quickly develop and test Smartphone applications without using a physical device. Details about the Smartphone emulator can be found in Chapter 3.

Figure 1-1 depicts a common layout of a Smartphone device. Although color doesn't appear in these screenshots, a color screen at the top and the number keys at the bottom are the most common elements. What separates a Smartphone from a PDA are the software keys. Notice the two soft keys (left and right) directly below the screen. These correspond to the menu bar and commands at the bottom of the screen. Depending on how applications define the functions of the menu bar, these soft keys may perform specific tasks. Also note the five-way (up, down, left, right, and "Select") navigation pad and the four fixed function keys: Call, Home, Hang Up, and Back. Pressing the Call key will bring you to the phone call screen, where you can enter a phone number. After you enter a phone number, pressing the Call key again will initiate the call. The Hang Up key, of course, is used when you want to end a call. The Home key always takes you to the home screen shown in this figure, and the Back key enables you to go back to the previous screen. At the home screen, pressing any number key will automatically bring you to the phone call screen.

Figure 1-1

A Smartphone usually has a power button, a record button (for voice recording), and two volume control buttons. Some Smartphones have a built-in camera, so there will be another button for it (some have a high-quality digital camera, such as Nokia N90, which features a Carl Zeiss lens, 2-megapixel resolution, and a 20x zoom).

The first-generation Smartphone applications are mostly clones of desktop Windows applications, enabling users familiar with those desktop applications to avoid learning a new one. Typical Smartphone applications are Pocket Outlook, calendar, contacts, Pocket Word, Pocket Excel, Windows Media Player, MSN messenger, games, and some accessories. The second-generation Smartphone applications are exclusively designed to leverage the advantages of mobility and ubiquitous wireless access. In the next several years we will see a whole new set of applications that utilize real-time location information in conjunction with always-on wireless data access.

Overall, the design of a Smartphone aims to take advantage of a user's prior experience with Windows desktop systems. After all, a Smartphone is a small computer running a stripped-down version of the Windows operating system. This greatly helps users familiar with Palm or Symbian cell phones because the learning curve is largely eliminated.

Summary

This chapter introduced the concept of the smartphone and the trend of convergence in the mobile computing and communication domain. As a converged mobile device, smartphone, in conjunction with supporting new mobile wireless services and applications, will gain widespread popularity worldwide in the foreseeable future. Microsoft Windows Mobile is a software platform that enables fast, efficient, and feature-rich application development for Smartphone devices. Many smartphone and cell phone manufacturers (including Palm, which traditionally uses Palm OS on their PDAs and smartphones) have started to use Windows Mobile as the underlying software platform on their products, largely because users can leverage their knowledge of Windows desktop systems in using a Smartphone.

Beginning with the next chapter, our discussion moves to the technical foundations of Windows Mobile. We will focus on the core components of the .NET Compact Framework, as well as the class libraries and type systems of .NET. In addition, a primer on the C# programming language will be provided.

2

.NET Compact Framework Fundamentals

The last chapter described Microsoft's commitment to smart device application development with the general Windows Mobile platform. As cell phones and smartphones continue to proliferate in people's daily lives, it is quite natural to think of a smartphone as a computing platform that supports mobility and wireless access on-the-go. Windows Mobile-based Smartphone application development can be divided into two categories: *managed code*, which runs on top of the .NET Compact Framework, and *unmanaged code* (also known as *native code*), which executes directly on top of the underlying operating system.

Microsoft's decision to incorporate Smartphone into the general .NET infrastructure can be seen as a strategy to promote Windows Mobile technology. With the .NET Compact Framework and supporting programming tools, developers can choose any .NET programming languages to write secure and high-quality code targeting mobile devices. This chapter discusses the .NET Compact Framework, and one of the .NET languages, C#. The following topics are discussed in this chapter:

- An overview of the .NET Framework and the .NET Compact Framework
- The components of the .NET Compact Framework
- A quick review of C#

Introducing the Microsoft .NET Framework

Before introducing the .NET Compact Framework, we'll first look at its "full" version, the Microsoft .NET Framework, which is key to Microsoft's .NET technology. In a more general sense, .NET represents Microsoft's software architecture, which provides the capability to quickly build, deploy, manage, and use feature-rich, security-enhanced software solutions. Both the .Net Framework and the .Net Compact Framework are implementations of the CLI (common language infrastructure). As an ECMA (European Computer Manufacturer's Association) specification (#335), the CLI defines platform-independent, unified runtime support for programs written in different languages. The CLI is composed of a file format, a common type system (CTS), an extensible metadata system, a common intermediate language specification, and a base class library. ECMA #334 defines the C# language specification submitted by Microsoft.

Microsoft provides another shared source implementation of CLI, code-named Rotor, that runs on Windows XP, FreeBSD, and Mac OS X. You can download Rotor from `www.microsoft.com/downloads/details.aspx?FamilyId=3A1C93FA-7462-47D0-8E56-8DD34C6292F0&displaylang=en`.

In addition to the .NET Framework, the .NET Compact Framework, and Rotor, are non-Microsoft .NET Framework implementations, such as the Mono project (`www.mono-project.com`) and the DotGNU project (`www.dotgnu.org`). These projects represent efforts at enabling .NET applications to run on Unix and Linux.

At of the heart of the .NET Framework and the .NET Compact Framework is the CLR (common language runtimes), which enables .NET applications to execute in a type-safe and secure environment. Note that the .NET Compact Framework is not a tailored version of the .NET Framework to Windows Mobile. It is built independently, but like the .NET Framework it exports similar interfaces in the form of class libraries. In most cases, the frameworks are shipped within the operating systems, or they can be deployed onto target systems.

For developers, the .NET Framework refers to a software platform for managed code development and execution. The latest version at the time of this writing is 2.0. Major components of the .NET Framework include the following:

- Common language runtime (CLR), in which managed code is JIT (just-in-time) compiled and executed
- .NET Framework class libraries, including a set of language-independent classes
- .NET tools and application development tools

The Common Language Runtime

The CLR is a software virtual machine similar to a Java Virtual Machine (JVM). It is a layer between the operating system and the managed code, providing services such as the following:

- Process and thread scheduling
- Memory management (including garbage collection)
- Type safety
- Code security
- Cross-language integration
- Cross-language exception handling
- Enhanced security

- Versioning and deployment support
- Component interaction
- Debugging and profiling

Note that the notion of *runtime* implies anything related to a programming execution environment, which may consist of class libraries, a class loader, an interpreter, a JIT compiler, and so on. The runtime environment provided by .NET is the CLR; therefore, the CLR is the core of the .NET Framework. Any .NET language code will be executed within the CLR, rather than directly by the underlying operating system.

The CLR provides a set of compilers that compile managed source code into intermediate code called *MSIL (Microsoft Intermediate Language)*. MSIL defines a set of platform-independent instructions. The compilers also generate metadata that describes types in the MSIL code, security-related data, and versioning data. MSIL code and the metadata are linked into an assembly, which could be a dynamic link library (DLL), an executable (EXE), or a binary module. An assembly will be loaded by a class loader at runtime, checked for type safety, JIT compiled into native code, and then sent to the processor for execution.

Application Domain

The CLR executes a .NET application within an application domain. An application domain hosting an application is under the control of the CLR. Application creation and termination in an application domain impose much less overhead than processes. The CLR ensures type-safe checking for applications in an application domain, and forbids any inter-application domain direct access (which can be done only via remoting or networking services). The CLR can run multiple application domains within a single process. The CLR itself is loaded and initialized by a special process called the *CLR host*. When the CLR is initialized within a process, a default application domain is created.

Common Type System

The term *type* refers to a class within the context of the .NET CLR. A type contains data fields, methods, properties, and events (see the section ".NET Compact Framework" for details). A fundamental building block of the .NET programming paradigm is the common type system (CTS), which defines a set of standard types that are common to any .NET language. Because different .NET language compilers must produce intermediate code that conforms to the CTS, managed code can interoperate across languages. Moreover, multiple source files in .NET-compliant languages can be compiled into the same assembly.

.NET Languages

Each .NET-compliant language requires a compiler to produce a program's MSIL code. Microsoft has provided these compilers for C#, Visual Basic .NET, J#, JScript, and managed C++. In addition, some other programming languages have been ported to .NET, including Cobol, Fortran, Caml, LISP, and Perl. You can find a complete list of .NET-compliant languages at `www.gotdotnet.com/team/lang`.

.NET Framework Class Libraries

.NET Framework class libraries provide more than 2,000 classes, interfaces, and value types organized into numerous namespaces. These classes enable developers to perform almost anything related to desktop standalone applications and network applications. These namespaces include the following:

- **Base classes** in the `System` namespace that implement strings, arrays, collections, math functions, time, values types, type conversion, events, event handlers, and so on
- **Language compilation and code-generation control classes** in the `Microsoft.*` namespaces, such as `Microsoft.CSharp`, `Microsoft.VisualBasic`
- **Data access classes** in the `System.Data` namespace that implement database access across multiple data providers
- **Networking classes** in the `System.Net` namespace that provide programming interfaces for network protocols, including a managed Winsock implementation
- **Remoting classes** in the `System.Runtime.Remoting` namespace that enable objects to interoperate with each other across application domains on the same machine or on remote machines
- **Web services and web application classes** in the `System.XML` namespace that provide XML web service support for distributed applications in a heterogeneous computing environment and HTTP application control
- **Security classes** in the `System.Security` namespace, including cryptographic classes and support for the .NET security system, such as policies, permissions, and so on
- **UI classes** in the `System.Windows.Forms` and `System.Web.UI` namespaces for Windows desktop UI applications and web applications
- **I/O classes** in the `System.IO` namespace that implement asynchronous and synchronous I/O functionality
- **Threading classes** in the `System.Threading` namespace that provide multithreading programming interface and thread synchronization classes, such as `Mutex`, `Monitor`, `Interlocked`, and so on

A similar namespace family for the .NET Compact Framework is presented later in the chapter.

Visual Studio 2005

Visual Studio 2005 is the major multi-language IDE (Integrated Development Environment) for Windows applications. Both managed and unmanaged code of Windows desktop applications can be developed using Visual Studio 2005. It allows managed and unmanaged Pocket PC and Smartphone application programming targeting Windows Mobile devices. With the .NET Framework on their target systems or devices, developers can use a single set of .NET Framework classes in different languages, with the same naming and calling convention, and similar syntax. In particular, mobile application developers can leverage their experience and programming techniques obtained from smart client application development with Visual Studio 2005, and the only difference is the class libraries: They will work with a subset of the desktop .NET Framework class libraries. Visual Studio 2005 (and the previous version, Visual Studio .NET) is the key element for rapid and high-performance .NET application development, testing, and deployment over various systems and devices.

Prior to Visual Studio 2005, developers had to use Visual eMbedded C++ for unmanaged C++ development for Windows Mobile devices.

You will see a step-by-step example of creating Smartphone applications in Visual Studio 2005 in the next chapter.

Introducing the .NET Compact Framework

The .NET Compact Framework represents the materialization of .NET for software platforms on mobile devices. Many developers believe the .NET Compact Framework to be a subset of the "full" version of the .NET Framework. This is only partially correct. Although the .NET Compact Framework maintains a high-level of consistency with the desktop framework, it has been heavily optimized for resource-constraint mobile devices, which usually have limited computing power and memory, and are battery powered; and it has added new classes that are unique to mobile application development. Resource-constrained devices require the operating system and applications to perform efficiently due to such factors such as CPU power, memory size, battery power, screen size, input methods, and so on, that may not be a problem for desktop application development at all. The .NET Compact Framework has been available on a number of Windows Mobile platforms, including Smartphone 2003, Pocket PC 2002, Pocket PC 2003, and Windows Mobile 5.0.

CLR of the .NET Compact Framework

The CLR of the .NET Compact Framework is built from the ground up, following the same rationale as the CLR of the .NET Framework. C# or Visual Basic.NET code is compiled into MSIL code that conforms to the CLI specification. The MSIL code, in turn, can be JIT-compiled into native machine code for the processor of the mobile device by the CLR. Type safety and memory management are done by the CLR as well. You may already realize that memory management of the .NET Compact Framework posed significant challenges to Microsoft's .NET Compact Framework team, because highly efficient memory management is of paramount importance to system and application performance. A Windows Mobile device usually has 32MB to 64MB of memory (flash or RAM). Thus, the footprint of the .NET Compact Framework and any .NET applications running on top of it must be small in memory, while at the same time the .NET Framework must provide support for a large number of classes (but not the entire list of .NET Framework classes) tailored for Windows Mobile applications.

.NET Compact Framework Class Libraries

The .NET Compact Framework class libraries are composed of a subset of the desktop .NET Framework classes, plus some new classes designed especially for mobile device applications and services. To reduce the CLR's size, some classes or functionality of classes in the .NET Framework are dropped. Figure 2-1 shows a comparison between .NET Framework classes and .NET Compact Framework classes. Notice the shaded components available only in the .NET Framework.

.NET Compact Framework libraries create a consistent namespace hierarchy similar to that of .NET Framework classes. Developers familiar with desktop Windows .NET application development can leverage their experience with .NET Framework classes to build .NET Compact Framework–based applications. There are, however, remarkable differences between some of the widely used classes in these two frameworks. For example, some events, methods, and properties in Windows Forms, some controls such as RichTextBox and CheckListBox, drag and drop operations, and .NET Remoting are not supported in the .NET Compact Framework. You will see how they differ for a specific functionality in the following chapters.

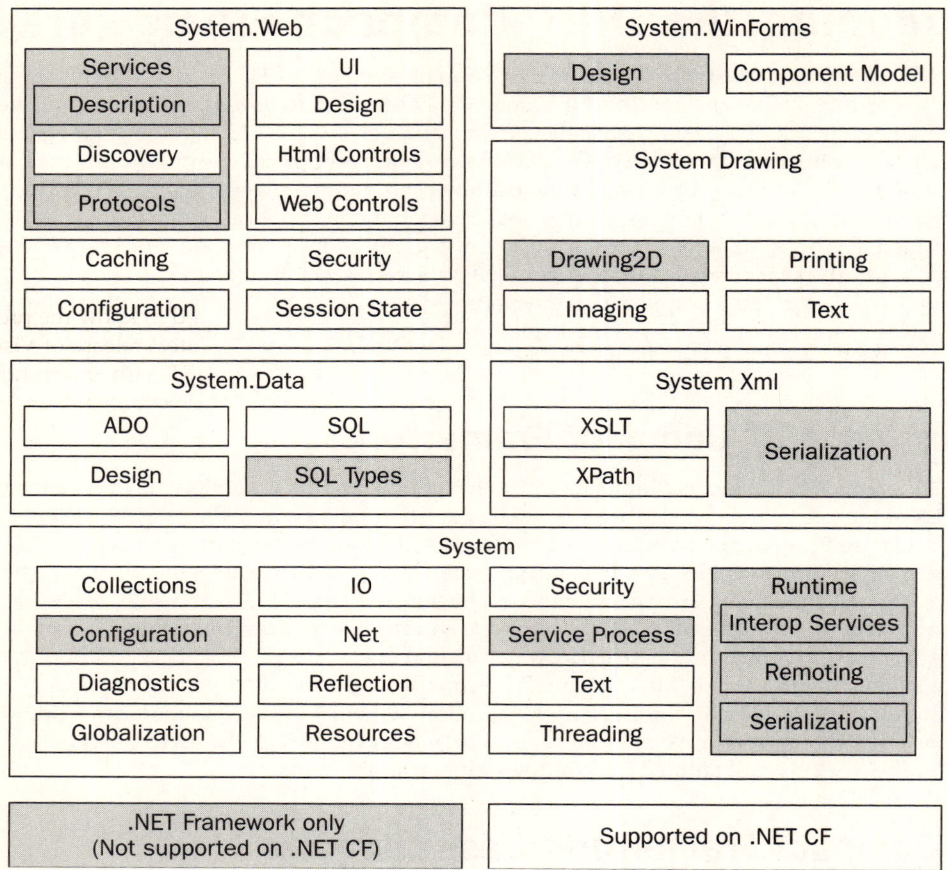

Figure 2-1

Platform Invoke

Both the .NET Framework and the .NET Compact Framework provide a way to access unmanaged code in native Win32 DLLs: Platform Invoke, or P/Invoke. In the case of the .NET Compact Framework, P/Invoke allows managed code to call methods in Windows CE native DLLs. P/Invoke performs marshalling of data types between the CLR and the underlying native runtime environment. Another major use of P/Invoke is to access COM objects from within a managed application. The .NET Compact Framework does *not* allow COM interop directly, but you can circumvent this by using P/Invoke to call a DLL wrapper of the COM object you want to access. P/Invoke is provided in the `System.Runtime.InteropServices` namespace.

There has been some confusion about the .NET Compact Framework and another Microsoft mobile application development technology called ASP.NET Mobile Controls. Keep in mind that the former technology is for developing applications running on mobile devices that have the .NET Compact Framework installed, whereas the latter technology is for developing web applications on the web server side (presumably Microsoft Internet Information Server), which has the .NET Framework full version installed, and it does not require mobile client devices to have the .NET Compact Framework. Most of the content of this book is devoted to software development for Smartphone devices with the .NET Compact Framework installed.

For a more detailed discussion of Platform Invoke, see Chapter 10, "Platform Invoke."

Smartphone Development Tools

If you are new to mobile software development, then you probably don't need to know the previous versions of software development tools provided by Microsoft. Because of the latest release of Visual Studio 2005, you can simply unify your work on a single IDE tool: Visual Studio .NET with Smartphone Device Programmability (SDP), using C# and Visual Basic .NET for managed code development, and C++ for native code development.

Prior to the release of Visual Studio 2005, two other tools were used by Smartphone application developers:

- Microsoft eMbedded Visual C++ 4.0 (for native code development)
- Microsoft eMbedded Visual Basic (for managed code development)

Although these tools were replaced by the built-in components of Visual Studio 2005, they can be used for application development for Windows Mobile 2002 and 2003 for Smartphone, and Windows Mobile 2003 for Smartphone and Pocket PC. Smartphone developers should be aware that Windows Mobile for Smartphone 2002 (also known as Smartphone 2002) does *not* support the .NET Compact Framework. Thus, you cannot use Visual Studio .NET with SDP for Smartphone 2002 software development.

Visual Studio 2005 with SDP

Smartphone application development has been integrated with Microsoft's unified IDE, Visual Studio .NET, in an effort to offer a universal software environment that supports both desktop and mobile .NET-based application development. It can be used to develop, debug, test, and deploy .NET Compact Framework–based Smartphone applications. In fact, the .NET Compact Framework, including the CLR and class libraries, can be directly accessed from within Visual Studio .NET. The Smartphone SDK provides further help for Smartphone application development. The latest Smartphone SDK, Windows Mobile 5.0 for Smartphone SDK, consists of a number of code samples, useful tools, sample security configuration files, emulator updates, and documentation. Very often a Smartphone developer needs to have both Visual Studio .NET and the latest Smartphone SDK installed on a development computer.

Unlike desktop application development, programming for a Smartphone device requires some specific procedures:

- You must choose a target platform when you create a project in Visual Studio .NET for Smartphone application development. At the time of this writing, the available Smartphone platforms supported by Visual Studio 2005 include Smartphone 2003 and Windows Mobile 5.0 for Smartphone.
- You don't need to have a physical Smartphone mobile device (a smartphone with Microsoft Smartphone 2003 or later versions) to develop applications for the target Smartphone platform.

Instead, you can use the software emulator of the target platform to test your application. An emulator is a software execution environment within a Windows process that is used to debug programs targeting the emulated platform. As a real Windows CE operating system tailored for the Smartphone platform, the Smartphone emulator can directly execute the same instruction set of the target platform's processor.

- You can connect a Smartphone device to the development desktop system by using either a USB, Ethernet, or serial port. Visual Studio .NET can detect the device via a program called *ActiveSync*, and deploy your application to the device. An interesting feature of Visual Studio .NET with Smartphone SDK is remote debugging, which enables you to debug code running on your mobile device from within the Visual Studio .NET environment.
- Online help (within the IDE) for Smartphone application development has been integrated into Visual Studio .NET's help system, so at any time you can access filtered help pages from the Help menu or press F1 for instant help in the current context.

Visual Studio .NET (with built-in Smart Device Programmability) enables you to leverage your .NET desktop Windows application development knowledge and programming skills for mobile software development. The .NET Compact Framework class libraries are mostly compatible with the full version of the .NET Framework, which makes it considerably easy to port your applications from the desktop to mobile devices.

Microsoft eMbedded Visual C++

During the non-.NET years, the typical way to develop mobile applications for Microsoft Windows CE operating systems was to use eMbedded Visual Tools, including eMbedded Visual Basic and eMbedded Visual C++. It is strongly suggested that developers who are familiar with eMbedded Visual Basic move to Visual Basic .NET on Visual Studio for .NET Compact Framework–based application development. For eMbedded Visual C++ developers, however, there are still some cases for which native C++ code is preferred to managed .NET code, such as small footprint device drivers and mobile games.

eMbedded Visual C++ is a standalone programming tool that combines the compiler, emulators of mobile devices, and IDE supporting tools. Smartphone software development with eMbedded Visual C++ is similar to the traditional Windows desktop software development with Visual C++. You are free to use the Win32 API for Windows CE, MFC (Microsoft Foundation Classes) library for Windows CE, ATL (Active Template Library) for Windows CE, and whatever libraries are specific to the Smartphone platform. eMbedded Visual C++ also provides an emulator of the Smartphone platform to assist in debugging native C++ applications.

.NET Compact Framework Type System

.NET languages such as Visual C# and Visual Basic are based on the same type system as the .NET Compact Framework and use the same class library. Thus, developers familiar with one .NET language (and therefore the .NET Compact Framework type system) can easily move to another .NET language.

Every class in the .NET Compact Framework, including those with built-in value types, is directly or indirectly inherited from the `System.Object` class. Only single inheritance is allowed in the .NET Framework and .NET Compact Framework; a class cannot have more than one base classes. Any user-defined classes are inherited from the `System.Object` class if no base class is explicitly specified.

Types

There are two kinds of objects in the .NET Compact Framework: built-in *value types* and *reference types*. A *value type* is a primitive data type that holds only values. They are actually structures allocated on the stack. *Reference types* are instances of classes and must be created using the `new` keyword. They are allocated on the heap and are subject to garbage collection. Reference types are understandably larger than value types because they encapsulate more data. On 32-bit systems, every object has a header of 8 bytes. The smallest object, an object of `System.Object`, is 12 bytes. Reference types are completely object-oriented, meaning that you can leverage encapsulation, inheritance, and polymorphism to develop your own types. If a reference type must be submitted, a value type can be "boxed" into a reference type, either explicitly or implicitly, by creating a reference type allocated on the heap and copying the value of the value type to the reference type. For example, an integer value type needs to be "boxed" in order to be placed into an `ArrayList` object.

Data Types

Table 2-1 summarizes the numeric types in the .NET Compact Framework. Note that for simplicity, an alias is often used for each built-in data type. For example, in C#, you can use `int` for the `System.Int32` type in the .NET Compact Framework. The decimal type has a greater precision but a smaller range than floating-point types and is used primarily for financial and monetary calculations.

Table 2-1 .NET Compact Framework Numeric Data Types

.NET CF Type (In System Namespace)	Number of Bytes	C# Keyword	VB Keyword	Description
Boolean	1	`bool`	`Boolean`	True or false
Sbyte	1	`byte`	`Byte`	A single byte
Char	2	`char`	`Char`	Unicode character
Int16	2	`short`	`Short`	Signed 2-byte integer
UInt16	2	`ushort`	-	Unsigned 2-byte integer
Int32	4	`int`	`Integer`	Signed 4-byte integer
UInt32	4	`uint`	-	Unsigned 4-byte integer
Int64	8	`long`	`Long`	Signed 8-byte integer
UInt64	8	`ulong`	-	Unsigned 8-byte integer
Single	4	`float`	`Single`	4-byte single-precision floating number
Double	8	`double`	`Double`	8-byte double-precision floating number
Decimal	16	`decimal`	`Decimal`	Greater precision but smaller range than double

Class Objects

Like C++ or Java, the .NET Framework and the .NET Compact Framework provide object-oriented design patterns centered around the concept of the class. You build applications by creating objects or instances of a class and by using their members to realize your programming logic. The `new` keyword is used to create an object, which will call one of the constructors of the class. Thanks to garbage collection, you don't need to explicitly delete the object when your code no longer references the object. The garbage collector will automatically detect unreferenced objects, delete them, and free up memory. In other words, in the .NET Framework and the .NET Compact Framework, no destructors are needed for a class.

However, there might be some cases for which you need to do something before an object is garbage collected. In this case, the .NET Framework and the .NET Compact Framework provide a way to inject some code in the form of a *finalizer*. You can put code that does resource cleanup into the finalizer to ensure that the object's resource will be released when the object is terminated by the garbage collector. Note that the finalization of an object is not deterministic: It is up to the garbage collector to decide when to finalize those unreferenced objects. Internally, the garbage collector does not delete objects that have the finalizer method overridden. Instead, these objects are added to a list called the *finalization queue*. A runtime thread will thus call the finalizer of these objects and drain the queue. The next time the garbage collector runs, these objects are eventually terminated and their memory is released. Unlike the `Dispose()` method, which is provided for developers to explicitly perform some cleanup tasks, the finalizer is called by the garbage collector. Here is an example of using a finalizer:

```
class MyClass
{
MyClass() //A constructor
{...}
~MyClass() //Finalizer
{
   //Some clean-up operations to perform before the object is garbage-collected
}
}
...
//In a method of another class
void Do()
{
  //To create an instance of the class
  MyClass obj = new MyClass();
  //Use the object
  ...
} //When object gets out of scope of the method, it will be put into the
finalization queue and a runtime thread will run its finalizer
```

In the preceding code example, the `MyClass` class has a finalizer method called `~MyClass()` that can contain some cleanup code for the object. This ensures that the garbage collector will perform tasks in the finalizer before the object is terminated.

Attributes and Reflection

Attributes are part of the metadata of your code. You can add metadata for classes, methods, properties, events, and so on, directly into your code. It is like annotation for your program. Readers familiar with Java should notice that attributes are similar to the general-purpose annotation in Java 5.0.

To add an attribute to your class, simply add the attribute's name enclosed within a pair of square brackets in front of the class definition, as follows:

```
[serializable] public class Myclass {...}
```

You can also define your own type of attributes by deriving a class from `System.Attribute`.

Both the .NET Framework and the .NET Compact Framework provide *reflection*—a feature that allows your code to inspect an assembly at runtime. You can get type information from an assembly and call methods of those types. Recall that every object has a header. Reflection allows your program to access object headers. Reflection APIs are organized into the `System.Reflection` namespace.

Generics

The .NET Compact Framework 2.0 and later support parametric polymorphism using generics. (So does Java 5.0, but .NET 2.0 is superior to Java in this regard, a topic beyond the scope of this book.) Generics, much like templates in C++ and generics in Java, enable to you generate container class objects with arbitrary types you've provided. For example, you can use generics such as `Stack` and `Queue` in the Generic collection `System.Collections.Generic` to manage your own types, or you can write your generic class. A unique feature of .NET generics is that it is not just a compilation time concept like a C++ template; instead, it encompasses runtime type checking, reflection, and debugging.

To create an instance of a generic `Stack` object that holds a `MyObj` type in C#, you can do the following:

```
System.Collections.Generic.Stack<MyObj> myObjStack = new
System.Collections.Generic.Stack<MyObj) ();
```

Subsequent operations of the stack `myObjStack` are the same as using other collection objects.

Exception Handling

Exceptions are runtime errors that need to be handled to avoid program crashes. The .NET Framework and the .NET Compact Framework provide a built-in mechanism to handle exceptions using the well-known `try...catch...finally` semantics. As with exceptions in C++ and Java, an exception is a facility to catch potential runtime errors and transfer control to a specific error-handling block: the `catch` block. A `try` block can be associated with multiple `catch` blocks, each processing a specified exception. The `finally` block encloses cleanup operations that must be performed for any error-handling process defined in preceding `catch` blocks. The following code snippet shows the basic structure of `try...catch...finally`:

```
try
{
  //Statements that can cause exception, such as I/O operations
}
catch(ExceptionType1 x)  //
{
  //Statements to handle exception type 1
}
catch(ExceptionType2 x)  //
{
  //Statements to handle exception type 2
}
finally
{
  //Clean up
}
```

The preceding code uses two `catch` blocks for a `try` block. In addition, there is a `finally` block that will be executed regardless of which `catch` block is executed. Each `try` block deals with a specific type of exception.

In addition to the exceptions that the CLR generates, you can also explicitly throw an exception using the `throw` statement. You can even rethrow an exception your `catch` block caught to other code, as follows:

```
try
{
  ...
}
catch(FileNotFoundException e) //Catch a file-not-found exception
{
  Console.WriteLine("[File Not Found] {0}", e);

  //Creates a new FileNotFoundException with additional information and throws it
  throw new FileNotFoundException("[File Not Found. Check its path",e);
}
```

In the preceding example, the `catch` block catches `FileNotFoundException`, displays a message in the console, and rethrows the exception with more information.

For a more detailed discussion of exception handling, see Chapter 11, "Exception Handling and Debugging."

A Quick Review of C#

If you have experience with C++ or Java, then C# should not be problematic. As an object-oriented programming language, C# uses classes to encapsulate data and operations of a specific entity. A C# program must contain at least one class, and programs that can be directly executed by the CLR must have a `Main` method as the entry point for the class loader of the CLR. C# syntax largely follows C++ and Java, with some enhancements described in this section.

Value Types

A variable is of a specific data type. The data types shown previously in Table 2-1 list the data types of C#. You can initialize a value type by assigning a value of that type to the variable, as shown here:

```
bool isVisited = true;
uint studentID = 41292922;
int myAge = 24;
float interestRate = 0.056;
char partyLine = 'D';
```

Conversions between value types in C# are similar to other programming languages. For example, an `integer` can be cast onto a `long`, and vice versa. Precision may be changed when casting a "wide" type (such as a `long`) to a "narrow" type (such as an `int`), in which case an explicit conversion is needed. If the number being converted is too large to fit into a small type, some data will be lost without being noticed. A better way to perform such potentially dangerous conversions is to use the `System.Convert` class, as shown in the following example:

```
uint myTotal;
ushort myNumber;
...
myTotal = myNumber; //implicit Implicit cast

byte = mySmallNumber;
mySmallNumber = (byte) myNumber; // Explicit conversion is required. Data loss is
possible in this case. For example, if myNumber is larger than the maximum of a
byte type

double dBigNumber = 123.45;
try {
    int iSmallInt = System.Convert.ToInt32(dBigNumber);
}
catch (System.OverflowException) {
       System.Console.WriteLine("Overflow in double to int conversion. Double is
too big!";
}
```

The `System.Convert` class will throw an `OverflowException` exception when it detects data loss during a type conversion. Therefore, when you use the `Convert` methods, you should always check whether an `OverflowException` is raised.

To convert a string to a number, simply use the numeric type's `Parse` static method. A *static method* is a method that you can call directly without creating an object of the class. To convert a number to a string, simply use the `ToString()` method of the numeric type. The following shows how to convert a string to a number:

```
string s1 = "123";
int x = int.Parse(s);
```

The following is an example of converting a number to a string:

```
int y = 321;
string s2 = y.ToString();
```

There are two other data types: enum and struct. enum is the keyword that declares a set of named constants. The default type of these constants is integer. By default, the first constant has a value of 0, the second has a value of 1, and so on. Here is an example of an enum type:

```
enum employeeStatus {fulltime = 1, part-time, contractor, retired};
int i = EmployeeStatus.fulltime;  //Access an enumerator
```

The enumerator list of employeeStatus starts from 1, rather than 0.

A struct is like a class but it is not a reference type but a value type, meaning it contains the value itself, not a reference to the data. As with any other value type, if you assign a struct variable to another struct, the second one will have a copy of the first one's data. For a reference type, however, an assignment between two reference types will result in assignment of the reference, rather than the object being referenced. A C# struct can have constructors, fields, methods, properties, indexers, operators, events, and nested types. It is used primarily to hold lightweight objects. The following is an example of a struct:

```
public struct Telephone
{
     private uint phoneNumber;
     private string userName;
     public Telephone(uint n, string s)
     {
        phoneNumber = n;
        userName = s;
  }
}
```

In this example, a Telephone struct is defined. It has two private data members and a constructor.

Reference Types

A variable of a reference type contains the reference to the actual data. Reference variables are also known as *objects* or *instances*. Assignments of reference variables of the same type, such as *a=b*, does not copy the data object of *b* to *a*. Instead, *a* will reference to the data object of *b*. This effect is sometimes called *shallow copy*.

The following are commonly used reference types:

- Objects of classes
- Interfaces, events, and delegates (explained below)
- Objects of Windows system components such as threads, graphic objects, and so on
- Collections such as arrays, stacks, queues, and so on
- Strings
- Boxed value types

Operators

An expression is composed of variables, literals, and operators. Literals can be any string literal or a number literal. Operators in C# are similar to those in C++ and Java; the same associativity and precedence rules apply. For operators of the same precedence, the associativity rule dictates the following:

- The assignment operators are right associative — that is, computation is performed from right to left.
- All binary operators are left associative.

Table 2-2 shows the operator precedence table.

Table 2-2 C# Operator Precedence

Category	Operators

Primary	`x.y`
	`f(x)`
	`a[x]`
	`x++`
	`x--`
	`new`
	`typeof`
	`checked`
	`unchecked`

Unary (applied to one operand)	`+`
	`-`
	`!`
	`~`
	`--x`
	`(T)x`

Multiplicative	`*`
	`/`
	`%`

Additive	`+` `-`
Shift	`<<` `>>`

Relational and type testing	`<`
	`>`
	`<=`
	`>=`
	`is`
	`as`

Equality	`==` `!=`

Table continued on following page

Category	Operators
Logical AND (bitwise)	`&`
Logical XOR (bitwise)	`^`
Logical OR (bitwise)	`|`
Conditional AND	`&&`
Conditional OR	`||`
Conditional Ternary	`?:`

Assignment	`=`
	`*=`
	`/=`
	`%=`
	`+=`
	`-=`
	`<<=`
	`>>=`
	`&=`
	`^=`
	`|=`

A few operators need some introduction. The `typeof` operator is used to obtain a `System.Type` object of a type you specified, as shown in the following example:

```
System.Type t = typeof(MyClass); //Here MyClass is the name of the class
```

You may be wondering how to get the type of an object at runtime. Use the `GetType()` method of the object. Every type supports this method.

The `checked` and `unchecked` operators are used to control the overflow-checking context for integer-type arithmetic operations and conversions. If `checked` is applied to an expression, and the arithmetic operation produces a value that is out of the range of a destination type, an exception will be raised. If the expression is marked as `unchecked`, the result will be truncated.

string and object

C# has two built-in reference types: `object` and `string`. The `object` reference type is an alias of the `System.Object` type in the .NET Compact Framework, whereas `string` is an alias of `System.String`. The following code provides some examples of `string` variables:

```
string productName = "Windows Mobile for Smartphone";  //Create a literal string
referenced by productName
string filePath = @"c:\windows\system32\";  //Use this form to avoid escaping
sequence
```

In the preceding example, two strings are defined. The second one is using the @ form to avoid escaping sequence.

Literal strings are stored as UTF-16 Unicode characters on the runtime heap. The CLR ensures that portion of a literal string may be shared by other string objects. As the example shows, the easiest way to create a string object is to assign a literal string to the object. The `System.Object` class has a `ToString()` method that any other classes can override to return a string object that provides meaningful information about the underlying object.

Another way to associate strings to a string object is to use a string resource, which can be added into an Assembly Resource File. Literal strings defined in an Assembly Resource File are saved in UTF-8 code. The .NET Compact Framework provides a `System.Resources` namespace for resource file access.

The `string` type supports common string operations such as searching a character or a substring, string concatenation, comparisons, and so on. Note that a string object is not mutable in that the sequence of characters referenced by the string object cannot be modified; you can't remove a character from the sequence or append a new character to it. When you use the `ToUpper` and `ToLower` methods to switch between lowercase and uppercase, respectively, the original string is not changed at all; instead, a new string is generated from the resulting conversion, whereas the original string stays intact and can be referenced by other objects or collected as garbage later. You can explicitly let a string object reference another literal string. This design rationale of immutable string objects simplifies multithreading applications that have multiple threads accessing the same string object — a read-only object is obviously easy to share. The downside, of course, is the new string allocation on the runtime heap while references are changed. In the following example, first a string object `productName` is created for the literal string `"Windows Mobile for Smartphone"`. Then a new literal string is allocated and referenced by `productName`. The original literal string stays on the heap and does not change at all until next garbage collection.

```
string productName = "Windows Mobile for Smartphone";  //Create a literal string
referenced by productName
string productName = "Windows CE"; //"Windows CE" is allocated and is referenced by
productName now
```

To have full control over a sequence string in your program, you need to use the `System.Text.StringBuilder` class, which provides methods such as `Append()`, `Insert()`, and `Remove()` for manipulating the string. The following are examples of some common string operations:

```
Using System.Text;
...
StringBuilder sb = new StringBuilder();
sb.append("abcde"); //Append a string to the current string
sb.insert(2,"xyz"); //The first parameter is the position for insertion. Now the
string is "abxyzcde"
sb.remove(4,2); //The first parameter is the starting index, and the second
parameter is the length of string to be removed. Now the string is "abxyde"
```

The preceding code shows examples of using the `append()` method, the `insert()` method, and the `remove()` method of the `StringBuilder` class. All these methods have several overloaded forms. The MSDN documentation provides a detailed introduction to each of these methods.

Classes and Interfaces

Any class, interface, event, and delegate is a reference type. A class type is declared using the keyword `class`. A class can contain fields, methods, properties, indexers, delegates, or other classes (see Table 2-3).

Table 2-3 Class Members

Class Member	Description
Fields	Data members of a class
Methods	An operation of a class (similar to member functions in C++)
Properties	Get and/or set a private field of a class using dot (.) notation (see the example below)
Indexers	Provide an array index notation to access a collection of a class (see example below)
Delegates	Provide a way to pass a method to other code (similar to function pointers in C and C++)

The following code is an example of these class members (except delegates):

```
// Code of class fields, properties, etc.
using System;

namespace ClassDemo1
{
    struct StudentDataEntry
    {
        public uint studentID;
        public string studentName;
        public StudentDataEntry(uint id, string name)
        {
            studentID = id;
            studentName = name;
        }
    }
    class Group
    {
        private StudentDataEntry[] studentData = new StudentDataEntry[4];
        private uint groupID;

        /*constructor. studentData has at most 4 elements*/
        public Group(uint assignedGroupID, StudentDataEntry[] inputData)
        {
            groupID = assignedGroupID;
            int i = 0;
            foreach(StudentDataEntry d in inputData)
            {
                studentData[i++] = d;
```

```
            }
        }
        /*GroupID Property*/
        public uint GroupID
        {
            get
            {
                return groupID;
            }
            set
            {
                groupID = value;
            }

        }

        /*indexer*/
        public StudentDataEntry this[int index]
        {
            get
            {
                return studentData[index];
            }
            set
            {
                studentData[index] = value;
            }
        }
        static void Main(string[] args)
        {
            StudentDataEntry[] inputStudentData = new StudentDataEntry[4];
            inputStudentData[0] = new StudentDataEntry(19, "John");
            inputStudentData[1] = new StudentDataEntry(23, "Joe");
            inputStudentData[2] = new StudentDataEntry(56, "Kevin");
            inputStudentData[3] = new StudentDataEntry(71, "Rachel");

            // Create a Group class object
            Group g = new Group(1, inputStudentData);
            Console.WriteLine("Group ID: {0}", g.GroupID); // Test GroupID property
"Get"
            g.GroupID = 100; //test Test GroupID property "Set"
            Console.WriteLine("Group ID: {0}", g.GroupID);
            Console.WriteLine("Student data index: 2: {0} {1}", g[2].studentID,
g[2].studentName);

        }
    }
}
```

This example defines a `StudentDataEntry` struct and a `Group` class in the `ClassDemo1` namespace. The `Group` class has a property of `GroupID` and an indexer to access the `studentData` array in the class. The static `Main()` method is the entry point of the program.

Parameter Passing

Parameter passing of value types in C# is *call-by-value* — that is, the value of a variable the caller submits as parameter is passed to the method, in which a local copy of the parameter is created on the call stack. Thus, any change to the local copy will not influence the variable the caller uses.

Parameter passing of reference types in C# is *call-by-reference* — that is, the reference type itself is passed to the method. Thus, any changes made to the referenced object will be reflected in the caller. For example, a reference to an array object can be passed to a method that directly modifies the array. In fact, this is also done in a call-by-value manner: The reference itself is copied to the method's stack space.

Another facility C# provides for passing multiple values to a method is the `ref` keyword. Using the `ref` keyword, you can force a value type to be passed call-by-reference, just like a reference type. In the following example, notice that the method returns nothing (`void`), but two parameters, `sum` and `diff`, are specified as `ref` parameters. Therefore, the computation results can be returned using these two parameters:

```
public void SumAndDifference(int x, int y, ref int sum, ref int diff)
{
    sum = x+y;
    diff = x >= y ? x-y : y-x;
}

//to To use the method
int Sum =0;
int Diff = 0;
SumAndDifference(5, 8, ref Sum, ref Diff);
```

You can also use the `out` keyword to replace the `ref` keyword so that you don't need to initialize the reference parameters in the caller method.

Delegates and Events

In the C# programming paradigm, it is always necessary to be able to allow callback functions that will be called when a specific event occurs. The event can be a GUI event, a timer timeout, I/O completion, and so on. A *delegate* in C# is a facility to enable type-safe callback functions. You can specify a method using the `delegate` keyword. Then the method will be encapsulated into a delegate type derived from `System .Delegate`. When you want to create an object of the delegate type you declared, you must use the `new` keyword just as you would when creating any other class objects. In the .NET Compact Framework, delegates are often used to handle events. An *event* is a member of a class that can be used to inform a client of the object that the state of the object has changed. An event delegate can be connected with the event to call a predefined event handler method. The following example shows how to combine an event with a delegate. First, a delegate is declared, which encapsulates an event handler that accepts an object that raises the event and the event object:

```
public delegate void EventHandler(Object sender, EventArgs e);
```

Then the event handler is defined within a class:

```
public class Notification
{
    ...
    DoNotification(Object sender, EventArgs e) { ...} // Notice the same signature
as the delegate
}
```

The next step is to bind the event handler to the delegate:

```
Notification nObj = new Notification();
EventHandler handler = new EventHandler(nObj.DoNotification);
```

Finally, you can use the delegate as shown in the following example. The class has a public event member and a method to raise the event. The actual event handling is passed to the event delegate, which in turn delegates to an event handler bound to it (`nObj.DoNotification()` method):

```
public class A
{
public event EventHandler MyEvent;
protected virtual void OnMyEvent (EventArgs e)
    {
        if(MyEvent!=null)
MyEvent(this, e)
}
}
```

This example doesn't define any specific data for the event; instead, it uses a plain `System.EventArgs` event. You certainly can define an event class that is derived from `EventArgs`, and let it carry some data to the handlers.

Interfaces

Because C# does not allow multiple inheritance, you cannot derive your class from more than one base class. However, you can derive your class from multiple interfaces and/or a single base class. An interface is a contract between the interface designer and the developer who will write a class that implements the interface. It specifies which methods, events, indexers, or properties need to be implemented in that class. An interface itself does not implement those things. In the following example, `ClassA` is derived from `BaseClass` and two interfaces, `Interface1` and `Interface2`; thus, `ClassA` must implement all members specified in those two interfaces:

```
class ClassA: BaseClass, Interface1, Interface2
{
   //Class members
}
```

Member Accessibility

A C# class uses the following five modifiers to specify the accessibility of class fields and methods:

- `public`—`public` modifiers are open to everyone.
- `protected`—`protected` fields and methods can be accessed from within the class and any derived classes.
- `private`—`private` fields and methods are available only to the underlying class. By default, class members are private.
- `internal`—`internal` fields and methods are available only to the underlying assembly. This is the default access level.
- `internal protected`—`internal protected` fields and methods can be accessed from within the assembly and derived classes outside the assembly.

Class Accessibility

A class can also have an access modifier, such as `public` and `internal`. The `public` modifier makes the class available to everyone, whereas the `internal` modifier makes the class available only within the underlying assembly. The default class access modifier is internal. Note that the member access modifiers are restricted by the class modifiers. For example:

```
internal class A
{
    public uint num;
    ......
}
```

The data field `num` will still be `internal` because the class modifier is applied first.

Polymorphism

As an object-oriented language, C# implements a special polymorphism mechanism. You can specify a method in a base class by using the `virtual` keyword, making it a virtual function to be overridden in a derived class. Virtual functions or methods are not bound to an object at compilation time, as with most method calls; rather, the binding is done at runtime, depending on the actual object. In the derived class, the method that overrides the base class's virtual method must be specified with the `override` keyword.

If this dynamic binding is not needed, you don't want to put `virtual` in front of the method in the base class. Then, in your derived class, if you write a method that has the same signature as the one in question in the base class, the C# compiler will issue a warning, requesting you to clarify whether you want the method "overriding" or "hiding." To use your new method in the derived class, you need to use the `new` keyword to hide a derived method in the derived class.

In the following example, a base class `A` has a virtual method called `do()`, and its derived class `B` has an overridden method called `do()`. Class `A` has another method, a nonvirtual method called `do2()`. In class `B`, a method of the same signature is provided, which hides `do2()` derived from `A`. The method `do2()` will be statically bound at compilation time. Notice that in the new `do2()`, you can call the derived `do2()` using the keyword `base`:

```
class A {
    virtual public void do() {Console.Println("In A's do method.");}
    public void do2() {Console.Println("In A's do2 method.");}
}
class B : A {
    override public void do() {Console.Println("In B's do method.");}
    new void do2() { base.do2();Console.Println("In B's do2 method.");}
}
...
A a;
B b = new B();

a=b; // This is fine because class B is derived from class A
a.do();  // B's do() will be called because the binding is resolved at run time
a.do2(); // A's do2 will be called because do2() is statically bound at compilation
time
Console.Println("----");

A a2 = new A();
```

```
A2.do();  // Call A's do
A2.do2(); // Call A's do2

Console.Println("----");
b.do(); // Call B's do
b.do2(); // B's do2
```

The output of the preceding code is as follows:

```
In B's do method.
In A's do2 method.
----
In A's do method.
In A's do2 method.
----
In B's do method.
In A's do2 method.
In B's do2 method.
```

Arrays and Collections

An *array* is a collection of elements of the same type. The `System.Array` class is the abstract base class for all arrays. The element type of an array can be any value type or reference type, even an array type. An array can be one-dimensional or of multiple dimensions. When creating an array, you either specify the size of the array explicitly or initialize the array element using an initialization list. Array elements are garbage-collected when no references to them exist. Setting an array object to `null` will force the garbage collector to reclaim the memory allocated to it.

Because arrays are derived from `System.Array`, they can use the following members defined in the `Array` class:

- The `Sort()` static method sorts a one-dimensional array in place (i.e., without requiring extra storage).
- The `BinarySearch()` method searches for a specific element in a sorted array.
- The `Length` property stores the number of elements currently in the array.
- The `IndexOf()` static method searches for a specific element and returns its index.
- The `CreateInstance()` method creates an array containing items of the specified `System.Type` with a given length, with indexing. Although the `Array` class is an abstract base class (an abstract class is a class that cannot be initiated), it provides this method to construct an array of specified element types.

The following code shows two different ways to initialize an array, and the use of `CreateInstance()` method:

```
using System.Array;
...
int[] myIntArray1 = { 1, 2, 3, 4, 5 };
int[] myIntArray2 = new int[3] { 1, 2, 3 };
Array.sort(myIntArray1);  //sort Sort myIntArray1
foreach ( int i in myIntArray1 )
```

```
{  //access Access each element in myIntArray1
    myIntArray1[i] *= 3;
}
Array.CreateInstance(typeof(uint), 5); // Create an array of 5 uint integers
```

The preceding example also demonstrates using an array indexer to access array elements. The `foreach` construct is quite handy to iterate through all array elements.

Aside from the `Array` construct, the .NET Compact Framework provides the following set of collection classes and interfaces in the `System.Collections` namespace:

- `ArrayList`—A linked list whose size is dynamically changed as needed. The `ArrayList` class implements the `IList` interface.
- `BitArray`—An array of bit values, either `true` or `false`.
- `Hashtable`—A hash table providing key-value mapping. Keys and values are both objects.
- `Queue`—A first-in-first-out (FIFO) collection of objects. The capacity of the queue is automatically increased when more objects are added.
- `SortedList`—A sorted collection of key-value pairs, which can be accessed using indexers (like an array) or keys (like a hash table).
- `Stack`—A last-in-first-out (LIFO) collection of objects. The capacity of the stack is increased automatically when more objects are pushed onto the stack.
- `Comparer and CaseInsensitiveComparer`—Two classes implement the `IComparer` interface.

Collection interfaces include the following:

- `ICollection`—A general interface for all collection classes. The `ICollection` interface specifies properties such as `count` and `isSynchronized`, and the `CopyTo()` method.
- `IList`—An interface for a collection of objects that can be accessed using an index. The `IList` interfaces specifies methods such as `Add()`, `Insert()`, `Remove()`, `RemoveAt()`, and `Contains()`. Classes that implement `IList` include `Array`, `ArrayList`, and a number of GUI control classes.
- `IComparer`—An interface specifying only a `Compare()` method. Classes implementing `IComparer` include `Comparer`, `CaseInsenstiveComparer`, and `KeysConverter`.
- `IDictionary`—An interface representing a collection of key-value pairs. Classes implementing `IDictionary` include `Hashtable` and `SortedList`.
- `IEnumerable`—An interface specifying a `GetEnumerator()` method over a collection of objects. Classes implementing `IEnumerable` include `Array`, `ArrayList`, `BitArray`, `Hashtable`, `Stack`, `Queue`, and `SortedList`.
- `IEnumerator`—An enumerator interface used in conjunction with the `IEnumerable` interface. The `IEnumerator` interface specifies the methods `MoveNext()` and `Reset()`, and a property of `Current`. Classes implementing `IEnumerator` include `CharEnumerator`, `DBEnumerator`, and `MessageEnumerator`.

Summary

This chapter discussed the core of managed Smartphone application development: the .NET Compact Framework. The .Net Compact Framework enables developers to utilize a rich set of unified classes and facilities to develop type-safe and high-performance mobile applications across different hardware and operating systems.

After reading this chapter, it is assumed that you understand the .NET Compact Framework type system and the rationale of language-independent programming on the CLR. You should also be familiar with a set of Smartphone development tools. In addition, you should become familiar with C#. If you are a developer working on C, C++, or Java, you will find C# is quite easy to learn.

In the next chapter, you will start to build your first Smartphone application. After learning how to set up your Smartphone application development environment, you will use Visual Studio .NET to write a simple application. You will also learn how to test, debug, package, and deploy a Smartphone application with the emulator or a Smartphone device.

Part II
Smartphone Application Development

3

Developing Your First Smartphone Application

The previous chapter talked about the core of the Microsoft Smartphone platform, the .NET Compact Framework and the C# programming language, which lay the foundation for managed Microsoft Smartphone application development. In this chapter, you will start to get hands-on experience with the Smartphone development environment, mainly using Visual Studio (2005 or .NET), the Windows Mobile 5.0 SDK, and some supporting tools. The chapter will walk you through the development, testing, debugging, and deployment of your first Smartphone application. Specifically, the chapter discusses Visual Studio's support of Smartphone security models and policies. This chapter is intended to give you a quick overview of the basic development stages of a Smartphone application. Topics covered in this chapter include the following:

- An introduction to Visual Studio
- Creating a simple Smartphone application
- Testing and debugging a Smartphone application
- Packaging and deploying a Smartphone application

Required Tools

To get started with Smartphone software development, you need to have a set of tools installed on your development computer. These tools form a basic Smartphone application development environment. The development computer must be running Windows Server 2003, Windows XP, or Windows Vista.

The download URLs provided in this section may change as Microsoft updates its website. For up-to-date links and tools, you can visit the Windows Mobile section at Mobile Developer Center at `http://msdn.microsoft.com/mobility/windowsmobile/`. *You will always find links to the latest tools and technical articles there.*

Visual Studio 2005

Visual Studio 2005 (code-named *Whidbey* while in development) is Microsoft's latest integrated development environment (IDE) for both desktop and device application development. The previous version of Visual Studio, Visual Studio .NET 2003, can also be used for application development targeting Windows Mobile 2003 devices. Both Visual Studio versions enable you to program, debug, test, and deploy an application targeting Windows Mobile devices. If you have don't the release version of Visual Studio 2005, you can download a trial version from `http://msdn.microsoft.com/vstudio/`. Note that the free Visual Studio Express Edition does *not* support application development with the .NET Compact Framework for Windows Mobile devices. You need to use Standard Edition, Professional Edition, or Team System for Smartphone application development.

Developers who are familiar with earlier versions of Visual Studio will find the 2005 version quite easy to get along with. For example, you will still see windows such as the control toolbox window, Solution Explorer window, and output window. The Form Designer enables you to drag and drop controls onto a form. The Class Designer simplifies class design by providing a list of toolbars for classes, interfaces, abstract class, struct, delegates, inheritance, and so on. Object Browser in Visual Studio enables developers to quickly browse objects and their members.

In addition to general IDE improvements such as code snippets, revision marks, refactoring, and so on, there are some new features that a Smartphone developer should know about:

- **.NET Framework 2.0 and .NET Compact Framework 2.0.** The new versions of both frameworks have been enhanced with new functionality and performance.
- **Native code development is supported.** Developers do not need to resort to eMbedded Visual C++ for native code development anymore.
- **MSBuild.** MSBuild is the new XML-based build system for managed application development. The build process is comprised of a number of atomic units of language-independent build tasks that developers can customize, augment, or even redefine. MSBuild is also a core component of the .NET Framework redistributable. Projects created in Visual Studio 2005 are now in MSBuild format (an XML file).
- **The new Device Emulator.** This is described in an upcoming section.
- **.NET Remote tools,** including the following:
    - *Remote File Viewer* — To browse and transfer files to a device or an emulator
    - *Remote Heap Walker* — To view memory usage on a device or an emulator
    - *Remote Process Viewer* — To view process information on a device or an emulator
    - *Remote Registry Editor* — To edit the registry of a device or an emulator
    - *Remote Zoom In* — To zoom in on a remote display on a local computer
    - *Remote Spy* — To view messages a window on a device receives

The MSDN library is also a must-have component that enables you to access online help while you program, test, and debug your application.

Windows Mobile 5.0 SDK for Smartphone

The Windows Mobile 5.0 SDK for Smartphone includes the Smartphone emulators (which are also in Visual Studio 2005), some command-lines tools, help files, header files, and libraries for native code development. The SDK also provides a set of sample configuration files and digital certificates for day-to-day application development with a physical device. You can download the SDK from `www.microsoft.com/downloads/details.aspx?familyid=DC6C00CB-738A-4B97-8910-5CD29AB5F8D9&displaylang=en`.

You can download optional tools for special-purpose Smartphone development from Microsoft's website. This includes localized emulator images for the Windows Mobile–based Smartphone in various languages, such as Spanish, French, Italian, Chinese, and so on. In addition, there is a popular application framework—namely, Smart Device Framework—developed by the open-source community OpenNETCF.org (`www.opennetcf.org`). The Smart Device Framework provides a rich extension to .NET Compact Framework libraries, offering a number of classes and controls.

Smartphone Device Emulator

Visual Studio 2005 comes with a new device emulator called *Microsoft Device Emulator*, which is a completely rewritten version of the Smart Device software emulator. Previously, the device emulator for Pocket PC and Smartphone was a Windows CE operating system image running in a virtual x86 emulation environment. However, in the mobile and embedded world, many devices use other processors, such as ARM, MIPS, and SH4. Among all these processors, including x86, ARM is far and away the market leader. In fact, ARM is not a single processor but a processor architecture designed by a British company called Acorn, which licenses the architecture to processor manufacturers such as Intel and Texas Instruments. The gap between an x86-based emulator and an ARM-based mobile device turns out to be a serious limitation of the Smartphone and Pocket PC programming support in Visual Studio .NET 2003. Even when an application is tested within an emulation environment of the x86 architecture, it may not function well on other processors. In addition, developers have to compile the application first for the x86 in order to debug it in the emulator, and then again for ARM or other processors to test it on a physical mobile device.

The new emulator in Visual Studio 2005, as shown in Figure 3-1, solves the problem. Now the emulator itself is a Windows Mobile operating system for a targeting processor (ARM), running within a Microsoft Virtual PC environment. (*Virtual PC* is a virtual machine technology that enables a guest operating system to run on top of another operating system.) Your application will be directly compiled against the ARM-based Windows Mobile operating system, thereby eliminating the gap between the x86 and target processor architecture. Here is a list of features that the new emulator offers:

- **Run code compiled for ARM processors, rather than x86 processors.** The emulator executes true ARM instructions. Because the underlying Windows Mobile operating system is exactly the same OS running on a physical device, developers can run the same binaries on the emulator and the physical device.
- **Support synchronizing with ActiveSync.** ActiveSync is a tool that can synchronize a device with a computer. Now, with the new emulator, you can establish a partnership between your computer and the emulator. A Device Emulator Manager tool is provided to work like a "cradle" for a physical device so that synchronization can be performed.

- **Configurable screen resolution and flexible display orientation.** You can easily rotate the emulator screen, zoom the display, and change the skin of the emulator.
- **Storage card emulation and serial port emulation.** You can share a folder on your development computer that will appear as a storage card to the emulator.
- **With appropriate images, the emulator supports Windows CE device emulation, Smartphone device emulation, and Pocket PC device emulation.** Figure 3-2 shows the Device Emulator Manager, which lists all available emulator images.

ActiveSync

You need to have Microsoft ActiveSync to connect your Smartphone device to your Windows PC. ActiveSync acts as the gateway between your PC and your Windows Mobile device so that you can easily transfer files or synchronize application data such as e-mail, a calendar, and so on. For Smartphone development, ActiveSync is needed by Visual Studio to transfer data between the development PC and the device or the emulator. You can download ActiveSync from `www.microsoft.com/windowsmobile/activesync/default.mspx`.

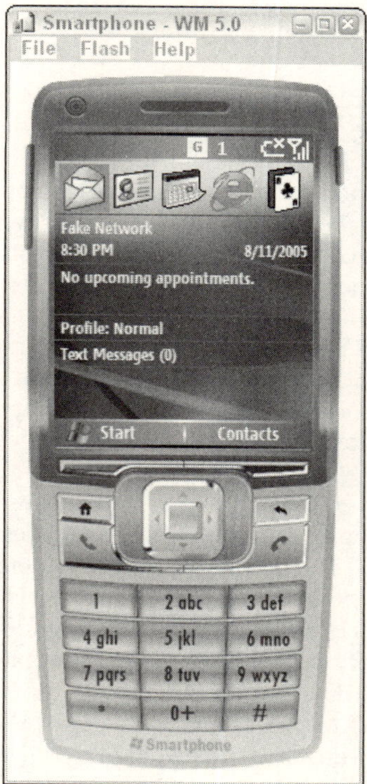

Figure 3-1

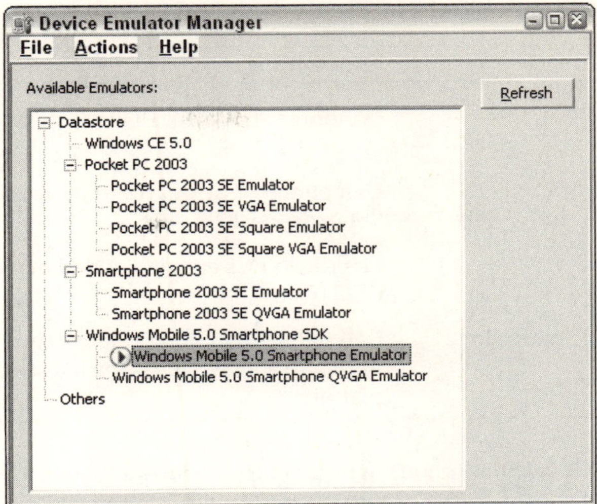

Figure 3-2

All-In-One Package

All the required tools for Smartphone application development can be purchased as an all-in-one package in a DVD, the Windows Mobile 5.0 Developer Resource Kit. You pay only the shipping and handling fee. It includes a 90-day trial version of Visual Studio 2005 Professional Edition, Windows Mobile 5.0 SDKs for Pocket PC and Smartphone, ActiveSync 4.1, .NET Compact Framework 2.0, localized emulator images and other useful developer tools, SQL Server 2005 Mobile Edition, plus developer resources such as links to technical whitepapers and webcasts, WeFly247-50 sample applications, hands-on labs and videos, and partnering opportunities. You can also download a subset of the package from `http://msdn.microsoft.com/mobility/windowsmobile/howto/resourcekit/`. The downloadable package does not include the trial version of Visual Studio 2005 or some emulator images.

Building Your First Smartphone Application

Now it's time to start building your first Smartphone application. You will create a Smartphone project, which is a collection of source code files, resource files, and other files. You will use the Form Designer to design your first Smartphone application, and then add some code to the form and test it on an emulator.

Creating a Smartphone Project

After you install the required tools listed in the previous section, you can start programming your very first Smartphone project. To begin, launch Visual Studio 2005 from the Start menu, and then create a new Smart Device project by clicking New➪Project. You will see the project wizard. Here you can choose from four types of targeting platforms, as described in Table 3-1.

Table 3-1 Smart Device Projects in Visual Studio 2005

Targeting Platform	Description
Pocket PC 2003	Windows Pocket PC PDAs running Windows Mobile for Pocket PC 2003
Smartphone 2003	Cell phones and Smartphones running Windows Mobile for Smartphone 2003
Windows CE 5.0	Handheld PCs and embedded devices running Windows CE 5.0
Windows Mobile 5.0 Smartphone	Cell phones and Smartphones running Windows Mobile 5.0 for Smartphone. This is the latest version of the Smartphone platform.

Choose Windows Mobile 5.0 Smartphone in order to try Windows Mobile 5.0 for Smartphone application development (see Figure 3-3).

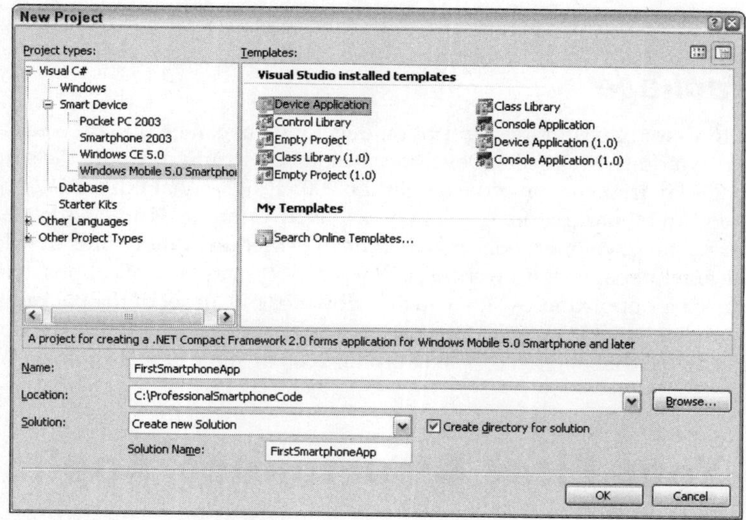

Figure 3-3

From the right side of this dialog box, choose a template for your application. A *template* is a framework consisting of the necessary source code files, resource files, and properly configured project references and properties. As shown in Figure 3-3, for a Smartphone application, you can choose from Device Application, Class Library, Control Library, Console Application, Empty Project, Device Application (1.0), Class Library (1.0), Console Application (1.0), and Empty Project (1.0). The version number 1.0 refers to .NET Compact Framework 1.0. For backward compatibility, Windows Mobile 5.0 for Smartphone supports .NET Compact Framework 1.0. You will build your first .NET Compact Framework 2.0 Smartphone application with has a simple GUI using forms, so choose Device Application. Rename the project to **FirstSmartphoneApp**. Optionally, you can select a directory in which to save the project. Finally, click OK. Visual Studio 2005 will create the framework of the application for you.

Figure 3-4 shows the Visual Studio main window when the project generation is complete. Visual Studio 2005 will put your project into a solution, which is simply a collection of projects, one of which is designated as the "Startup" project. Note that you may see a slightly different layout of windows if you have a specific IDE workspace configuration. (For example, the Solution Explorer windows may appear on the left side of the window.) The Form Designer shows an image of a conceptual Smartphone with Form1 loaded into its screen. The form appears to be empty; no controls have been placed onto it. However, there is already a control named "mainMenu1" at the bottom of the form. Usually, all forms in Smartphone applications should have a Main Menu control for soft keys immediately under the screen of a Smartphone. The automatically generated project now has only two files: `Form1.cs` and `Program.cs`. `Form1.cs` is the GUI interface with which a Smartphone user will interact, whereas `Program.cs` is the place where the .NET Compact Framework will find the main entry point and load the application.

Now, let's bring up the properties window of the form, and make some modifications to the form. Move the cursor onto the design window of Form1 (the window titled "form1.cs [Design]", where you see the form on a Smartphone image), and click the right mouse button. Then click the Properties context menu. Every project, as well as every file in the project, has a properties page where you can change the configuration of the underlying project. A properties page is also available for every single control on a form.

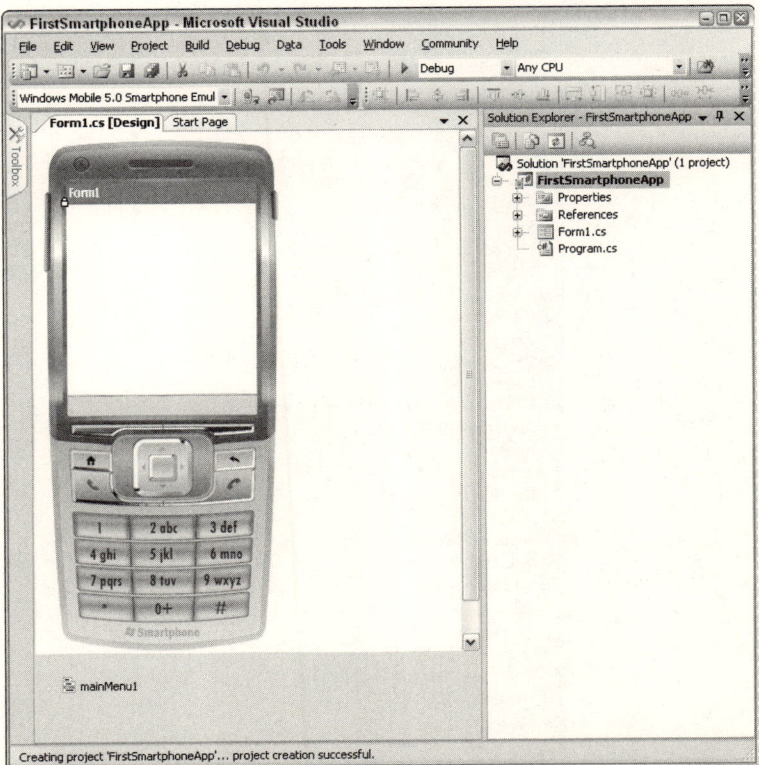

Figure 3-4

Figure 3-5 shows the Properties window of the form. It looks exactly like the regular properties window you probably have seen while developing desktop Windows applications. However, the .NET Compact Framework form controls don't have all those properties for controls in the full .NET Framework library. For example, the `Tab Index` property, which is available in a form in the .NET Framework, does not appear in the list of the Properties window here. The Design section of the Properties page lists some interesting mobile-related properties, such as `Name`, `FormFactor`, `Locked`, and `Skin`. The `Name` property refers to the form name (that is, the object name used in your program). The `FormFactor` property specifies the platform's display capability in terms of screen resolution, including 176 × 220 (the default for Windows Mobile 5.0) and 240 × 320 (also known as QVGA, or Quarter Video Graphics Array). The `Locked` property indicates whether the form or control can be moved or resized. Usually you will lock the form or control. Use the `Skin` property to indicate whether you want to see the phone image in the Form Designer. You can view all the properties in alphabetical order by clicking the second small icon, AZ, at the top of the window.

Now let's modify the `Text` property of the form to be My First App, which will appear as the title of the form on the title bar. It is strongly suggested that the form text be very short, because the title bar cannot contain too many letters.

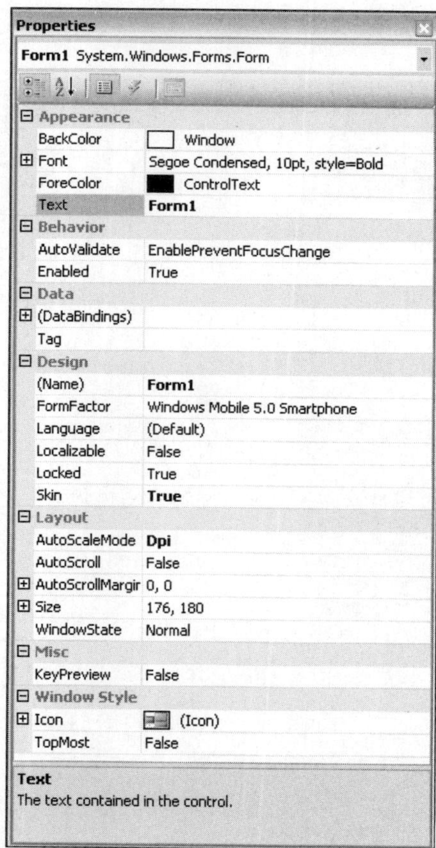

Figure 3-5

The next step is to add some controls to the form. First close or dock the Properties window of the form, and bring up the control toolbox. You are going to add the following five controls to the form:

- A label with the text "Employee Name"
- A label with the text "Department"
- A text box
- A combo box
- A checkbox with the text "Contractor"

To add a control to the form, simply drag and drop the icon of the control from the control toolbox window onto the form.

In Visual Studio .NET 2003, the order in which you add these controls to the form must be the reverse of the tab order because the control-adding code generated by the Form Designer in Visual Studio .NET 2003 follows a stacking pattern; that is, the last control dropped onto the form will be the first control added in the code, and the order in which the controls are added to the form is the tab order. This limitation has been removed in Visual Studio 2005.

For the two Label controls and the CheckBox control, only the `Text` property will be modified. The TextBox control has a default value of the `Text` property, which you must delete so that you can have an empty text box when the form shows up on the screen of an emulator or a device. The ComboBox also control requires some attention. You need to populate the list of items in its drop-down list. To do this, bring up its Properties window, find Items, and click the small button on the right. Then enter the following strings into the String Collection Editor window, one per each line: **IT, Human Resource**, **Marketing,** and **Product Dev.** Use the Align tool (Tools⇨Align) to align your controls. After that, modify the `Text` property of the main menu control to be `OK`. Feel free to adjust the size of these controls by dragging the border handles on the frame. The form with the newly added controls is shown in Figure 3-6.

So far you have added some controls to the automatically generated Windows form; you have not entered any code. Even without a single line of code, this application can be compiled without generating any errors or warnings.

To build the application, click Build Solution in the Project menu, and the project will be compiled and linked. You can view the build information in the Build Output window at the bottom of the workspace. Now click Debug⇨Start Without Debugging and a window will appear, prompting you to choose a device or a Smartphone emulator from a list of emulators or devices. Select the Windows Mobile 5.0 Smartphone emulator and click OK. A window with a Smartphone image will then appear; and, after a short delay for emulator initialization, you will see the application running in the emulator, as shown in Figure 3-7. Note that the initial focus is on the text box, where you can enter text for Employee Name. To move to the next tabbed control, click the down arrow button on the keypad of the emulator. The combo box is not really a "combo"; instead, it should be named "spin box," as it can only be spun horizontally, unlike a typical drop-down list used in a desktop ComboBox control. At any time, you can always click the soft key located right below the OK button.

Figure 3-6

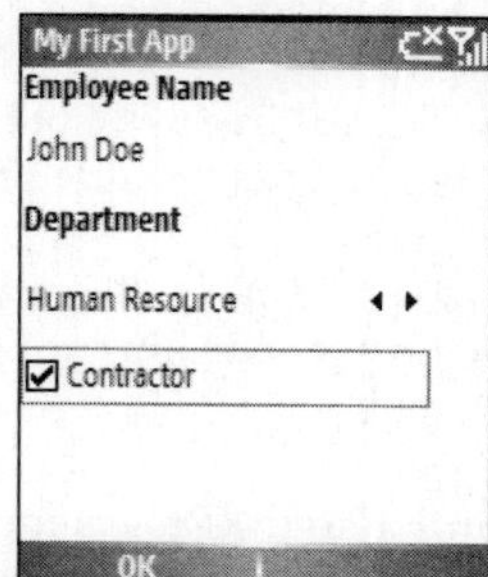

Figure 3-7

What happens when you select Start Without Debugging? Visual Studio 2005 performs the following steps behind the scenes:

1. It connects to the physical device or the emulator you selected.
2. As needed, it downloads the .NET Compact Framework onto the device or the emulator.
3. It copies the project files to the device or the emulator. The files are .NET assemblies, along with digital certificates or other security provisioning files (see "Signing Applications" later in this chapter).
4. It launches the application on the device or the emulator.

Because you did not add any exit code to close the application, there is no way to terminate it in the emulator. Thus, if you want to run it again, you have to perform a soft reset on the emulator or device.

As you might have noticed, Smartphone application development does not look very different from desktop application development; you place controls on a form and adjust their properties. Other than the particular UI characteristics of some controls (such as the horizontal spin box), you can leverage your desktop development experience for Smartphone development. In fact, there is a difference between these two programming paradigms; and although you are using the .NET Compact Framework rather than the .NET Framework, Visual Studio 2005 does a good job of making this divergence largely transparent to developers, ensuring a consistent programming experience.

Adding Code to the Form

The next step is to add some code to the form. We want to display a message box after a user enters some information and clicks the OK button. The message box will contain text for the employee's name, the department, and the employee's status.

Go back to the Form Designer and double-click OK at the bottom-left corner of the form. You are now in the event handler of menuItem1_Click() in the design view of the form, as shown in Figure 3-8. You will enter some code to display a message box with the information a user just entered in the aforementioned controls.

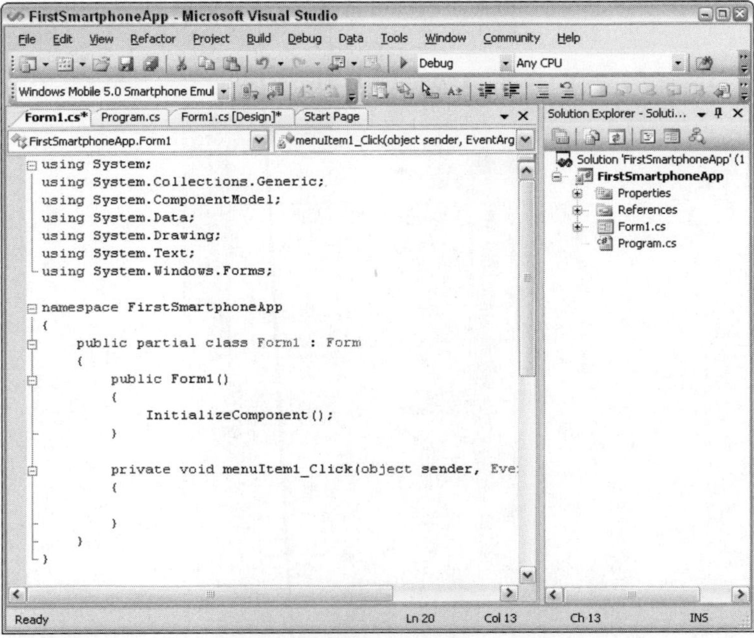

Figure 3-8

Enter the following code into the event handler `menuItem1_Click()`:

```
string status;
if(this.checkBox1.Checked)
status = "contractor";
else
status = "permanent employee";

MessageBox.Show("You are " + textBox1.Text +
                ", working at " + this.comboBox1.SelectedItem +
                " as a " + status,
                this.Text,
                MessageBoxButtons.OK,
                MessageBoxIcon.Exclamation,
                MessageBoxDefaultButton.Button1
                );
Application.Exit();
```

Don't worry about the classes and methods used here; they will be discussed extensively in later chapters. For now, simply run the program by selecting Debug⇨Start Without Debugging. Once the main form appears in the emulator, enter a name (**John Doe**), select a department (**Product Dev**), check the Contractor checkbox, and then click OK. You will see the screen shown in Figure 3-9. Congratulations! You have successfully finished your first Smartphone application.

Figure 3-9

When testing an application in the emulator, use the mouse to click the navigation keypad to move between controls, and click the soft keys (directly below OK) to launch the corresponding operation (OK in our example). The preceding example added a statement to exit the application once the message box has been displayed. Therefore, when the message box appears on the screen, clicking the OK soft key will terminate the application.

Project Files

You may wonder which files have been automatically generated by Visual Studio 2005 so far; you did not specify any filenames yet. In fact, Visual Studio 2005 has created a number of files for the FirstSmartphoneApp project. Knowing the details of these files will be crucial for further development. Admittedly, there is no big difference between the project files of a Smartphone project and those of a regular C# project, so feel free to skip this section if you are familiar with programming C# in Visual Studio.

When you use the project wizard to create a project, you have the option to create a solution for the project as well. Table 3-2 explains the files generated by Visual Studio 2005 for the FirstSmartphoneApp project (the only project in the solution).

Table 3-2 FirstSmartphoneApp Files

Filename	Directory	Description
SolutionName.sln	Solution folder	A plain-text solution file that lists all the projects in the solution and the configuration of the target platforms
ProjectName.csproj and ProjectName.csproj.user	Project folder	An XML project file (C# project) that describes the project's configurations and all the information needed by MSBuild, the Visual Studio 2005's built-in compiler
Source files (.cs files)	Project folder	Program.cs has the `main()` method; Forms files are partial classes derived from `System.Windows.Forms.Form`.
Form designer files (FormName.designer.cs)	Project folder	Files that contain automatically generated code in a partial class and will be combined with user's form code
Resource file of a form (FormName.resx)	Project folder	An XML file describing resources defined in the form
Assemblies (ProjectName.exe) and program debug database file (ProjectName.pdb) generated in "Debug" mode	Project folder\bin\debug	These are files generated and used under the configuration of "Debug"

Table continued on following page

Filename	Directory	Description
Assemblies (ProjectName.exe) and a program debug database file (ProjectName.pdb) generated in "Release" mode	Project folder\bin \Release	These are files generated and used under the configuration of "Release". The Release version is production code optimized for performance.
Assemblies, program debug database files, resource files, Property resource files, etc.	Project folder\ obj\Debug and Project folder\obj\ Release	Intermediate object-code files, including assemblies
AssemblyInfo.cs	Project folder\ Properties	A C# file containing attributes of the assembly
Resources.Designer.cs	Project folder\ Properties	A strongly typed resource C# class automatically generated by Visual Studio 2005
Resources.resx	Project folder\ Properties	A .NET-managed resource file (XML) that aggregates form resource files

Only the form files and `Program.cs` file are directly exposed to you; the other files are generated and managed by Visual Studio 2005; you don't even need to know where they are physically located.

Note that the `form` *class is composed of two "partial" classes located in two different files: one you can directly modify for event handling and business logic, etc., and one describing GUI components of the form generated by the Form Designer, which you don't and should not modify directly in most cases. The feature of "partial" class is introduced in .NET Framework 2.0 and .NET Compact Framework 2.0.*

If you choose to create a *strong name key* file (`.snk` file; explained below), it will appear in the project file list in the Solution Explorer. If you want to generate a package for your project that can be deployed to a device, you must add a Smart Device CAB project to your solution. The details of application packaging and deployment are discussed later in this chapter.

Testing and Debugging Applications

The process for testing and debugging a Smartphone application is similar to the process for a desktop .NET application. When testing with the emulator, you are free to add breakpoints to your code, step into or over the code, add watches for variables and expressions, or view call stacks and local variables when the program pauses at the breakpoint.

To begin, move your cursor to the line of `MessageBox.show()`, right-click to bring up the context menu, and then click Insert Breakpoint. (Alternatively, you can press F9 to create a breakpoint.) Next, select Debug⇨Start from the main menu. When the main form appears on the emulator or your device, enter the name (John Doe) again, select a department (Product Dev), check the Contractor checkbox, and then click OK. You will see the program pause before displaying the message box, as shown in Figure 3-10.

At this point, if you move the mouse onto some variables, such as `status`, you will see a small pop-up window with the value of the variable. You can run step by step (line by line of your source code) to debug your program, or select Debug➪Continue from the main menu to run to the next breakpoint (if any) or to the end of the program.

Figure 3-11 shows the Debug menu of Visual Studio 2005 after you start debugging an application. These operations can be divided into three categories: debug control, exception, and breakpoint configuration. Debug control enables you to choose what to do when a certain breakpoint has been reached or a monitored exception occurs. Table 3-3 lists the debug controls you can apply.

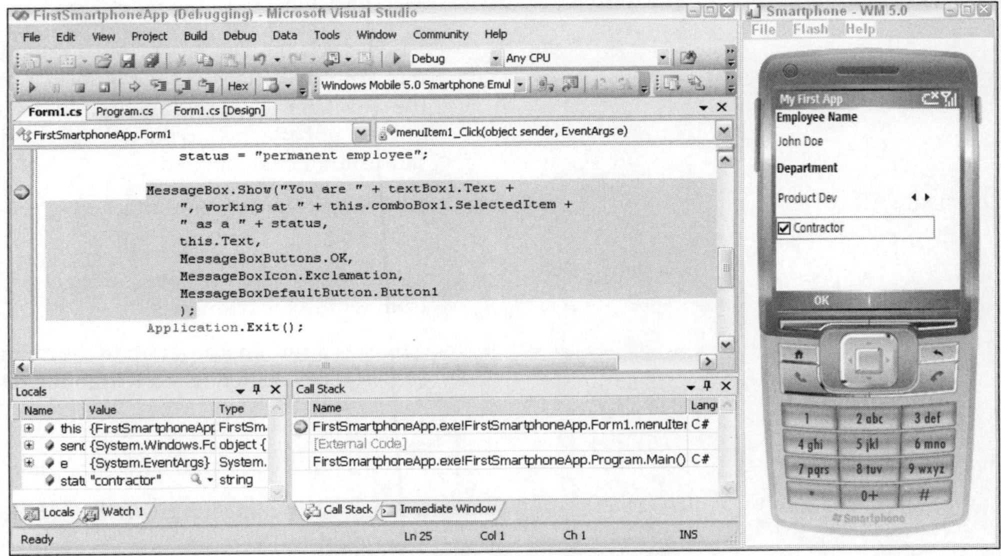

Figure 3-10

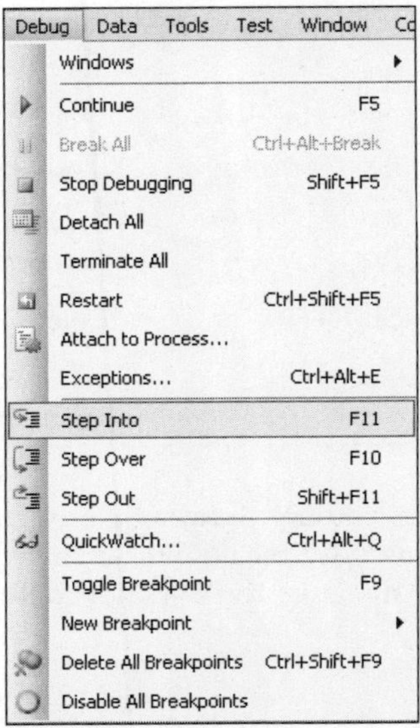

Figure 3-11

Table 3-3 Debug Controls

Debug Control	Keyboard Shortcut	Description
Continue	F5	Continues to run the program until another breakpoint is reached or the program terminates.
Break All	Ctrl+Alt+Break	Breaks the running application.
Stop Debugging	Shift+F5	Exits from the program immediately without executing the remaining part of the program.
Detach All		Disconnects the debugger from a program that you have attached to or launched from the debugger. The program is still running but it is not attached to the debugger, and thus cannot be debugged.
Terminate All		Terminates all programs attached to the debugger.
Restart	Ctrl+Shift+F5	Restarts the program from the beginning from within the debugger.
Attach to Process		Attaches a running program to the debugger so that it can be debugged within the IDE.
Exceptions	Ctrl+D, E	Defines breaks on certain exceptions.
Step Into	F11	Debugs line by line into a function call.
Step Over	F10	Debugs line by line but treats a function call as a single entity.
Step Out	Shift+F11	Gets out of the current function call and returns to its caller.
Quick Watch	Ctrl+Alt+Q	Brings up a window in which you can evaluate a variable or an expression (but not a method call).
Toggle Breakpoint	F9	Disables or enables a breakpoint.
New Breakpoint		Creates a new breakpoint that breaks at a function.
Delete All Breakpoints	Ctrl+Shift+F9	Deletes all breakpoints.
Disable All Breakpoints		Temporarily disables all breakpoints (they are not deleted).

Aside from these debugging controls, you can also drag the execution point to change execution sequence of statements when the program pauses at a breakpoint, as shown in Figure 3-12. This is useful when you want to test a specific code path that a regular run does not go through.

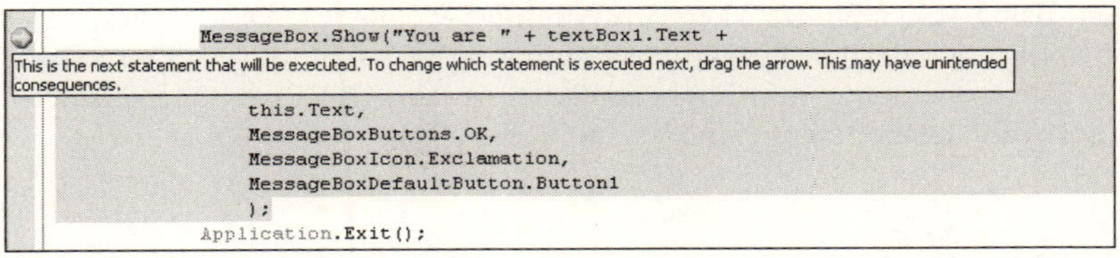

```
MessageBox.Show("You are " + textBox1.Text +
This is the next statement that will be executed. To change which statement is executed next, drag the arrow. This may have unintended consequences.
        this.Text,
        MessageBoxButtons.OK,
        MessageBoxIcon.Exclamation,
        MessageBoxDefaultButton.Button1
        );
Application.Exit();
```

Figure 3-12

The Exceptions command enables you to specify which exceptions to monitor while debugging a program. The program will pause at the `throw` code of the specified exception so that you can quickly identify which part of the code raises the exception. This feature is indispensable because when an exception occurs, you want to see the context in which the exception is raised, rather than the place it was caught. With this feature, you probably must add numerous breakpoints to the `catch` blocks. You can also define your own exception class and monitor it while executing the program.

Breakpoint configuration enables you to set new breakpoints and delete or disable existing breakpoints. When a program pauses at a breakpoint, the IDE presents a number of windows. You can view auto variables in the Auto window, local variables in the Local window, watches in the Watch window, call stacks in the Call Stack windows, and threads in the Threads window. *Auto variables* are variables that have been accessed recently while the program is being executed and debugged. *Local variables* are temporary variables used in a method. *Watched variables* are variables or expressions you want to monitor as a program executes. You can also use the Immediate window to obtain the value of a variable or an expression, or to evaluate a method call. The Call Stack window shows the current call stack, including every method call's parameter type and line number. The Thread window shows a running thread's ID, priority, status, and execution location (which method it is executing). These debugging information windows collectively give you a comprehensive view of the runtime information of a program.

Up to version 2005, Visual Studio does not support the Edit and Continue feature for .NET Compact Framework applications, including Smartphone and Pocket PC applications. The Edit and Continue feature, which is available for .NET Framework applications, enables you to edit the program in a debugging session. The changed program will be automatically recompiled and linked, and you can continue to debug the updated program from where it was.

Debugging a project running on a Smartphone device is also supported by Visual Studio 2005. In fact, you may not see big differences between debugging an application on a Smartphone emulator and on a real Smartphone device, except that you must specify the Smartphone device connected to the development computer via ActiveSync as the deployment device, rather than the emulator.

It is worth noting that debugging on a Smartphone device implies that code is running on the device's processor, not the desktop system's processor.

Device debuggers include the following limitations:

- Function or method evaluation is not supported in the native device debugger (the debugger for native code, rather than managed code). The feature is available for managed code device debugging using the managed device debugger.
- The Edit and Continue feature is not supported on either device debugger (native or managed).
- For applications with a mix of managed code and native code, two different instances of Visual Studio are needed: one for the native device debugger and another for the managed debugger.

As you can see, using Visual Studio 2005 with Windows Mobile 5.0 SDK, you can effectively leverage your desktop application design skills and techniques on Visual Studio 2005 for Smartphone application development. This is one of the most important advantages of .NET Compact Framework–based Smartphone software development using Visual Studio 2005.

Packaging and Deploying Applications

After debugging and testing an application on a real Smartphone device, you may deploy the application on a user's device. There are three stages in the entire deployment procedure:

1. Package the application using a Visual Studio 2005 Smart Device CAB project.
2. Deliver the package to a user's device via a web server, an ActiveSync copy, an e-mail attachment, or a storage card.
3. Install the package onto the device.

The following section focuses on the first stage, packaging the application, as this is a core functionality of Smart Device application support in Visual Studio 2005. You will notice the significance of application security in the domain of Smartphone application development. We then briefly introduce several methods for delivering and installing packages.

Packaging Applications

To package a Smartphone application, you must add a Smart Device CAB (cabinet) project to your Visual Studio 2005 solution. From the main menu, select File➪Add➪New Project. In the Other Project Types category, select Setup and Deployment, and then choose Smart Device CAB Project (*not* CAB Project) from the Visual Studio installed templates, as shown in Figure 3-13. A folder with the name of the Smart Device CAB project will be generated under the solution directory. The IDE also presents you with a File System Editor window that shows the filesystem entries you can create on a target device. A *CAB file* is a type of compressed executable archive file that can contain your application assemblies (EXEs and DLLs), dependencies such as DLLs, resources, help files, and so on, as well as any files related to application security. Most important, an INF (information) file will be generated to describe the destination directory of each installation file on a target device, versions of Windows Mobile for Smartphone on which the application is intended to run, and versions of required .NET Compact Frameworks.

To link the Smart Device CAB project with the FirstSmartphoneApp project, first click Application Folder in the File System Editor window. Then, from the Visual Studio main menu, select Action➪Add➪Project Output, choose the FirstSmartphoneApp project in the Add Project Output Group dialog box, and choose Primary Output from a list of packaging options; thus, only assembly files will be packaged. You can also right-click Application Folder in the File System Editor window to link the Smart Device CAB project with your application project, as shown in Figure 3-14.

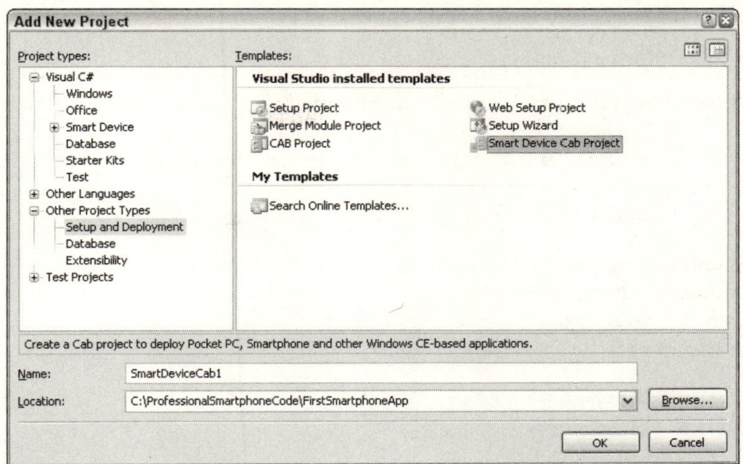

Figure 3-13

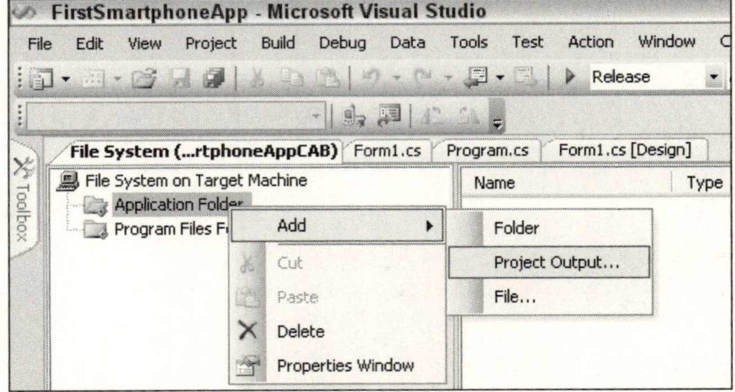

Figure 3-14

Once you add the application project to the Smart Device CAB project, you may choose to add a registry entry to a target device as part of the application's installation. In fact, many software products for Windows will create some registry keys for saving configuration settings. To add a registry key for your company named CoolMobile, open the Registry Editor by selecting View⇨Editors in the main menu. Then, in the Registry Editor window, browse to `HKEY_CURRENT_USER\SOFTWARE\%Manufacturer%` and add a new key with a string value of `CoolMobile`. Now you are ready to build the Smart Device CAB project. This will generate either a Debug folder or a Release folder in the Smart Device CAB project folder, depending on the active configuration of the project. In addition, three files are created in the Smart Device CAB project folder:

- A CAB file (`CABProjectName.cab`)
- An INF file (`CABProjectName.inf`)
- A Cab Wizard log file (`CabWiz.log`) that contains logging information about the packaging process

The following code shows an example of the INF file. When the package is installed on a Windows PC, the information in the INF file will be used.

```
[Version]
Signature="$Windows NT$"
Provider="Default Company Name"
CESignature="$Windows CE$"

[CEStrings]
AppName="FirstSmartphoneAppCAB"
InstallDir=%CE1%\%AppName%

[Strings]
Manufacturer="Default Company Name"

[CEDevice]
VersionMin=4.0
VersionMax=5.99

[DefaultInstall]
CEShortcuts=Shortcuts
AddReg=RegKeys
CopyFiles=Files.Common1

[SourceDisksNames]
1=,"Common1",,"C:\ProfessionalSmartphoneCode\FirstSmartphoneApp\FirstSmartphoneApp\
obj\Debug\"

[SourceDisksFiles]
"FirstSmartphoneApp.exe"=1

[DestinationDirs]
Shortcuts=0,%CE2%\Start Menu
Files.Common1=0,"%InstallDir%"

[Files.Common1]
"FirstSmartphoneApp.exe","FirstSmartphoneApp.exe",,0


[Shortcuts]

[RegKeys]
"HKCU","Software\%Manufacturer%\CoolMobile","","0x00000000","
"HKLM","Software\%Manufacturer%","","0x00000000","
```

If you want to deploy your application onto the end user's device without using Visual Studio 2005, you must manually write an INF file, and then use the command-line tool `CabWizSP.exe` to generate the CAB file. As shown in the preceding code, an INF file is made up of a number of sections, beginning with a string enclosed in a pair of square brackets (`[]`). The following sections are required:

- `Version` — The application provider's name and the application's version
- `CEStrings` — Some system strings
- `CEDevice` — The target platform
- `DefaultInstall` — Sections that define the installation
- `SourceDisksNames` — The path of the source files
- `SourceDisksFiles` — The source files
- `DestinationDirs` — The destination directory on the device

If the application must modify the device registry, a `RegKeys` section is also needed (see the RegKeys section in the preceding INF file where CoolMobile is added).

Signing Applications

Any Smartphone CAB files, application executables, and libraries *must* be digitally signed before being deployed to a user's device. The privileges of both signed and unsigned applications are restricted by the device's security policy. The reason for application signing is to ensure application security. Otherwise, how can you trust an application published on the Internet? Signing solves this problem.

- You can ensure that the program code comes from the entity it claims.
- You can ensure that no one has tampered with the program code.

The security technology behind digital signing is quite sophisticated and is discussed thoroughly in Chapter 12. Here you just need to know that in Visual Studio 2005, you can sign your applications or packages using digital certificates. During the development stage of a Smartphone application, you probably won't want to bother with getting a certificate from the cryptographic service providers (CSP). The Windows Mobile 5.0 SDK provides two sample certificates for day-to-day development: `SDKSamplePrivDeveloper.cer` and `SDKSampleUnprivDeveloper.cer`, each packaged with three other related files. `SDKSamplePrivDeveloper.cer` is a sample of a privileged certificate, which is used to enable the underlying application to run in Privileged mode, in which any APIs can be called. Unprivileged certificates such as the second sample certificate are used to verify applications running in Unprivileged mode, which only allows limited access to APIs.

You need to install the SDK sample certificate onto your development computer and your device; otherwise, you will not be able to use the sample certificate in Visual Studio 2005. To install a sample certificate, on your development computer, browse to the directory of the Windows Mobile SDK (by default, `Program Files\WinCE tools\WCE500\Windows Mobile 5.0 Smartphone SDK\Tools`*). Then double-click the pfx file of each certificate. To install the certificate onto a device or an emulator, you need to use the tool Rapiconfig with the XML provisioning file* `SDKCerts.xml` *in the same directory. The command is* `rapiconfig -p SDKCerts.xml`.

Digital certificates are saved in certificate stores on a Smartphone device.

Smartphone Security Model and Security Policies

Windows Mobile for Smartphone supports two types of security models: one-tier security and two-tier security. The one-tier model dictates that signed applications with a certificate are generally trusted, can run on the device, and have privileged access to the device (Privileged mode). For unsigned applications, the device's security policies will be checked to determine two things: 1) whether it can run on the device, with or without prompting the user; 2) if it can run, then in which mode (Privileged or Unprivileged).

The two-tier model further differentiates the access levels of signed applications. Unlike the one-tier model, not all signed applications can run in Privileged mode. Only those signed applications with a certificate chain that maps to a root certificate in the privileged store can run in Privileged mode. Signed applications with a certificate chain that maps to a root certificate in the unprivileged store can only run in Unprivileged mode; thus, they have only limited access to device resources and system APIs.

All security policies can be defined in a security policy provisioning XML file and copied to the device for security policy configuration. For development and testing purposes, the Windows Mobile SDK provides a number of XML files for device security configuration, including `OneTierLocked.xml`, `OneTirePrompt.xml`, `SecurityOff.xml`, `TwoTeirLocked.xml`, and `TwoTierPrompt.xml`, in the directory `Program Files\Windows CE Tools\wce500\Windows Mobile 5.0 Smartphone SDK\Tools\Securityconfiguration`. These files can be used to generate corresponding cpf (CAB Provisioning Format) files, which are compressed and signed using two command-line tools: `makecab.exe` and `Cabsigner.exe`. Each of these files contains numerous policy settings you can modify. For details about setting security policies, see Chapter 12.

Project Signing and Assembly Signing

After you have installed the certificates onto your development computer, you will be able to use them in Visual Studio 2005. It is recommended that a Smartphone application should have its assemblies (if it is a managed project), EXEs, DLLs, CABs, and MUI (Multilingual User Interface) files signed before they are delivered to the end user.

To sign the application files in a project, open the project's Properties page (Project⇨*(project name)* Properties), click the Devices page, and then check the Authenticode Signature box. If you have already selected some certificates for the project, you should see them in the text box below the checkbox. Otherwise, click Select Certificate and choose the certificates you installed on your development computer. Figure 3-15 shows the details of the two sample certificates, `SDKSamplePrivDeveloper.cer` and `SDKSampleUnprivDeveloper.cer`. You can also import, export, or remove installed certificates by clicking the Manage Certificates button. The selected certificate will appear on the Devices tab of the Properties page.

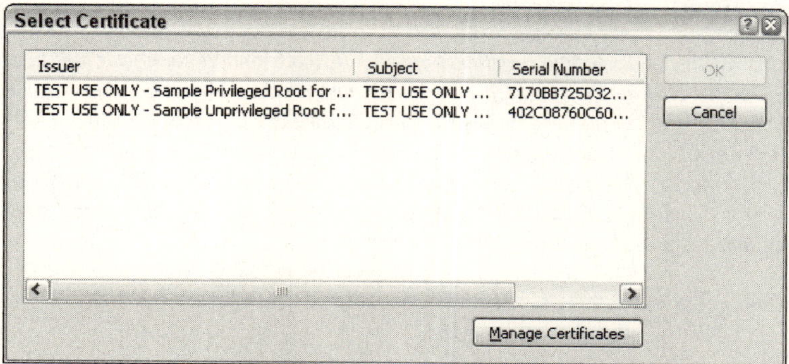

Figure 3-15

On the same Devices tab, you may also specify whether the selected certificate will be added to the target device, a procedure known as *provisioning*. After that, when you build your project, the application files will be signed using the selected certificate. CAB files can also be signed by changing the Smart Device CAB project properties.

A single assembly file can also be signed using a strong name key file. A key file contains a public key and the corresponding private key. A strong name consists of the assembly's identity, a public key, and a digital signature. To generate a strong name for an assembly, you must choose a key file. In Visual Studio 2005, you can create a key file in the project's Properties page (in the Signing page). Alternatively, you can use the command-line tool `sn.exe` (in `Program Files\Microsoft Visual Studio 8\SDK\v2.0\Bin directory`) to create a key file. Once you create a key file in Visual Studio 2005, it will appear in the file list of Solution Explorer.

Delivering and Installing Applications

You can deliver an application package to a user's Smartphone device in a number of ways, including ActiveSync, a web download, an e-mail attachment, Service Indication (SI) SMS (Short Messaging Service) or Service Loading (SL) SMS, or storage cards. The SMS methods for application package delivery take advantage of commonly used text messaging services to send a download URL as a text message to a user. Visual Studio 2005 does not provide a simple solution to application delivery. It is up to the developer or the mobile network operator to select which method to use. The following is a summary of these methods:

- **ActiveSync.** A user's Smartphone can be connected (docked in a cradle) to a desktop computer via a USB or serial port. Once a partnership relationship is established via the ActiveSync program, the Smartphone can synchronize e-mail, contacts, calendars, pictures, and so on with the desktop computer. Usually an application's CAB file is further packaged using an installation setup tool such as InstallShield. To transfer the setup package of an application to the Smartphone, a user needs to run a setup program of the application on a desktop computer, which performs some compatibility checking and copies the CAB file to the Smartphone device.
- **Web download.** Users can use the Pocket Internet Explorer on the Smartphone to visit a website and download the application. The URL of the site can be obtained from e-mails SMS, or other means.
- **E-mail attachment.** Mobile network operators can send an e-mail to users with an attachment of the application. When a user opens the attachment, the application is installed on the device.
- **Service Indication (SI) and Service Loading (SL) messages.** Mobile network operators can send these special SMSs to users, which contain a link to an application download website.
- **Storage cards.** The CAB file can be put onto a storage card such as a CompactFlash card or a MultiMedia Card (MMC). When these cards are inserted into a Smartphone, the application package will be installed automatically, provided that there is an Autorun file and that the device's security allows you to run it.

Regardless of which method is used for application delivery, internally the application installation procedure always starts with an on-device program called `wceload.exe`. Once a user clicks the downloaded package on a Smartphone, `wceload.exe` will be launched automatically. It checks the package being loaded on the device against the security polices and determines whether the package can be installed or whether the user needs to be prompted for the installation.

Summary

This chapter demonstrated how you can use Visual Studio 2005 in association with the Windows Mobile 5.0 SDK to develop a simple Smartphone application in a Smart Device project. The Visual Studio IDE provides an easy-to-use Form Designer that enables you to drag and drop .NET Compact Framework windows form controls. Event handling of these controls is similar to desktop Windows form application development. The debugging features have been made available to Smart Device development using either a Smartphone emulator or a Smartphone device. Moreover, the packaging and deployment of a Smartphone application in Visual Studio 2005 has been enhanced with Authenticode signing for CAB files and individual assemblies.

You now have a general overview of managed Smartphone application development using Visual Studio 2005. This concludes the first part of "Smartphone and .NET." Beginning with the next chapter, we will move on to Part II "Smartphone Application Development," which discusses a number of Smartphone development topics, including UI handling, data storage and file I/O, data access with SQL Server Mobile, networking, e-mail and messaging, XML and web services, Platform Invoke, and error handling and debugging.

4

User Interface and Input

In the previous chapter, you learned how to write, debug, and deploy a simple Smartphone program using Visual Studio 2005. You might think that developing a Smartphone application differs little from coding a PDA application or even a desktop program. Well, it depends on how you look at it. On the surface, all you have to do is drag a control to the form, modify the control's properties, and then add classes, methods, and events — a procedure familiar to veteran Windows programmers. Sooner or later, however, you will reach a point where conventional programming techniques do not apply directly to the Smartphone platform. For example, some controls are not supported, and there is no keyboard, mouse, or stylus for user input. You then have to examine what is available in the Windows Mobile 5.0 SDK and find a solution to work around any awkward situations.

This chapter discusses the following:

- Controls in the Smartphone SDK and their impact on UI design
- Smartphone UI design
- A detailed explanation of user input
- UI-related topics such as auto-save mode, DPI, and performance

UI Design with Forms and Controls

The term *forms* refers to the windows of .NET-based applications. The responsibilities of forms include displaying information and receiving user input through events. To create a Smartphone form in Visual Studio 2005, choose the New Project Wizard and then choose the C# application type. The template you are going to use is Device Application. The Form Designer in Visual Studio 2005 enables you to add controls to the form simply by dragging them from the toolbox and dropping them on the forms at design time.

Supported Controls

When a Windows Mobile–based Smartphone device application project is open in Visual Studio 2005, a simple way to see the available controls is to examine the Toolbox window in the Visual Studio IDE, as shown in Figure 4-1. If for any reason the Toolbox window is not shown on the IDE, you can open it by clicking View⇨Toolbox from Visual Studio 2005.

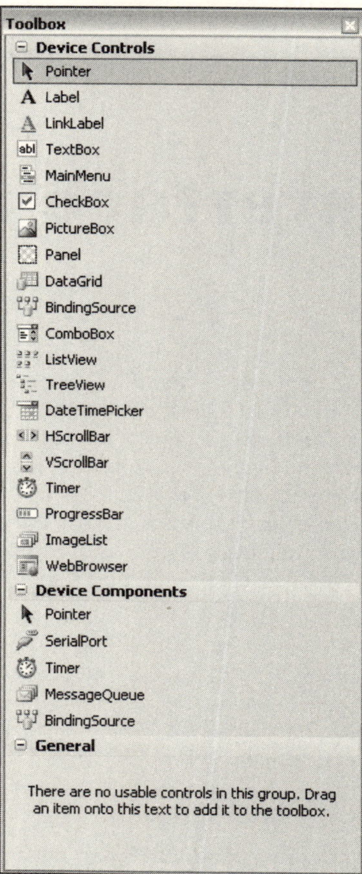

Figure 4-1

A number of popular controls are not available in the Smartphone SDK, such as Buttons, ListBox, and OpenFileDialog, which could be a little inconvenient for programmers. However, don't assume by looking at the Toolbox that these are all the controls Visual Studio 2005 provides. You can also use the Project menu to add other controls, including Form, MessageBox, and Cursor. In addition, compared to Smartphone 2003, Windows Mobile 5.0 supports four more controls: LinkLabel, BindingSource, DateTimePicker, and WebBrowser. To avoid any confusion, Table 4-1 lists all the supported controls in Windows Mobile 5.0.

Table 4-1 Controls in Windows Mobile 5.0-based Smartphone

Control	Shown in Toolbox	Supported in Smartphone 2003
BindingSource		
CheckBox	✓	✓
ComboBox	✓	✓
Cursor		✓
DataGrid	✓	
DateTimePicker	✓	
Form		✓
HScrollBar	✓	✓
ImageList	✓	✓
Label	✓	✓
LinkLabel	✓	
Panel	✓	✓
Menu	✓	✓
MessageBox (with 1 or 2 buttons only)		✓
PictureBox	✓	✓
ProgressBar	✓	✓
Screen		✓
TextBox	✓	✓
Timer	✓	✓
VScrollBar	✓	✓
WebBrowser	✓	

For detailed information about the controls in the .NET Compact Framework 2.0, visit the Microsoft MSDN website at `http://msdn2.microsoft.com/en-us/library/hf2k718k.aspx`.

Control Behaviors

Even though Windows Mobile 5.0 controls inherit their properties, methods, and events (PME) from the .NET System.Windows.Forms.Control class, not all the PMEs defined in the base class are implemented because of the limited computing power and memory space available to the Smartphone devices. For example, the Maximum and Minimize buttons of a form are not supported in the Smartphone SDK, which makes sense for Smartphone devices because the size of the screen is small and pointing devices are not available to manipulate the windows.

Conversely, Windows Mobile 5.0 for Smartphone supports more control features than its predecessors. Tab order and auto scroll, which are missing in Smartphone 2003, are now available in Windows Mobile 5.0 for Smartphone. Windows Mobile 5.0 also introduces new features that enable applications to change screen orientations.

Tab Order

Two important properties, `TabStop` and `TabIndex`, are not supported prior to Windows Mobile–based Smartphone forms. That means you cannot change the tab order during design time and runtime. Technically, tab order does not exist in Smartphone applications because there is no Tab key on Smartphone devices. However, when users navigate the different controls using the navigation keys, such as up arrow key and down arrow key, there should be a way to determine the order in which controls gain focus. In this book, this order is denoted as *tab order*.

You can set the tab order when you design the Smartphone forms. An intuitive way to do so is to choose View➪Tab Order from Visual Studio 2005. Note, however, that the Tab Order option is enabled only when the active window is the Form Designer.

For example, you can start a new Windows Mobile–based Smartphone device application by choosing File➪New➪Project. Then, in the Form Designer, add three text boxes to the form by dragging and dropping three TextBox controls from the Toolbox to the form. By selecting View➪Tab Order, you can see that three TextBox controls are marked with numbers in the tab order editor, as shown in Figure 4-2.

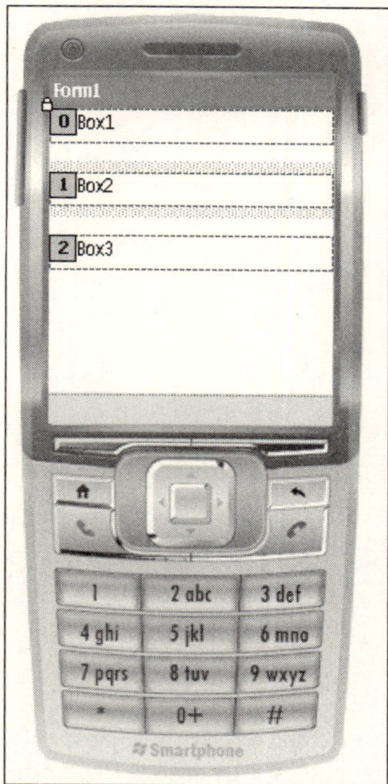

Figure 4-2

The numbers you see in the tab order editor are the values of the `TabIndex` properties of those three `TextBox` controls. In Figure 4-2, the `TabIndex` property of `textBox1` is value 0, which is the smallest among all three controls. It means that at runtime `textBox1` will be the first to receive focus, followed by `textBox2` and `textBox3`.

To change the tab order, simply click the numbers in the tab order editor. For example, if you click the number 1 on the left side of `textBox2`, then the `TabIndex` of `textBox2` is changed to value 2. It is then changed to value 0 in the next click. If you click again, the value will be changed back to value 1.

Alternatively, you can change the value of the `TabIndex` property from the Properties window by clicking View⇨Properties Window or by pressing Ctrl+W+P. Figure 4-3 shows the properties of `textBox2`, and you can set the value of the `TabIndex` property from there.

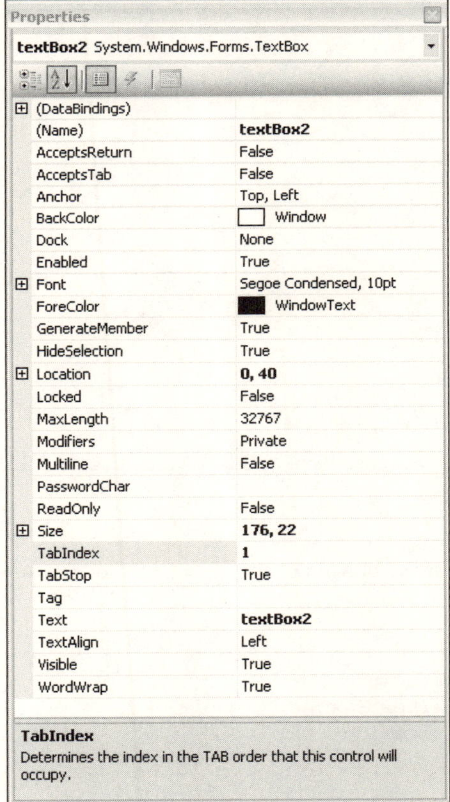

Figure 4-3

What if all the controls in a form have the same `TabIndex`? What determines which control gets the focus first and which control gets the focus last? For a Windows Mobile–based Smartphone, it depends on how the control is added to the form's control collections.

Continue from the previous example and change the `TabIndex` value of all three text box controls to 0. Then choose View➪Solution Explorer. Next, expand the `Form1.cs` from the Solution Explorer and right-click `Form1.designer.cs`, which is the code automatically generated by the Form Designer, and choose View Code. Note that part of the code may be hidden in the Visual Studio IDE; you need to click the plus (+) sign on the left of the Windows Form Designer–generated code to expand the code and make it visible, as indicated in Figure 4-4.

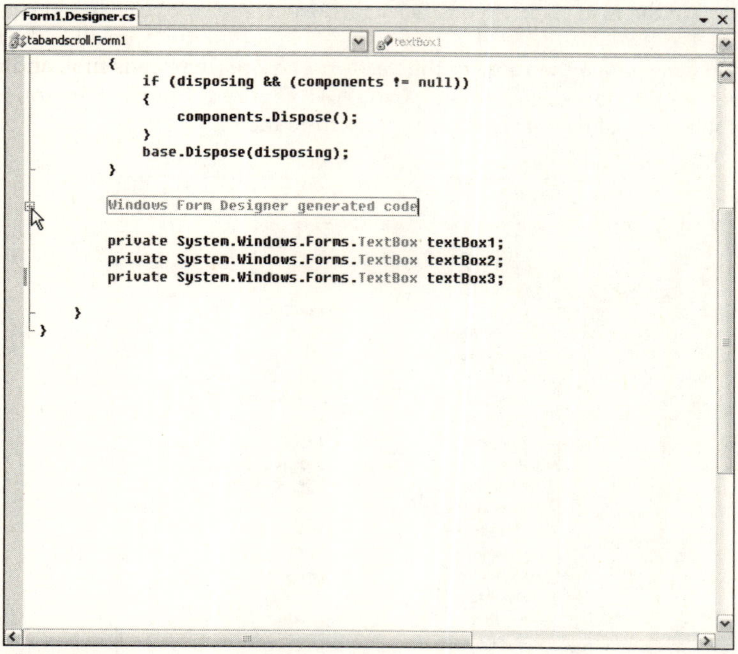

```
{
    if (disposing && (components != null))
    {
        components.Dispose();
    }
    base.Dispose(disposing);
}

Windows Form Designer generated code

private System.Windows.Forms.TextBox textBox1;
private System.Windows.Forms.TextBox textBox2;
private System.Windows.Forms.TextBox textBox3;

    }
}
```

Figure 4-4

When you examine the code at the bottom of the `InitializeComponent()` function, you find the following:

```
...
this.Controls.Add(this.textBox3);
this.Controls.Add(this.textBox2);
this.Controls.Add(this.textBox1);
...
```

If you build the sample code by pressing F6 and then press Ctrl+F5 to start the application, then you will notice that at run time `textBox3` is the first to receive the focus. If you press the down arrow key, then `textBox2` is the next to get focus, and `textBox1` is the last to get focus. Indeed, when controls have the same `TabIndex` value, the tab order is exactly the same as the order in which they are added to the control collections, which is `this.Controls` in the example.

Note that by default, the From Designer in Visual Studio adds controls to the control collection in reverse order. In the preceding example, `textBox1` is the first added to the form from Form Designer, but it is the last added to the control collection of the form. That explains why in Smartphone 2003, in which `TabIndex` is not supported, the tab order is the reverse of the order in which you add controls to the form from the Form Designer.

You can change the tab order of controls that have the same `TabIndex` values (or do not have `TabIndex` values as in Smartphone 2003) by reordering them in the `InitializeComponent()` function. For instance, if the `TabIndex` properties of those three TextBox controls are all `0`, the following code in the `InitializeComponent()` function will make `textBox2` receive focus first, and `textBox1` receive focus last:

```
...
this.Controls.Add(this.textBox2);
this.Controls.Add(this.textBox3);
this.Controls.Add(this.textBox1);
...
```

Auto Scroll

Prior to Windows Mobile 5.0, Smartphone forms do not have the `AutoScroll` property. As a result, when the size of a form is bigger than the screen, the out-of-area controls are never displayed, regardless of how a user navigates the form. The `AutoScroll` property is now supported in Windows Mobile 5.0 Smartphone, albeit with a minor problem: If a control on the top of the form does not receive cursor focus — for example, a Label control — it will not show up on the screen again after it is navigated out of the screen.

To verify this problem, start a new Smartphone device application. Then, from the Properties window, change the size of the form to (176, 320) and make sure that the `AutoScroll` property is enabled (see Figure 4-5). Now add four Label controls and four TextBox controls to the form. From the Properties window, set the `Location` properties of those four labels to (0, 0), (0, 80), (0, 160), and (0, 240), respectively. Then set the `Location` properties of the four text boxes to (0, 40), (0, 120), (0, 200), and (0, 280), respectively. The `Location` property sets or gets the coordinates of the coordinates of the upper-left corner of the control. Figure 4-6 shows this simple UI interface.

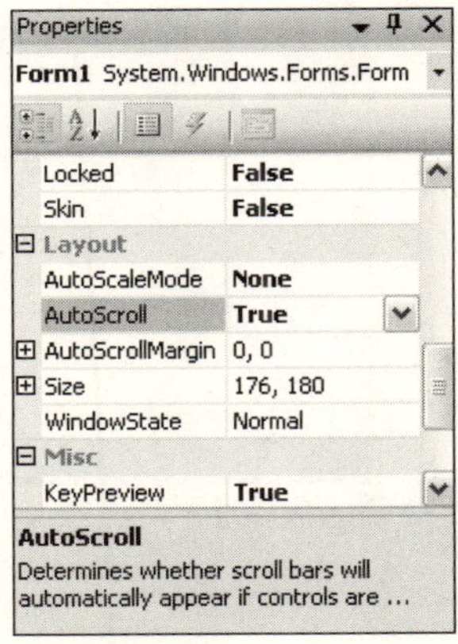

Figure 4-5

Now you are ready to test the default auto scroll behavior. Press F6 to build the solution and press Ctrl+F5 to launch the application. The form is displayed nicely when the program starts and `label1` is on the top of the screen, followed by the `textBox1`. However, when you navigate the form to `textBox4` by pressing the down arrow key, and then navigate back to `textBox1` by pressing the up arrow key, `label1` is now missing and `textBox1` appears on the top, as indicated in Figure 4-7.

This is a problem because a Label control is usually needed right before a TextBox control to give users hints as to what to input. In addition, because there are no pointing devices in Smartphone to move the scroll bar, once `label1` is navigated out of the screen users cannot get it back on the screen by pressing the navigation keys. This means that users may not be able to recall what `textBox1` refers to. As an example, imagine that the `Text` property of `label1` is a question like 1+1 = ?, and users are supposed to input their answers in `textBox1`. If `label1` is missing, how could users check their answers?

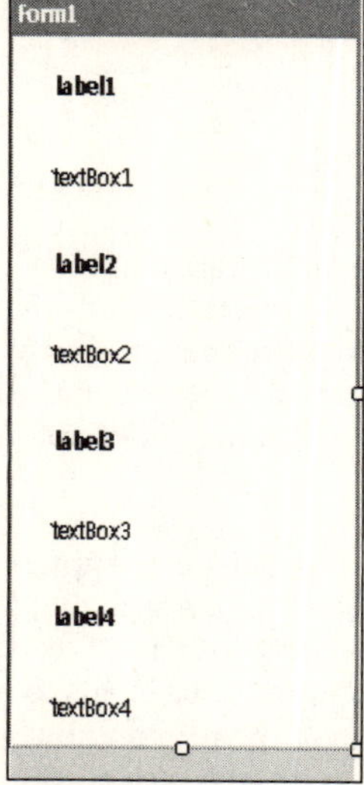

Figure 4-6

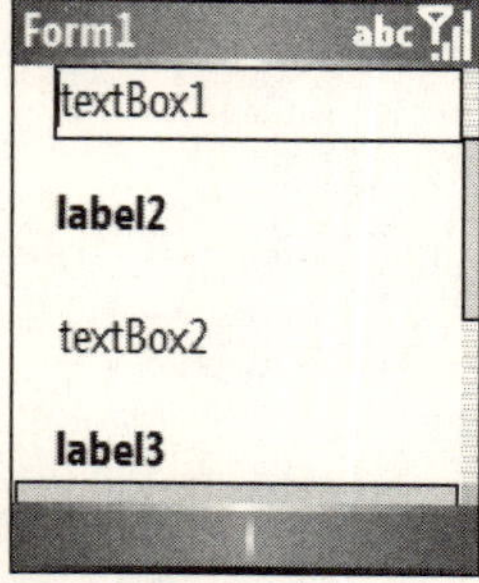

Figure 4-7

Because the auto scroll feature is not perfect on Windows Mobile 5.0, and it is missing on Smartphone 2003, we are going to present a solution that works on both platforms.

The idea is to add a Panel control and a VScrollBar control to the form and have the Panel contain all the other controls, such as Labels and TextBoxes. Whenever a control gets focus, change the `value` property of the VScrollBar and the `Top` property of the Panel accordingly to ensure that the active controls appear in the display area. For example, suppose a Panel control has a height of 320 pixels and the viewable area of a Smartphone device has a height of 180 pixels; if the `Top` property of the Panel is 0, the area of the Panel from 0 to 180 will be displayed. If the `Top` property of the Panel becomes –100, the area displayed on the screen will be from 100 to 280.

Start a new Smartphone device application from Visual Studio 2005 by choosing File⇨New⇨Project, and name the project **FormScroll**. Then click the Form1.cs[Design] tab design UI. On the Properties window, change the `AutoScroll` property of `Form1` to `false` and set the `Size` property to (`176, 320`). Next, add a VScrollbar control `vScrollbar1` to `Form1`. You don't need to set any properties of `vScrollbar1` at this point. You can adjust the settings later in your code so that settings can be applied during the application initialization process. Then add a Panel control `Panel1` to `Form1`. Set the `AutoScroll` property of `Panel1` to `false`, the `Location` property to (`0,0`), and the `Size` property to (`176,320`). Then add four pairs of Labels and Textboxes to `Panel1` and set the `Location` properties of the Labels and Textboxes the same way you did in the previous example. The resulting UI is very similar to what is shown in Figure 4-6. The difference is that whereas Labels and Textboxes are added directly to `Form1` in the previous example, they are now nested in `Panel1`, and `Panel1` is directly contained in `Form1`. Another difference is that a VScrollbar control is added to the UI.

From the Solution Explorer (choose View⇨Solution Explorer), right-click `Form1.cs` and choose View Code. You can now set the necessary properties of `vScrollBar1` in the `Form1` constructor, as follows:

```
public Form1()
{
    InitializeComponent();

    //Set the initial size, position, and value of VScrollbar1
    this.vScrollBar1.Height = this.ClientSize.Height;
    this.vScrollBar1.Left = this.Width - vScrollBar1.Width;
    this.vScrollBar1.Minimum = 0;
    this.vScrollBar1.Maximum = this.panel1.Height - this.ClientSize.Height + 5;
    this.vScrollBar1.Value = 0;

}
```

The preceding code makes the `Height` of `vScrollbar1` the same as the `ClientSize` of a form. By doing so, your application will behave correctly on both regular Smartphone devices and QVGA-enabled devices (QVGA stands for Quarter VGA and has an image size of 240 × 320 pixels). The code also defines the `Maximum` property of `vScrollBar1` as the height difference of `panel1` and `ClientSize` plus `5`. Therefore, when the screen scrolls, the `Top` property of `panel1` can be calculated as the `Value` property of `vScrollBar1` multiplied by `-1`.

Assuming the height of the ClientSize of Form1 is 180 pixels, the preceding code enables the Value of vScrollbar1 to change from 0 to 145. Therefore, the Top of panel1 can go from 0 to –145, which means you can move panel1 up by 145 pixels to show the area of panel1 from the height of 145 pixel to the height of 325 pixel. An extra 5 pixels are added here to provide a bit of room.

When users press the navigation keys and move the focus to the next control, the desired behavior is that the Textbox that receives the cursor focus and its preceding description Label are both displayed on the screen. To this end, you need to know the positions of the Label and Textbox, and the current scroll position. If the desirable display area is outside the current scroll position, change the Top of Panel1 to cause panel1 to move according. If the desirable display area is inside the current scroll position, simply do nothing.

Assuming topCtrl represents the Label and bottomCtrl represents the Textbox, the following code inside the function SetPos() can find the desirable display area:

```
//Set the position of vScrollbar1 and panel1
private void SetPos(Control topCtrl, Control bottomCtrl)
{
    //Get the desired displaying area
    int top    = topCtrl.Top;
    int bottom = bottomCtrl.Bottom;
...
}
```

For example, the Location of label3 is (0,160), the Location of textBox3 is (0,200), and the Size of textBox3 is (172,22). If you pass label3 and textBox3 to SetPos(), top will be 160, and bottom will be 222 (which is 200+22). This indicates you need to ensure that the area of panel1 from 160 to 222 appears on the screen.

The current scroll position, or the top of current viewable area, is indicated by the Value property of vScrollbar1. You can find the viewable area by using the following code:

```
    //Get current viewable area
    int screenTop    = this.vScrollBar1.Value;
    int screenBottom = ScreenTop + this.ClientSize.Height;
```

If the desirable display area falls within the current scroll position, then pos should be less than screenTop, and bottom should be less than screenBottom. For instance, if the current scroll position is 100, then the viewing area is from 100 to 280. In order to display label3 and textBox3, which is from 160 to 222, you do not have to do anything because 100 is less than 160, and 222 is less than 280. The following code illustrates how to the handle this situation:

```
    //If the bounds are within current view, do nothing
    if (screenTop < top && bottom < screenBottom ) return;
```

If screenTop is less than top, the viewing area is below the desirable display area, and the screen needs to be scrolled up. The can be done by setting the Top of panel1 to the value of top multiplied by –1 as follows:

```
    //If the bounds are above view, scroll up
    if (top < screenTop)
```

```
    {
        this.vScrollBar1.Value = top;
        this.panel1.Top = -this.vScrollBar1.Value;
        return;
    }
```

Assuming the height of the desirable viewing area is less than the height of the `ClientSize` of `Form1`, the only case left is when the desirable viewing area is below the view. The following code shows how to scroll down:

```
    //The bounds are below view, scroll down
    this.vScrollBar1.Value = bottom - this.ClientSize.Height + 5;
    this.panel1.Top = -this.vScrollBar1.Value;
    return;
```

Putting the previous example code together, you now have a `SetPos()` function that can adjust `panel1` to ensure that the two controls `topCtrl` and `bottomCtrl` are both displayed on the screen:

```
//Set the position of VScroll bar and Panel
private void SetPos(Control topCtrl, Control bottomCtrl)
{
    //Get the desired displaying area
    int top = topCtrl.Top;
    int bottom =bottomCtrl.Bottom;

    //Get current viewable area
    int screenTop = this.vScrollBar1.Value;
    int screenBottom = screenTop + this.ClientSize.Height;

    //If the bounds are within current view, do nothing
    if (screenTop < top && bottom < screenBottom ) return;

    //If the bounds are above view, scroll up
    if (top < screenTop)
    {
        this.vScrollBar1.Value = top;
        this.panel1.Top = -this.vScrollBar1.Value;
        return;
    }

    //The bounds are below view, scroll down
    this.vScrollBar1.Value = bottom - this.ClientSize.Height + 5;
    this.panel1.Top = -this.vScrollBar1.Value;
    return;
}
```

The next thing you need to consider is when to call this `SetPos()` method. At runtime, when the cursor moves into a control, the Windows Mobile–based Smartphone operating system will raise a `GotFocus` event. Therefore, a good place to call the `SetPos()` method is inside the `GotFocus` event handler.

The following steps describe how to create a `GetFocus` event handler for `textBox1` in Visual Studio 2005:

1. Make the Form Designer the current active window by clicking the Form1.cs[design] tab in Visual Studio 2005.
2. From the drop-down list of the Properties window, choose textBox1 and click the Events button, as illustrated in Figure 4-8.
3. Double-click the `GotFocus` event (see Figure 4-9). Visual Studio 2005 will automatically create a `textBox1_GotFocus()` event handler.

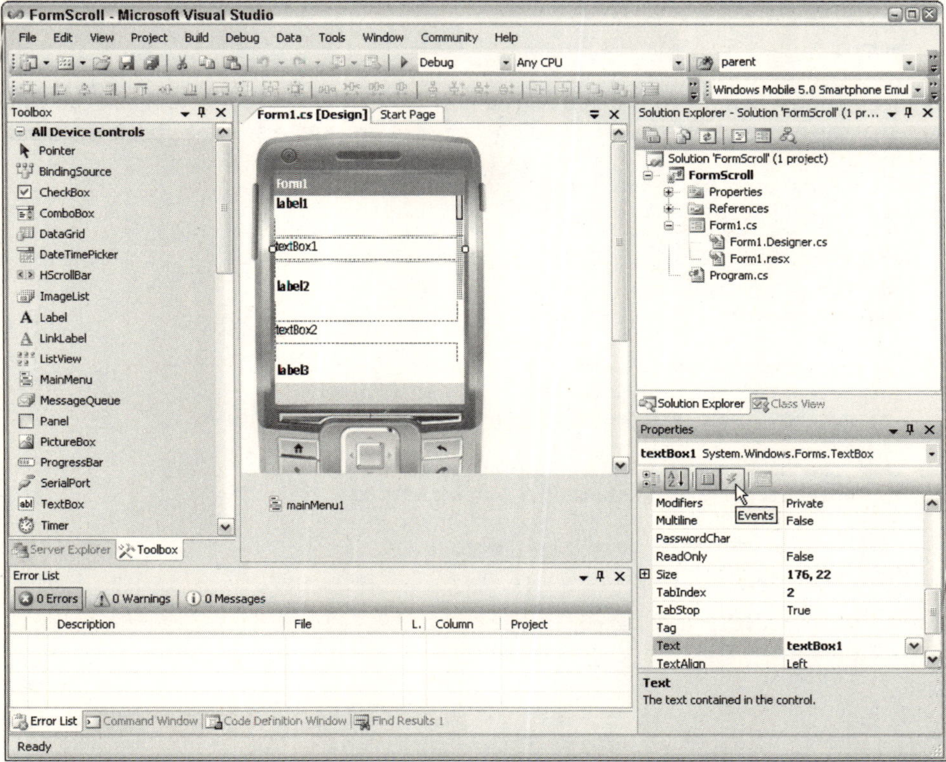

Figure 4-8

The code generated by Visual Studio 2005 is separated in two files. In the `Form1.designer.cs` file, the following code is added in the `InitializeComponent()` method:

```
this.textBox1.GotFocus += new System.EventHandler(this.textBox1_GotFocus);
```

The preceding code means the event `this.textBox1.GotFoucs` is registered to the system event delegate `System.EventHanlder`, which links the event to the event handler `this.textBox1_GotFocus`.

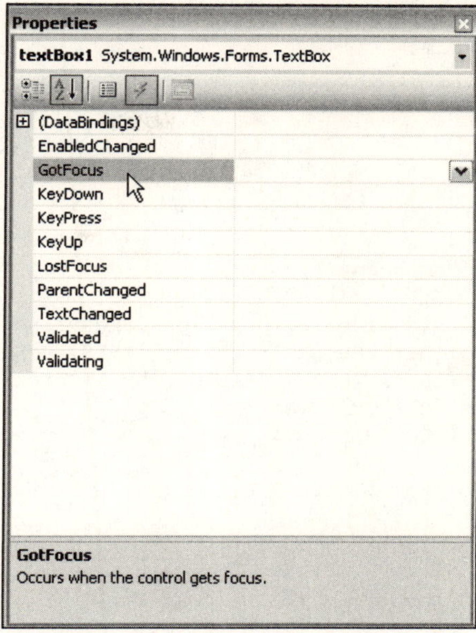

Figure 4-9

In the file `Form1.cs`, the event handler is automatically generated and defined as follows:

```
private void textBox1_GotFocus(object sender, EventArgs e)
{
}
```

The parameter `sender` represents the source of the event, and parameter `e` is an instance of `EventArgs`, which can be used to contain event data. For the sample application, when `textBox1` receives the cursor focus, you need to simply call the `SetPos()` method and pass `label1` and `textBox1` as the two arguments. The following code illustrates the event handler when `textBox1` gets the focus:

```
private void textBox1_GotFocus(object sender, EventArgs e)
{
    SetPos(label1,textBox1);
}
```

Following the same approach, add `GotFocus` event handlers for `textBox2`, `textBox3`, and `textBox4`, respectively.

The following is the complete code of `Form1.cs`:

```
using System;
using System.Collections.Generic;
using System.ComponentModel;
using System.Data;
using System.Drawing;
```

```
using System.Text;
using System.Windows.Forms;

namespace FormScroll
{
    public partial class Form1 : Form
    {
        public Form1()
        {
            InitializeComponent();

            //Set the initial size, position, and value of VScrollbar1
            this.vScrollBar1.Height = this.ClientSize.Height;
            this.vScrollBar1.Left = this.Width - vScrollBar1.Width;
            this.vScrollBar1.Minimum = 0;
            this.vScrollBar1.Maximum = this.panel1.Height - this.ClientSize.Height
+ 5;
            this.vScrollBar1.Value = 0;

        }

        //Set the position of VScrollbar and Panel
        private void SetPos(object topCtrl, object bottomCtrl)
        {
            //Get the desired displaying area
            int top = topCtrl.Top;
            int bottom =bottomCtrl.Bottom;

            //Get the current viewable area
            int screenTop = this.vScrollBar1.Value;
            int screenBottom = screenTop + this.ClientSize.Height;

            //If the bounds are within current view, do nothing
            if (screenTop < top && bottom < screenBottom ) return;

            //If the bounds are above view, scroll up
            if (top < screenTop)
            {
                this.vScrollBar1.Value = top;
                this.panel1.Top = -this.vScrollBar1.Value;
                return;
            }

            //If the bounds are below view, scroll down
            this.vScrollBar1.Value = bottom - this.ClientSize.Height + 5;
            this.panel1.Top = -this.vScrollBar1.Value;
            return;
        }

        //Scroll for textBox1
        private void textBox1_GotFocus(object sender, EventArgs e)
        {
            SetPos(label1, textBox1);
        }

        //Scroll for textBox2
```

```
        private void textBox2_GotFocus(object sender, EventArgs e)
        {
            SetPos(label2, textBox2);
        }

        //Scroll for textBox3
        private void textBox3_GotFocus(object sender, EventArgs e)
        {
            SetPos(label3, textBox3);
        }

        //Scroll for textBox4
        private void textBox4_GotFocus(object sender, EventArgs e)
        {
            SetPos(label4, textBox4);
        }

    }
}
```

Now build the application by pressing F6, and then start the application by pressing Ctrl+F5. The scroll bar is now positioned in the correct place when you scroll down to the bottom of the form and then up to `textBox1` again, as displayed in Figure 4-10.

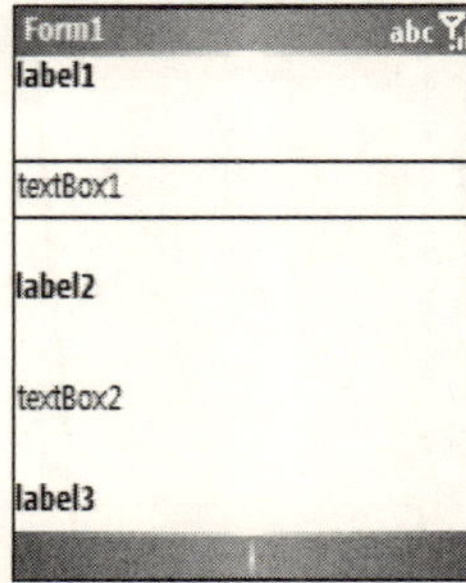

Figure 4-10

In the .NET Compact Framework, controls that have user interaction can receive focus. In Smartphone applications, the following controls can receive focus:

- CheckBox
- ComboBox
- Form
- ListView
- TextBox
- TreeView

For a Smartphone application, the first form you created is the first focusable control in `Controls.Collection`. However, the form is *not* the first control to receive focus because a form is a container control and the .NET Compact Framework will recursively search for focusable controls inside a container until one is found. A control with its `Enabled` property set to `false` will not be considered in this process.

The vertical scroll bar in this example is not as useful as it could be in .NET desktop editions or PDA applications because users are not likely to be able to manipulate the scroll bar directly from Smartphone devices. We added this control in the example mainly for two reasons. First, it is consistent with the UI style of Windows applications. Second, it indicates to users that the form is longer than the length of the screen.

Menus

The MainMenu control is also available in the .NET Compact Framework to enable you to easily create menus for your mobile applications. As you may have noticed, you can add only two menu items at the top level, which is largely due to the small display size of Smartphone devices. All other features look very similar to their full .NET counterparts.

By default, a MainMenu control is automatically added to your project when you create a new Smartphone device application in Visual Studio 2005. You can then add MenuItem controls from the top-left of the menu, as shown in Figure 4-11.

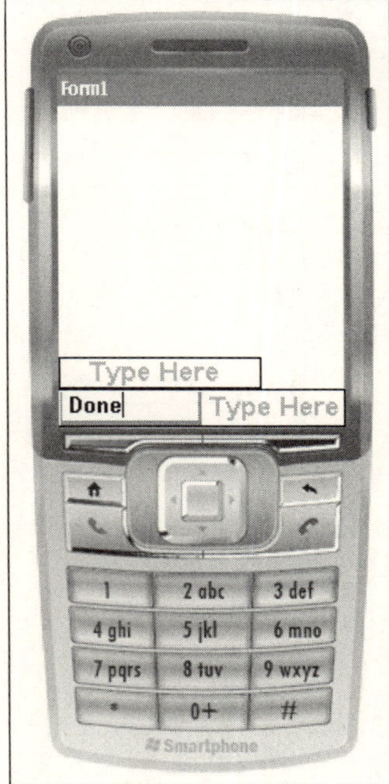

Figure 4-11

Historically, Smartphone devices didn't support submenus to the top-left menu and supported up to two levels of submenus to the top-right menu. Even though such restrictions no longer apply in Windows Mobile 5.0, we believe it is still a good idea to add submenus only to the top-right menu so that the menu style is consistent with Smartphone devices prior to Windows Mobile 5.0. Therefore, we recommend that you create the top-left menu first, followed by the top-right menu, and then stack submenu items to the top-right menu.

For Smartphone applications, submenu items are numbered based on the order in which they are added to the parent MenuItem control. For instance, from the Form Designer, if you add three MenuItem controls (R1, R2, and R3) to the top-right menu, the numbers 1, 2, and 3 are automatically assigned to them, as shown in Figure 4-12.

At runtime, users can select a menu item by first clicking the right soft key followed by clicking the navigation keys to pick a specific menu item. Alternately, users can simply use the number to the left of a submenu item to invoke it. For example, if users want to choose menu item R2, they can simply press 2 on the keypad.

The number assigned to a menu item serves as a shortcut key. This is different from how a shortcut key is constructed in the full .NET Framework, in which menu items are never numbered. To create a shortcut key for a menu item in the full .NET Framework, put an ampersand (&) to the left of the shortcut key from the Form Designer. For example, if the `Caption` of a menu item is `&Write`, then the menu item will be presented as `Write` during runtime and can be invoked when users press the W key. For Smartphone applications, it is still possible to use the & symbol to create a shortcut key for a menu item, but it is not practical to use and is unnecessary.

Just like other controls in the .NET Compact Framework, a MenuItem control can be added to the form during design time as well as runtime. The next example illustrates some basic operations related to the menu operation in a Smartphone application.

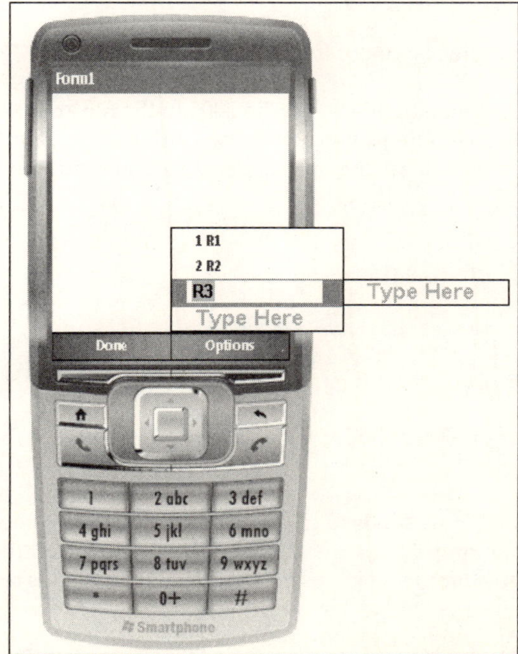

Figure 4-12

Start a new Windows Mobile 5.0–based device application and name the project **menuBehavior**. From the Form Designer, add two Labels to the form and rename them **lbCntTitle** and **lbCnt**, respectively. Set the `Location` of `lbCntTitle` to (`0,40`), `Size` to (`80,22`), and the `Text` to `'Counter ='`. Then set the `Location` of `lbCnt` to (`90,40`), `Size` to (`80,22`), and the `Text` to an empty string. Next, add menu items to the main menu. Table 4-2 lists the order in which MenuItems are added to the form and summarizes the settings of each MenuItem added to the example application.

Table 4-2 MenuItems in the menuBehavior Example

Name	Parent	Text
mnuDone	mainMenu1	Done
mnuOptions	mainMenu1	Options
mnuR1	mnuOptions	R1
mnuR2	mnuOptions	Counter
mnuSep	mnuOptions	-
mnuR3	mnuOptions	R3
mnuR4	mnuOptions	Remove R5
mnuR5	mnuOptions	R5
mnuCntInc	mnuR2	Inc
mnuCntDec	mnuR2	Dec

Figure 14-13 presents the graphic layout of the UI and menu items.

If you examine the code generated by Visual Studio 2005 in the file `Form1.Designer.cs`, then you will see that the hierarchy of the menu items is achieved by adding each menu item to the `MenuItems` collection of its parent. For example, `mnuCntInc` and its parents and grandparents are constructed as follows:

```
...
this.mainMenu1.MenuItems.Add(this.mnuOptions);
...
this.mnuOptions.MenuItems.Add(this.mnuR2);
...
this.mnuR2.MenuItems.Add(this.mnuCntInc);
```

As you can see, the top-right menu `mnuOptions` is added to `mainMenu1`, and it contains `mnuR2`, which in turn contains `mnuCntInc`.

In the example, the function of `mnuSep` is to create a separator between menu items. This is easily achieved by setting the `Text` property of `mnuSep` to `-`. Next, you will learn how to dynamically change menu items at runtime. For example, when users select `mnuR4`, `mnuR5` is removed and `mnuR4` will therefore be disabled.

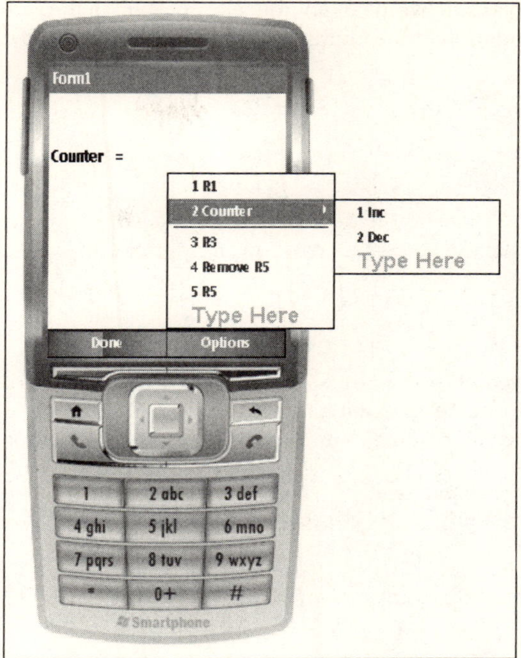

Figure 4-13

From the Form Designer, click `mnuOptions` to show the `MenuItems` directly attached to it. Then double-click `mnuR4`. Visual Studio 2005 will then automatically generate a stub of the event handler in the file `Form1.cs` for `mnuR4` when it is chosen at runtime:

```
private void mnuR4_Click(object sender, EventArgs e)
{

}
```

Visual Studio 2005 will also register an event handler to the `System.EventHandler` in the `InitializeComponent` method in the file `Form1.Designer.cs`:

```
this.mnuR4.Click += new System.EventHandler(this.mnuR4_Click);
```

To delete a MenuItem control at runtime, first remove it from the `MenuItems` collection of its parent, and then call the `Dispose()` method to release the resources. To disable a menu item, you can simply set the `Enabled` property to `false`. The following code shows the event handler for the `Click` event of `mnuR4`:

```
private void mnuR4_Click(object sender, EventArgs e)
{
    this.mnuOptions.MenuItems.Remove(this.mnuR5);
    this.mnuR5.Dispose();
    this.mnuR4.Enabled = false;
}
```

The sample code also displays the value of an integer `counter` on the screen. This can be achieved by passing the string representation of the `counter` to the `Text` property of `lbCnt`, as follows:

```
private int counter;

public Form1()
{
    InitializeComponent();

    //Reset counter and send the string representation to the text of lbCnt
    this.counter = 0;
    this.lbCnt.Text = counter.ToString();
}
```

When `mnuCntInc` is selected, the `counter` will be increased by one. To do so, double-click the `mnuCntInc` from the Form Designer to create an event handler to respond to the `Click` event of `mnuCntInc`. In the event handler, increase the value of the `counter` and pass the information to `lbCnt`, as follows:

```
//Increase counter by one
private void mnuCntInc_Click(object sender, EventArgs e)
{
    this.counter++;
    this.lbCnt.Text = this.counter.ToString();
}
```

Then add an event handler for `mnuCntDec` by double-clicking `mnuCntDec` from the Form Designer. Inside the event handler, decrease the value of `counter` by one:

```
  //Decrease counter by one
private void mnuCntDec_Click(object sender, EventArgs e)
{
    this.counter--;
    this.lbCnt.Text = this.counter.ToString();
}
```

Finally, an event handler is necessary to terminate the application when users press `mnuDone`. Double-click `mnuDone` from the Form Designer and add the following line to the `click` event handler of `mnuDone` to quit the application:

```
Application.Exit();
```

The following shows the full listing of the code in the file `Form1.cs`:

```
using System;
using System.Collections.Generic;
using System.ComponentModel;
using System.Data;
using System.Drawing;
using System.Text;
using System.Windows.Forms;

namespace menu
```

```
{
    public partial class Form1 : Form
    {
        private int counter;

        public Form1()
        {
            InitializeComponent();

            //Reset counter and send the string representation to the text of lbCnt
            this.counter = 0;
            this.lbCnt.Text = counter.ToString();
        }

        //Remove MenuItem mnuR5
        private void mnuR4_Click(object sender, EventArgs e)
        {
            this.mnuOptions.MenuItems.Remove(this.mnuR5);
            this.mnuR5.Dispose();
            this.mnuR4.Enabled = false;
        }

        //Terminate the application
        private void mnuDone_Click(object sender, EventArgs e)
        {
            Application.Exit();
        }

        //Increase counter by one
        private void mnuCntInc_Click(object sender, EventArgs e)
        {
            this.counter++;
            this.lbCnt.Text = this.counter.ToString();
        }

        //Decrease counter by one
        private void mnuCntDec_Click(object sender, EventArgs e)
        {
            this.counter--;
            this.lbCnt.Text = this.counter.ToString();
        }

    }
}
```

Build the sample application by pressing F6, and then launch it by pressing Ctrl+F5. Figure 4-14 shows the menu items when a user presses the right soft key.

After `Remove R5` is selected, `R5` disappears from the menu list and `Remove R5` is grayed out, which indicates the menu item is disabled. Figure 4-15 illustrates how the menu looks after users choose `Remove R5`.

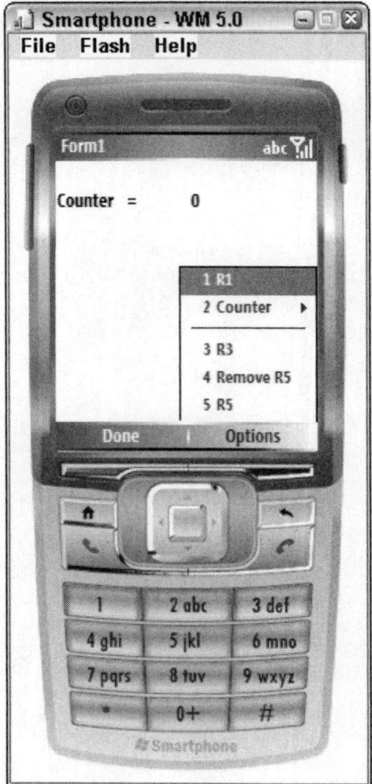

Figure 4-14

Focus Behavior for ListView, TreeView, and Panel

The nature of a Smartphone device's small screen size determines how you want to use container controls such as ListView, TreeView, and Panel. In full .NET applications, you can resize those controls as you like and use tab keys to navigate in and out of them. Things are a little different for Smartphone applications. The ListView, TreeView, and Panel controls are expected to expand to fill the entire screen, and users can use tab keys to navigate the child controls inside those containers, but not between those container controls.

To make a container control expand to the size of the form, set the `Bounds` property of that control to the `ClientRectangle` property of the form, as follows:

```
listView1.Bounds = this.ClientRectangle;
```

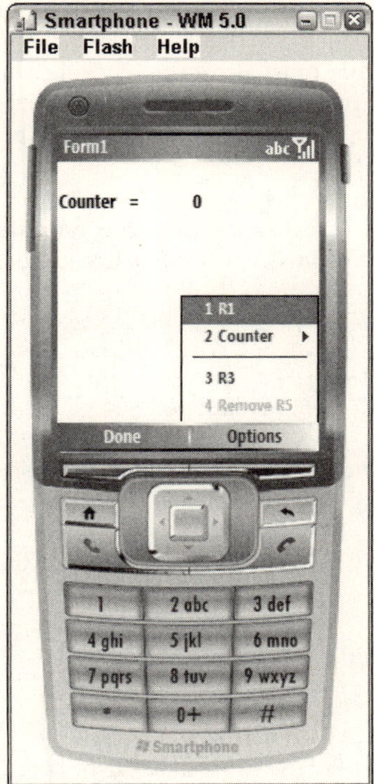

Figure 4-15

Smartphone UI Design

You need to consider several issues when designing a UI. A user-friendly UI is simple, clear, and consistent, and is normally optimized for performance. This section first introduces several Smartphone-specific issues that affect UI design. A typical UI flow for a Smartphone application is then explained, followed by a multi-form sample application to illustrate how this can be achieved programmatically.

Microsoft provides detailed design guidelines on MSDN. It is not our intention to illustrate all the UI details in this book. For further information, please refer to the website at `http://msdn.microsoft.com/library/default.asp?url=/library/en-us/uiguidesp/html/SPUser_Interface_Guidelines_SKLK.asp`.

Soft Keys

The Smartphone-specific features that most affect the UI design are likely the two soft keys: The top-level menu items are invoked only by the pressing the soft keys. As a direct result, a MessageBox control for Smartphone applications supports only two buttons; there are simply no additional soft keys to respond to a third button.

For instance, the following code will create a message box with three buttons (Yes, No, and Cancel) in the full .NET Framework and the Windows Mobile 5.0–based Pocket PC:

```
MessageBox.Show ("Do you really want to quit?", "Important",
                  MessageBoxButtons.YesNoCancel,
                  MessageBoxIcon.Question,
                  MessageBoxDefaultButton.Button1  );
```

In a Windows Mobile 5.0–based Smartphone, even though the preceding code can still pass the compile and build stages, an error indicating that the value is out of the range will be thrown during runtime.

The fact that a Smartphone has only two soft keys also dictates that only two top-level menus are supported.

The following are the UI design guidelines for the soft keys:

- The left soft key:
    - Should always display the most likely user task
    - Normally is the Done soft key that closes the window
- The right soft key:
    - Is blank if it is not needed
    - Should display the second most likely user task if there is no menu
    - Displays the Menu soft key when there is a menu

The Home and Back Keys

Besides the soft keys, two other keys are special in Smartphone devices: Home and Back. As its name implies, the Home key returns users to the Home screen from anywhere at any time. The Back key performs different functions depending on the state of the application. It is generally designed to return users to the previous screen, with the following exceptions:

- Applications use a hierarchical structure with regard to the Back key. For example, if a user navigates to a child form from a main form, pressing the Back key should take the user back to the parent form before quitting the application.
- The Back key performs a global back function, navigating out of the current application.
- When a user visits one screen several times in one session, only one instance of the screen is saved. Duplicates should be removed from the back stack.

- If a menu is open, the Back key should close the menu.
- Pressing the Back key in an edit control performs a backspace. For text boxes on edit screens with one or more lines, pressing and holding the Back key clears an entire box. For full-screen edits, pressing and holding the Back key performs repeated backspaces.
- Pressing the Back key when viewing a message box closes the message box and cancels the action.

General UI Flow of Smartphone Applications

A Smartphone application may require more than one form. How, then, do different forms relate to each other, and how does UI flow from one form to another form? A good example that demonstrates UI flow is the Smartphone native Contacts application.

As shown in Figure 4-16, the Contacts program begins with a list of all the contacts, which is referred to as *list view*.

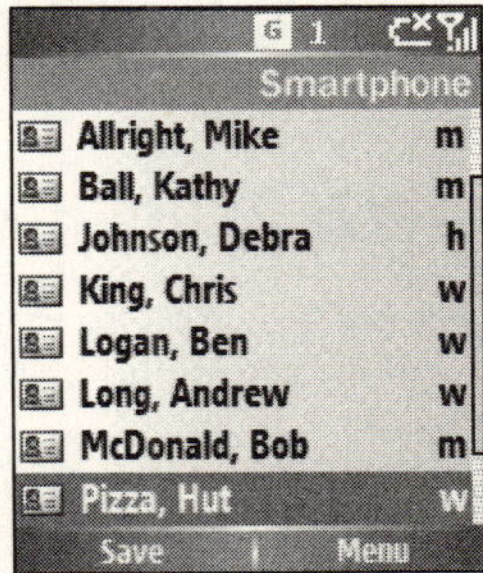

Figure 4-16

When users select an item, detailed information will be shown on the next screen, which is referred to as *card view*. For example, in Figure 4-17, when a contact is selected, detailed information about that contact is displayed, such as a picture, a work phone number, an e-mail address, a work address, and other information.

Figure 4-17

From the card view, users can also choose the edit option from the menu to further edit the detailed information of a contact. Figure 4-18 shows the edit view for the Contacts program.

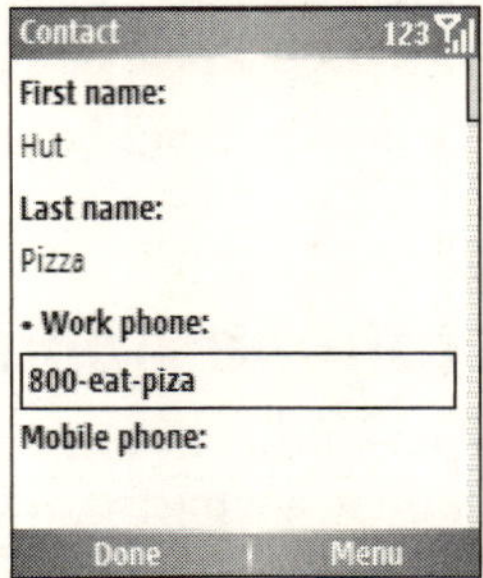

Figure 4-18

Depending on the needs of an application, you do not have to design the list view, card view, and edit view. For instance, the list view can go directly to the edit view without showing the card view. It is also possible for an application to start from a card view and flow to an edit view.

Creating an Application with Multiple Forms

One of the questions you may have at this point is how to programmatically control the UI flow from one view to another view. In the next example, you will learn how to add more forms to an application and how to the control the flow of the forms.

The example is a fairly simply one. When users click the left soft key, a second form with a combo box will appear to enable users to modify certain data. When users finish editing and press the Done key, the selected item will be printed on the first form and the second form will be closed. Of course, using two forms for this simple application is overkill, but the concept presented in this application is important.

Start a new Windows Mobile 5.0 Smartphone application and name it **MultiForm**. Add two labels to the main form `Form1` and rename the Labels **lbTitle** and **lbSname**, respectively. Change the `Text` property of lbTitle to `'Main: Chose a School'`, and clear the `Text` property of lbSname. Then add a top-left menu item to `Form1` and rename the menu item to **mnuDone**. Next, add a top-right menu item to `Form1` and rename it **mnuEdit**. Figure 4-19 shows the simple UI of main form `Form1`.

To add a new form in Visual Studio 2005, choose Project➪Add Windows Form. Name the new form **subForm** and add a combo box named **comboBox1**. Then add a MenuItem `mnuDone` to the top-left of the menu and a MenuItem `mnuCancel` to the top-right of the menu, as shown in Figure 4-20.

You can edit the items in a combo box during design time. From the Properties window, choose comboBox1 from the drop-down list and then go to the Items property and click the ellipsis (...) button (see Figure 4-21). This will open the String Collection Editor for `comboBox1`. Figure 4-22 shows four public universities in Indiana added to the `Items` collection of `comboBox1`.

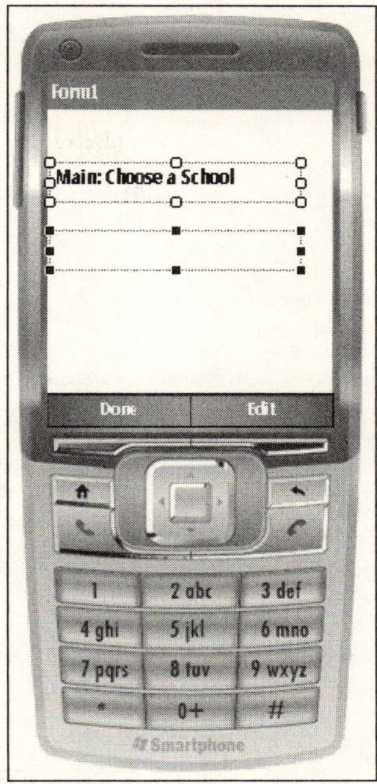

Figure 4-19

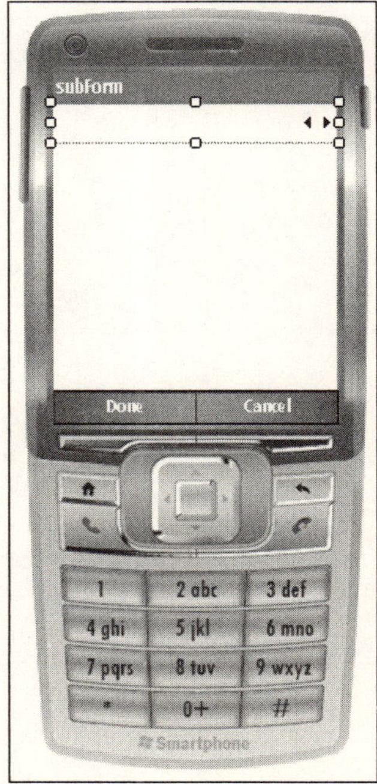

Figure 4-20

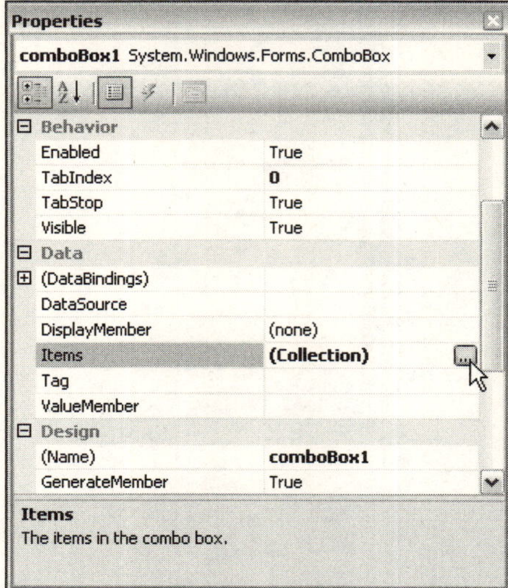

Figure 4-21

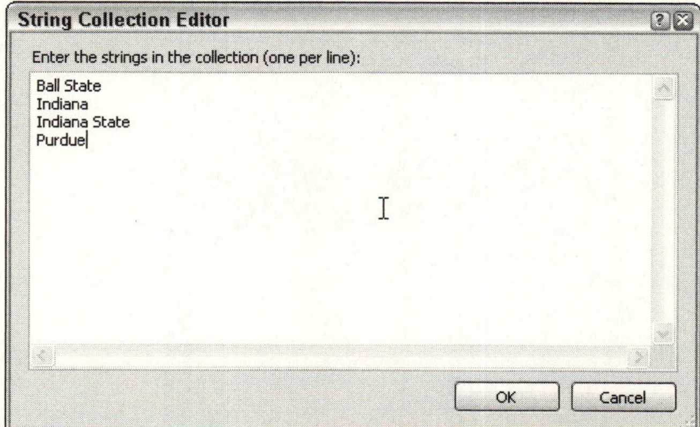

Figure 4-22

Because subForm will become a child form of Form1, you need to add a property in the subForm class to track its parent form:

```
//Add a parentForm property to refer to the parent form
private Form1 parentForm;
```

The parentForm info can be passed through the subForm constructor:

```
        //Set the parentForm property from constructor
        public subForm(Form1 parentForm)
        {
            InitializeComponent();
            this.parentForm = parentForm;
            this.Size = new System.Drawing.Size(parentForm.Size.Width,
parentForm.Size.Height);
             ...
        }
```

Also in the constructor of `subForm`, you can set the initial value for the combo box. The following code will mark the first item in the `Items` collection as the selected item and set the focus to `comboBox1`:

```
    //Set the initial value of ComboBox1
    comboBox1.SelectedIndex = 0;
    comboBox1.Focus();
```

On `Form1`, add an event handler to the `mnuEdit` click event so that the second form will be created and shown on the screen:

```
        //Show child form
        private void mnuEdit_Click(object sender, EventArgs e)
        {
            //Show waiting cursor
            Cursor.Current = Cursors.WaitCursor;

            //Create a new subForm and show it
            subForm subform1 = new subForm(this);
            subform1.ShowDialog();
        }
```

In this example, the transformation from `Form1` to `subForm` is similar to switching from card view to edit view. Two methods are available in the `Forms` class to display a form: `ShowDialog()` and `Show()`. In the example, `subForm1.ShowDialog()` is called so the statements following `showDialog()` will not be executed until after the form is shown, used, and closed. This is termed a *modal dialog* in Windows programming. If you use `subForm1.Show()` in this implementation, then the execution of the statements after the `Show()` method proceed even as the form is being displayed, which is known as *modeless dialog*. The preceding code also changed the style of the cursor to waiting style. This serves as a cue for users that some operations are happening in the background.

Once the child form `subForm` is called and is to be loaded, you will need to hide the parent form and change the type of cursor to normal. This can be achieved from the `Load` event hander of `subForm`:

```
        private void subForm_Load(object sender, EventArgs e)
        {
            this.parentForm.Hide();
            Cursor.Current = Cursors.Default;
        }
```

How do you pass the results back to the parent form? One solution is to add a public setter method in the parent form so that you can modify the properties of certain controls. In the example presented here, declare a method `SetLbSname` in the `Form1` class, as follows:

```
public void SetLb_SName (String name) {
    this.lbSname.Text = name;
}
```

Before the child form is closed, you should change the `Text` value of `lbSname` and show the parent form. The best place to do so is in the click event handler of the `mnuDone` menu item in the `subForm` class:

```
private void mnuDone_Click(object sender, EventArgs e)
{
    //Set the label text in parent form
    String SName = (String) this.comboBox1.SelectedItem;
    this.parentForm.setLbSname(SName);

    //Show parent form and close current form
    this.parentForm.Show();
    this.Close();
}
```

If users decide to quit the `subForm` without choosing a name from the combo box, you can simply close the form and show the parent form, as follows:

```
//Close out the edit windows and bring the parent window back to screen
private void mnuCancel_Click(object sender, EventArgs e)
{
    //Show parent form and close current form
    this.parentForm.Show();
    this.Close();
}
```

Of course, you should also add an event handler to respond to the click event of `mnuDone` to terminate the application. Then you are ready to test this simple application with two forms.

The following is the full code for `Form1.cs` and `subForm.cs`:

Form1.cs (parent form)

```
using System;
using System.Collections.Generic;
using System.ComponentModel;
using System.Data;
using System.Drawing;
using System.Text;
using System.Windows.Forms;

namespace MultiForm
```

```
{
    public partial class Form1 : Form
    {
        public Form1()
        {
            InitializeComponent();
        }

        // Add this public method so the child form
        // can change the text value of Lb_Sname
        public void setLbSname (String name) {
            this.lbSname.Text = name;
        }

        //Show child form
        private void mnuEdit_Click(object sender, EventArgs e)
        {
            //Show waiting cursor
            Cursor.Current = Cursors.WaitCursor;

            //Create a new subForm and show it
            subForm subform1 = new subForm(this);
            subform1.ShowDialog();
        }

        //Close the application
        private void mnuCancel_Click(object sender, EventArgs e)
        {
            Application.Exit();
        }
    }
}
```

subForm.cs (child form)

```
using System;
using System.Collections.Generic;
using System.ComponentModel;
using System.Data;
using System.Drawing;
using System.Text;
using System.Windows.Forms;

namespace MultiForm
{
    public partial class subForm : Form
    {
        //Add a parentForm property to refer the parent form
        private Form1 parentForm;

        //Set the parentForm property from constructor
```

```
        public subForm(Form1 parentForm)
        {
            InitializeComponent();
            this.parentForm = parentForm;
            this.Size = new System.Drawing.Size(176, 180);

            //Set the initial value of ComboBox1
            comboBox1.SelectedIndex = 0;
            comboBox1.Focus();
        }

        private void mnuDone_Click(object sender, EventArgs e)
        {
            //Set the label text in parent form
            String SName = (String) this.comboBox1.SelectedItem;
            this.parentForm.setLbSname(SName);

            //Show parent form and close current form
            this.parentForm.Show();
            this.Close();
        }

        //When called up, hide the parent form
        //and change the cursor type to default
        private void subForm_Load(object sender, EventArgs e)
        {
            this.parentForm.Hide();
            Cursor.Current = Cursors.Default;
        }
        //Close the current window and bring the parent window back to screen
        private void mnuCancel_Click(object sender, EventArgs e)
        {
            //Show parent form and close current form
            this.parentForm.Show();
            this.Close();
        }

    }
}
```

After `MultiForm` is compiled and built, it can be deployed to a Smartphone device or an emulator. Figure 4-23 shows the initial UI when the application is just loaded.

When users press the right soft key, the `subForm` will be displayed on the screen, as shown in Figure 4-24.

If users pick the first item (Ball State) from the combo box and then choose the child form by pressing the left soft key, the parent form will be displayed with the name of the school set to Ball State, as illustrated in Figure 4-25.

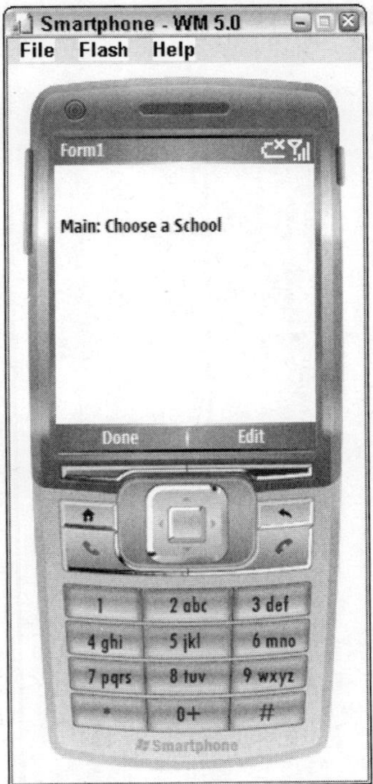

Figure 4-23

Figure 4-24

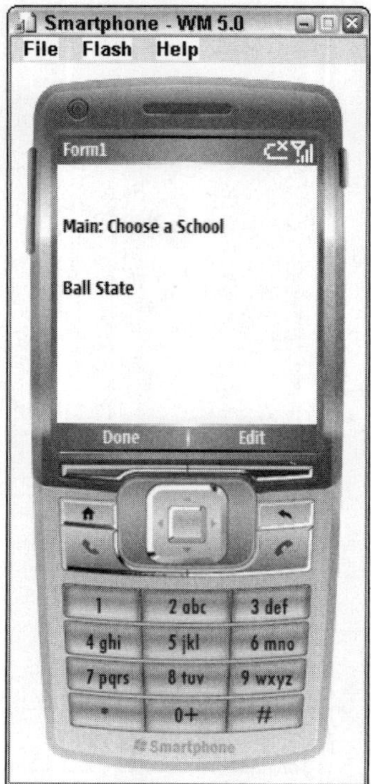

Figure 4-25

Keyboard Input and Input Mode

In the previous section you learned how to design a UI for Smartphone applications. One important aspect of dealing with the UI is responding to user input. This section discusses how to handle keyboard input in the .NET Compact Framework.

Input Mode

Unlike a PC's keyboard, a Smartphone device typically has only a number pad and a few other function keys. For Smartphone applications requiring user input, this is a little awkward. The most popular solution to this problem is to use the multi-tap function. For example, when a user presses the 2 key repeatedly, the letters A, B, C and the number 2 will cycle around until either the user clicks the forward navigation key or a short timeout occurs.

Another way to handle user input is by using predictive text technologies, such as T9, also known as *ambiguous word input*. In T9 input mode, to get a word "Hello," users can directly punch in the corresponding numbers one at a time (4, 3, 5, 5, 6), whereas the conventional multi-tap input would require a user to press the 4 key twice, the 3 key twice, the 5 key three times, and then another 5 key three times, followed by the 6 key three times. The T9 input, however, does not guarantee that the key combinations result in the desired word. If this is the case, users will have to select the word from a list. Nonetheless, it can save users a lot of time sending text messages. It should be noted, however, that the T9 input method is not supported on many models of Smartphone devices on today's market, nor does it work on Smartphone 2003 and Windows Mobile 5.0 emulators.

The default input behavior for a Smartphone device application uses mixed alphabetic characters and numeric characters. This is not desirable for some applications, however. For example, for a TextBox that expects users to input a zip code, it is more efficient to take the user input directly as a numeric number. For a TextBox that expects users to input names, it is probably better to take input as an alphabetic string.

Fortunately, the Windows CE .NET provides a number of APIs that enable you to call into the native code to change the way input is interpreted at runtime. Windows Mobile 5.0 also introduces a set of managed code to enable you to change the input method on-the-fly. This chapter describes how to change the input method using managed code. Chapter 10 describes how to use the native Windows CE APIs in your application.

The `Microsoft.WindowsCE.Forms` namespace provides a number of managed classes for programming device applications using the .NET Compact Framework. The `InputMode` enumeration supports the following input modes:

- `AlphaABC` — The conventional multi-tap input mode
- `AlphaCurrent` — A mix of `AlphaABC` and `AlphaT9`, which can be switched by pressing the star (*) key
- `AlphaT9` — The T9 predictive input mode
- `Default` — The user's preferred input mode, which is usually the user's last input mode selection
- `Numeric` — The mode that accepts numeric characters only

The `InputModeEditor` class in the `Microsoft.WindowsCE.Forms` namespace has an important static method, `SetInputMode()`, which can set the input mode for a TextBox control. For example, the following code sets the input mode of a `textBox1` to T9:

```
InputModeEditor.SetInputMode ( textBox1, InputMode.AlphaT9);
```

In the next example, you will learn how to preset an input mode for a TextBox control and how to change the input mode during runtime.

Start a new Windows Mobile device application for Smartphone, and name the project **myInputMode**. From the Solution Explorer, right-click `Form1.cs` and rename the file to **myInputMode.cs**. Then, from the Designer Window, change the name of Form1 to **FmInputMode** and set the `Text` of the form to `Input Form`.

Add a Label named **lbPhoneNo**, a TextBox named **txtPhoneNo**, another Label named **lbComments**, and another TextBox named **txtComments** to `FmInputMode`. The `Text` of `lbPhoneNo` is `Phone Number`, and the `Text` of `lbComments` is `Comments`. Next, add a menu item `mnuQuit` to the top-left menu, and add another menu item `mnuInputMode` to the top-right menu. Add three menu items — `mnuNumeric`, `mnuText`, and `mnuT9` — to `mnuInputMode`. Figure 4-26 shows the UI of this application.

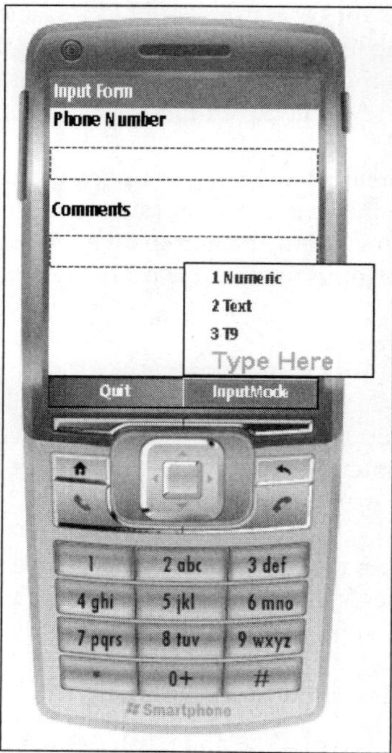

Figure 4-26

Because txtPhoneNo expects users to input numeric characters, it makes sense to change the input mode to InputMode.Numeric when the application starts. Note, however, that the Microsoft.WindowsCE.Forms namespace is not part of the .NET Compact Framework, so you must add a reference to the namespace in your project. Click Project➪Add Reference, and then from the .NET tab choose Microsoft.WindowsCE.Forms and click OK (see Figure 4-27). Add the following line in your code in order to use the InputMode class:

```
using Microsoft.WindowsCE.Forms;
```

You can preset the input mode of txtPhoneNo in the constructor of FmInputMode() as follows:

```
public FmInputMode()
{
    InitializeComponent();
    //Preset the input mode of txtPhoneNo to Numeric
    InputModeEditor.SetInputMode(txtPhoneNo, InputMode.Numeric);
}
```

For txtComments, users may want to pick their most convenient input method to type in the message, and they may need to change the input mode from time to time by clicking one of the three input modes from the right menu.

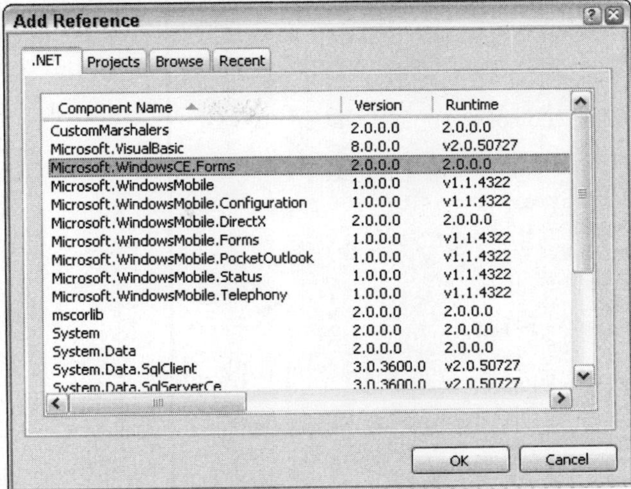

Figure 4-27

If you define the click event handler of `mnuT9` to change the current `TextBox` control to T9 input mode, you need to first determine which control currently has the cursor focus. The following code implements a `getFocusedCtrl()` method that will loop through all the controls in a control collection and return the one that has focus (or return `null` if no control has focus):

```
// Find the focused control in current form
private Control getFocusedCtrl(Form form)
{
    foreach (Control c in form.Controls)
    {
        if (c.Focused) return c;
    }
    //Return null if no control is focused
    return null;
}
```

After implementing the `getFocused()` method, you can now call this method in the click event handlers for `mnuNumeric`, `mnuText`, and `mnuT9`. For example, from the Form Designer, you can double-click `mnuText` to get the click event handler registered. The following code shows that the input mode of the current focused control will be set to `InputMode.Text`:

```
private void mnuText_Click(object sender, EventArgs e)
{
    //Set the input mode of txtComments to T9
    Control curCtrl = getFocusedCtrl(this);
    if (curCtrl == null) curCtrl = txtComments;

    InputModeEditor.SetInputMode(curCtrl, InputMode.AlphaABC);

}
```

If a focused control cannot be found, then the sample application will change the input mode for `txtComments`. Apply the same approach to the click event of `mnuNumeric` and `mnuT9`. In addition, add a click event hander for `mnuQuit` to terminate the application by calling `Application.Exit()`.

Following is the complete code of `myInputMode.cs`:

```
using System;
using System.Collections.Generic;
using System.ComponentModel;
using System.Data;
using System.Drawing;
using System.Text;
using System.Windows.Forms;

using Microsoft.WindowsCE.Forms;

namespace myInputMode
{
    public partial class FmInputMode : Form
    {
        public FmInputMode()
        {
            InitializeComponent();
            //Preset the input mode of txtPhoneNo to Numeric
            InputModeEditor.SetInputMode(txtPhoneNo, InputMode.Numeric);
        }

        // Find the focused control in current form
        private Control getFocusedCtrl(Form form)
        {
            foreach (Control c in form.Controls)
            {
                if (c.Focused) return c;
            }
            //Return null if no control is focused
            return null;
        }

        private void mnuQuit_Click(object sender, EventArgs e)
        {
            Application.Exit();
        }

        private void mnuT9_Click(object sender, EventArgs e)
        {
            //Set the input mode of txtComments to T9
            Control curCtrl = getFocusedCtrl(this);
            if (curCtrl == null) curCtrl = txtComments;

            InputModeEditor.SetInputMode(curCtrl, InputMode.AlphaT9);
        }

        private void mnuText_Click(object sender, EventArgs e)
        {
```

```
            //Set the input mode of txtComments to T9
            Control curCtrl = getFocusedCtrl(this);
            if (curCtrl == null) curCtrl = txtComments;

            InputModeEditor.SetInputMode(curCtrl, InputMode.AlphaABC);

        }

        private void mnuNumeric_Click(object sender, EventArgs e)
        {
            //Set the input mode of txtComments to T9
            Control curCtrl = getFocusedCtrl(this);
            if (curCtrl == null) curCtrl = txtComments;

            InputModeEditor.SetInputMode(curCtrl, InputMode.Numeric
                );

        }

    }
}
```

Soft Key Functionality

Typically, Smartphone applications use two soft keys to pop up menu items, but you may want to provide your own implementations to respond to those two keys. This is particularly true for gaming applications. Note that the customized soft key function works only when there is no MainMenu control in the form.

For example, you can start a new Smartphone device application and delete the auto-generated `mainMenu1` from `Form1`. Add the following line to the constructor of `Form1` to register a `KeyDown` event:

```
this.KeyDown += new System.Windows.Forms.KeyEventHandler(this.Form1_KeyDown);
```

The stub of the event handler is as follows:

```
        private void Form1_KeyDown(object sender, KeyEventArgs e)
        {

        }
```

When a key is pressed, the information of the key is passed to the event handler through `KeyEventArgs`. For example, if the `KeyCode` equals `System.Windows.Forms.Keys.F1`, it means the left soft key is pressed. Similarly, when the right soft key is pressed, the `KeyCode` equals `System.Windows.Forms.Keys.F2`.

The following code will display a message box that indicates which soft key is pressed:

```
        private void Form1_KeyDown(object sender, KeyEventArgs e)
        {
```

```
            if ((e.KeyCode == System.Windows.Forms.Keys.F1))
            {
                // Soft Key 1
                MessageBox.Show("The left soft key is hit");
            }
            if ((e.KeyCode == System.Windows.Forms.Keys.F2))
            {
                // Soft Key 2
                MessageBox.Show("The right soft key is hit");
            }
        }
```

To find out more about the `Keys` *enumeration in the .NET Compact Framework, visit the MSDN website at* `http://msdn2.microsoft.com/en-us/library/system.windows.forms.keys.aspx`.

Additional UI Considerations

Now that you have learned the fundamentals of UI design and key input, this section introduces some additional factors you should consider when developing Smartphone applications.

Auto-Save

Microsoft Windows Smartphones normally don't follow the desktop computer model of the File➪Save or Save As commands. Instead, the devices normally work in a so-called *auto-save mode* in which data is saved as soon as the user enters it.

If a user mistakenly changes any data, an Edit➪Undo command is available on a per-control basis. For PIM or other database items, an Undo command is supported to the extent that the view can be reverted to the way it was when it was last opened. In the case of file-based applications, until the user closes the file, the Undo stack remains. This enables users to select the Save As command at any time to keep their original documents intact and to keep any changes made since the files were last opened.

Sometimes auto-save is not desirable. Consider an application that enables users to change the operating system settings or configuration of the network or system. In the middle of operations, users may want to cancel changes and roll back to previous settings. You should provide a Cancel menu to enable users to call off any changes.

DPI and QVGA Issues

Traditionally, Smartphone devices have a resolution of 176 × 220 pixels and a DPI (dots per inch) value of 96. With the advances of display technology, QVGA (Quarter VGA) mode is now available in newer Smartphone devices whose screen can hold 240 × 320 pixels. Some Smartphone devices can now even support VGA mode, which has a resolution of 640 × 480 pixels. This change will surely bring sharper text on the screen and present more detailed images. However, it requires you to consider whether your application is targeting only conventional Smartphone device platforms. For Windows Mobile 5.0 Smartphones, an additional `AutoScaleMode` property is available for the Windows `Forms` class. You can set this property to `DPI` mode so that applications designed for regular Smartphone devices can still be displayed nicely on a higher-resolution device.

Performance

Developers have tested a number of scenarios to determine how to improve the performance of loading a form. Their findings suggest reducing the number of method calls and creating controls using the top-down technique.

For example, when setting the `Location` and the `Size` of a TextBox control, you can certainly use code as follows:

```
this.textBox1.Location = new Point(10, 20);
this.textBox1.Size = new Size(100, 22);
```

However, the preceding will not be as efficient as using one method call:

```
this.textBox1.Bounds = new Rectangle(10, 20, 100, 22);
```

Initializing the controls using the top-down technique means creating the parent control first and then the child control. In addition, instead of adding the child controls to the parent control, set the `Parent` property of a child control to its parent. For instance, the following code is generated automatically by Form Designer when you add a Panel to the form and a Textbox to the Panel:

```
// Before optimization
Panel panel1 = new Panel();
TextBox textBox1 = new TextBox();
textBox1.Text = "My Text";
panel1.Controls.Add(this.textBox1);
// Add the Panel to the Form's control collection
this.Controls.Add(panel1);
```

Optimizing this code snippet using the top-down technique results in the following snippet:

```
// After optimization
// Create a new panel and TextBox control
Panel panel1 = new Panel();
TextBox textBox1 = new TextBox();
// Parent the Panel to the current Form
this.panel1.Parent = this;
// Parent the TextBox to the Panel
this.textBox1.Parent=this.panel1;
// Set the Text property of the TextBox control
textBox1.Text = "My Text";
```

Summary

This chapter introduced various controls and their behaviors, followed by guidelines for UI design. Also presented were issues related to input. As you can see from reading this chapter, the controls available in the Smartphone .NET platform are a trimmed-down version of the .NET Compact Framework. It is very important to understand how those constraints change the way you program on thin-client mobile devices.

You should always design a simple UI that follows the Smartphone conventions. It is highly recommended that you go through the native Smartphone applications carefully to familiarize yourself with its design philosophies.

In the next chapter, you will learn how to access local data and get information about local files and directories.

5

Data Storage and File I/O

Unlike desktop PCs, for which battery life is not an issue, Smartphone devices prefer hardware and software designs that consume less power. The constraints that led to the data storage architecture of the Smartphone devices are fundamentally different from that of PCs. This chapter examines how data is stored in a Windows Mobile Smartphone device. You will also learn how the unique Windows Mobile filesystem can affect you when writing a program that accesses local data.

A good understanding of the I/O classes defined in the `System.IO` namespace is extremely important for a Smartphone programmer because knowing how to access local data is the foundation for accessing data from a database, over a network, and through web services.

This chapter includes the following:

- ❑ An overview of Smartphone data storage
- ❑ System.IO classes
- ❑ A sample file and directory browser application
- ❑ A sample memo application

Accessing external files — specifically, retrieving data from an SQL Server database — is discussed in Chapter 6; and accessing an XML file, whether internal or external, is presented in Chapter 9.

Overview of Smartphone Data Storage

Data storage is an area where battery-powered mobile devices differ significantly from their desktop peers. It is not suitable to save data on power-consuming, rotating media, such as hard drives. As a result, mobile devices rely primarily on ROM, RAM, and removable Flash memory to store data. The size of the data storage has to be small, and the access speed is expected to be fast.

Prior to Windows Mobile 5.0, a typical Pocket PC device put the operating system, protected file storage, some device drivers, and application updates on ROM, and split RAM in two portions: *program memory* and the *object store*. Program memory provides space for applications to load and run, which is pretty much the same way that RAM is used in desktop operating systems. The object store includes a filesystem, the system registry, and property databases. For example, a user's personal information, data (including e-mail, calendars, and contacts), and user-installed software all reside in a portion of RAM. Figure 5-1 shows how data is allocated in a RAM-based filesystem.

ROM

Operating system
Protected file storage
Drivers

RAM

Program memory

Object store
– User info
– User-installed applications
– System registry
– Property store

Figure 5-1

The advantage of a RAM-based filesystem is obvious: fast access. This is particularly true when most of the operating is handled over a network. RAM, however, is volatile in nature. Retaining the data saved on RAM requires a constant power supply, which is why a backup battery is still powering the RAM after you turn off a PDA. Normally, the more RAM you have installed on a mobile device, the more power it will consume to retain the information stored in RAM. If both the main battery and the backup battery are drained, you are out of luck: All the applications installed on RAM and all the data saved on RAM are gone. To avoid such a devastating situation, the industry recommends a "72-hour rule," which mandates that a Pocket PC device must be able to preserve data for at least 72 hours after the critical battery level is reached. The 72-hour rule enables users to recharge the device without losing data.

Smartphone devices, conversely, take a different approach and are designed to work with persistent storage-based filesystems. The concept of the object store is non-existent in the context of a Smartphone device. By adopting a persistent storage architecture, Smartphone devices, therefore, require less RAM, and less RAM entails a lower cost and less power consumption. Figure 5-2 shows how data is allocated in a persistent storage-based filesystem.

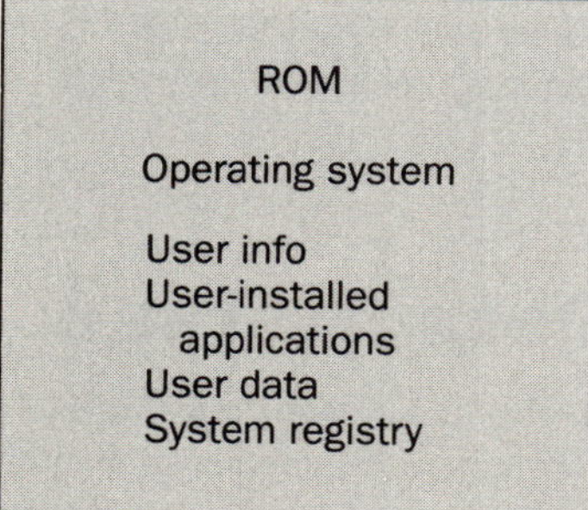

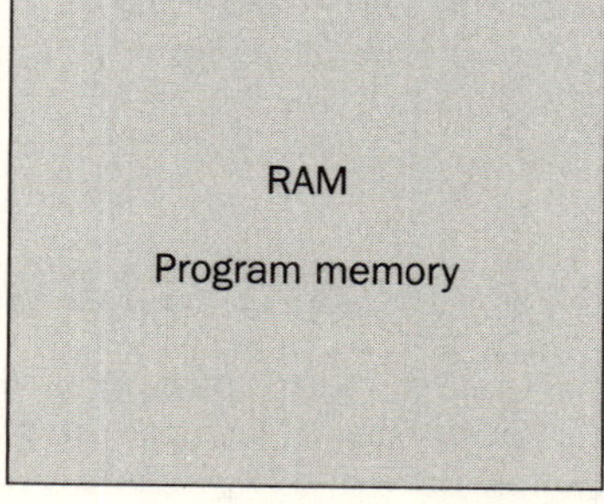

Figure 5-2

There is, of course, no free lunch. Using persistent storage filesystems results in slower access speed; it takes more than ten times longer to access Flash ROM than it does to access RAM, and the storage performance heavily affects the entire system's performance. Windows Mobile 5.0 optimized storage performance by buffering a block of data before writing it back to persistent storage. Software developers should also try to minimize any negative impacts your application may have on the entire system. The following are the best-practice guidelines:

- Minimize the size of your application.
- Minimize the amount of persisted data your application requires.
- Minimize the frequency of altering data.
- Minimize any writing to the registry.
- Minimize polling and always try to use an event-driven design, if possible.

Note that beginning with Windows Mobile 5.0, Pocket PCs a persistent storage filesystem is preferred even though a RAM-based filesystem is still supported. Another thing worth noting is that Windows Mobile adopts a *unified directory view*, which eliminates the drive letter in the traditional Windows directory structure. All the files, regardless of whether they are saved in RAM, ROM, or a storage card, have a centralized logical view, with the root of the directory beginning with the backslash (\).

Although mobile devices store data differently from desktop PCs, accessing Smartphone data programmatically still has a lot in common with accessing data in a desktop environment. The following section discusses which classes are supported and how to access local files in the Windows Mobile 5.0 Smartphone platform.

The System.IO Namespace

The `System.IO` namespace contains classes that are essential for accessing local files. Logically, the classes can be grouped into the following three categories:

- File-manipulation
- Byte-level I/O
- Higher-level I/O

File-Manipulation Classes

A number of classes have been defined and implemented in the `System.IO` namespace to interact with the filesystem. These classes enable you to manipulate files and directories, such as creating, deleting, copying, and moving files.

In the .NET Compact Framework, two classes are available for accessing information pertaining to directories: `Directory` and `DirectoryInfo`. Likewise, the `File` and `FileInfo` classes can be used to access file-based information.

The `Directory` and `File` classes are *static* classes, which means you do not need to create an instance to use them. For example, if you need to create a new file called `myfile1`, you can simply call the `Create()` method of the `File` class, as follows:

```
File.Create("myfile1")
```

Static classes such as `Directory` and `File` are particularly useful if you want to perform some quick directory-related operations.

The info classes `FileInfo` and `DirectoryInfo` provide similar functions to the `File` and `Directory` classes; however, they are not static classes. To create a new file named `myfile1` using the `FileInfo` class, you must first create an instance of `FileInfo` and then call the `Create()` method:

```
FileInfo myFileInfo = new FileInfo ("myfile1");
myfileInfo.Create();
```

Another difference between the nonstatic and the static file/directory classes is the level and format of the information they return. For instance, the following examples demonstrate how to search for all the `.txt` files in the My Documents folder using the `Directory` and `DirectoryInfo` classes, respectively:

```
//Search *.txt file with Directory class
string[] fileList = Directory.GetFiles(@"\My Documents", "*.txt");

//Search *.txt file with DirectoryInfo class
DirectoryInfo myDirInfo = new DirectoryInfo(@"\My Documents");
FileInfo[] fInfoList = myDirInfo.GetFiles("*.txt");
```

As you can see, the `Directory` class needs to take both the path name and search pattern, whereas the `DirectoryInfo` class requires only the search pattern because the path name is already known to the `myDirInfo` object when you create it. When the `Directory` class finishes searching, it returns an array of strings, whereas the `DirectoryInfo` class returns an array of `FileInfo` objects.

The `@` symbol indicates that the strings will be escaped. This is handy when you are dealing with path names. Without the `@` symbol, you have to use a string such as `"\\My documents\\subfolder"` rather than `@"\My Documents\subfolder"`.

Byte-Level I/O Classes

The `Stream` class is the base class of `System.IO` objects and is the foundation of all file access I/O classes. It provides a basic data transfer capability: moving bytes from one data unit to another, independent of data storage media. It includes methods such as `Read()` and `Write()` to perform I/O synchronously, and it includes asynchronous I/O methods, such as `BeginRead()` and `BeginWrite()`. The `Stream` class is designed as an abstract class (or interface) in the .NET Compact Framework, which means the base class only defines the functionalities, without really doing the actual work. You can perform the actual I/O tasks by using the derived `FileStream` class and `MemoryStream` class. The `FileStream` class supports stream access to physical files, whereas the derived `MemoryStream` class allows stream operations on physical memory.

The following is a typical procedure to access a physical file with the `FileStream` class:

1. Create a `FileStream` object and specify the filename and `FileMode`. The `FileMode` enumeration is a parameter to indicate whether to open a file or to create a file. The four common `FileMode` values are as follows:

- `FileMode.Open`—Opens an existing file. If the file does not exist, a `System.IO.FileNotFoundException` is thrown.
- `FileMode.Create`—Creates a new file. If the file already exists, it overwrites the existing file.
- `FileMode.OpenOrCreate` —Opens a file if the file exists; otherwise, a new file will be created.
- `FileMode.Append`—Opens or creates a file in the same way as `FileMode.OpenOrCreate`. Then it seeks the end of the file to read or to write.

2. Optionally, specify file access mode and file sharing mode. The `FileAccess` enumeration parameter has three values: `Read`, `Write`, and `ReadWrite`. Specifying this optional parameter can make your program safer. For instance, if you set the `FileAccess` enumeration to `Read`, an opened file stream can avoid accidental writing operations. Note that not every combination of `FileMode` parameter and `FileAccess` parameter is valid. For example, it is not applicable to create a new file with `FileMode.Create` and set the `FileAccess` to `Read`. An `ArgumentException` will be thrown because creating a file requires either `FileAccess.Write` or `FileAccess.ReadWrite`. The `FileShare` enumeration denotes how others can open the file. The typical enumeration values are `FileShare.Read`, `FileShare.Write`, `FileShare.ReadWrite`, and `FileShare.None`.
3. Read or write file with the `Read()`, `ReadByte()`, `Write()` or `WriteByte()` methods.
4. Close the file stream with the `Close()` method.

The following code snippet opens an existing file `myfile1` in the root directory and reads the first 100 bytes into the integer array `buff[]`:

```
using System.IO;
using System.Windows.Forms;
...

try
{
    //Create a filestream fsRead with Open mode and Read access
    FileStream fsRead = new FileStream(@"\myfile1", FileMode.Open,
FileAccess.Read);

    //Create the integer array buff[]
    int[] buff = new int[100];

    //Read the first 100 bytes from the fsRead to buff[]
    for (int i = 0; i < 100; i++)
        buff[i] = fsRead.ReadByte();

    //close the filestream fsRead
    fsRead.Close();
}

//Error handling:
catch (IOException e)
{
    //Show the error information.
    MessageBox.Show("I/O Error: " + e.ToString());
}
```

When you are dealing with file I/O, it is always recommended to put the file I/O access call inside the `try...catch` *block so that you can catch the file I/O-related exceptions and write your own error-handling functions. For more information about error handling, refer to Chapter 11.*

Writing data using the `FileStream` class is very similar to reading data from the file stream. The following code illustrates how to write ASCII-coded data into a `FileStream` object `fsWrite`. The example writes the lowercase letters `a` through `z` to file `myfile2` with the integer value of each letter's ASCII code:

```
using System.IO;
using System.Windows.Forms;
...

try
{
    //Create a filestream fsWrite with Write access, and Share Read mode
    FileStream fsWrite = new FileStream (@"\myfile2",
                FileMode.Create, FileAccess.Write, FileShare.Read);

    // Write to the stream with ascii value of each letter
    for (int i = 0; i < 26; i++)
        fsWrite.WriteByte( (byte) ('a'+i) );

    //Close the filestream fsWrite
    fsWrite.Close();
}
//Error handling:
catch (IOException e)
{
    //Show the error information
    MessageBox.Show("I/O Error: " + e.ToString());
}
```

Byte-level I/O classes provide access to data stored in various storage media in a byte-oriented fashion. If you are interested in reading or writing data in a form other than the native byte format (such as number, strings, or text with specific coding), you should familiarize yourself with the higher-level I/O classes.

Higher-Level I/O Classes

The higher-level I/O components include a variety of reader/writer objects to perform data-specific I/O operations. This is achieved by wrapping the higher-level I/O classes around the basic byte-level classes.

Two major reader/writer classes can be found in the `System.IO` namespace: `BinaryReader` and `BinaryWriter` and `TextReader` and `TextWriter`.

As their names imply, the `BinaryReader` and `BinaryWriter` classes read or write data in binary format, respectively. To use them correctly, you need to call the data type–specific methods for either reading or writing. For example, if you want to read a string from a stream using a `BinaryReader` object, use the `ReadString()` method instead of the `Read()` method. The `BinaryReader` and `BinaryWriter` classes also enable you to specify how the data is encoded. This can be done by specifying the optional `Encoding` enumeration type in the `System.Text` namespace. The following example shows how to read the first string from file `myfile1` using the default system encoding:

```
using System.IO;
using System.Windows.Forms;
using System.Text;
...

try
{
    //Create a filestream fsRead with Open mode and Read access
    FileStream fsRead = new FileStream(@"\myfile1", FileMode.Open,
FileAccess.Read);

    //Create a new BinaryReader bReader from fsRead
    //using the system default decoding
    BinaryReader bReader = new BinaryReader(fsRead, Encoding.Default);

    //Read the first string in the file to string str
    string str = bReader.ReadString();

    //Close the filestream fsRead
    fsRead.Close();
}

//Error handling:
catch (IOException e)
{
    //Show the error information
    MessageBox.Show("I/O Error: " + e.ToString());
}
```

You can see from this example that the `BinaryReader` class and the `BinaryWriter` class rely on the lower-level I/O classes to actually open or create the file. Once the file is opened or created, these two classes provide a number ways for programmers to read and write data using the desired data types, such as `string`, `double`, `short`, `long`, etc.

The `System.IO` namespace also provides the `TextReader` and `TextWriter` classes, which enable you to read or write text characters from/to a stream, respectively. In the .NET Compact Framework, both `TextReader` and `TextWriter` are abstract classes. For that reason, you should use the derived `StreamReader` and `StreamWriter` classes and the derived `StringReader` and `StringWriter` classes to access the data in `Stream` objects. The `StreamReader` and `StreamWriter` classes can read and write a number of data types, respectively, and support a variety of encoding methods. `StringReader` and `StringWriter` objects, conversely, support only read and write in string format, which is defined in the `StringBuilder` class from the `System.Text` namespace.

You can create a `StreamReader` or `StreamWriter` object either from an opened file stream or directly from the name of the file. The following code snippet reads the first line from file `\myfile1` to a string `str`:

```
streamReader sReader = new StreamReader (@"\myfile1");
string str = sReader.ReadLine();
```

A Summary of I/O Classes

Table 5-1 summarizes the commonly used the I/O class in the `System.IO` namespace of the .NET Compact Framework.

Table 5-1 Common Classes in the System.IO Namespace

Class	Description
`BinaryReader`	Reads data from an I/O stream as binary values in a specific coding.
`BinaryWriter`	Writes data to an I/O stream as binary values in a specific coding.
`Directory`	Exposes static methods for directory-related operations. Cannot be inherited.
`DirectoryInfo`	Exposes instance methods for directory-related operations. Cannot be inherited.
`File`	Exposes static methods for file-related operations. Cannot be inherited.
`FileInfo`	Exposes instance methods for file-related operations. Cannot be inherited.
`FileStream`	Provides a stream around a file. Supports both synchronous and asynchronous read and write operations.
`FileStreamInfo`	The base abstract class of `FileInfo` and `Directory` info classes.
`IOException`	The exceptions that are raised when I/O errors occur.
`MemoryStream`	Creates a `stream` around a block of memory.
`Path`	Exposes static methods to retrieve the path information of files or directories.
`Stream`	Provides a generic view of a sequence of bytes. This abstract class is the base class of the `FileStream` class and the `MemoryStream` class.
`StreamReader`	Implements a `TextReader` class for reading characters from a stream in a specific coding.
`StreamWriter`	Implements a `TextWriter` class for writing characters to a stream in a specific coding.
`StringReader`	Implements a `TextReader` class to read from a string.

Class	Description
`StringWriter`	Implements a `TextWriter` class to write to a string.
`TextReader`	Represents a generic reader that reads a sequence of characters. It is the base class of the `StreamReader` class and the `StringReader` class.
`TextWriter`	Represents a generic writer that writes a sequence of characters. It is the base class of the `StreamWriter` class and the `StringWriter` class.

The `System.IO` namespace in the .NET Compact Framework supports a subset of the classes and structures in the full .NET Framework. The following classes and structures are currently not supported in the .NET Compact Framework. In addition, file attributes, such as `Hidden`, `Archive`, and `Read-Only`, are not supported in the .NET Compact Framework. Therefore, methods such as `GetAttributes` are also not supported.

- `BufferedStream` class
- `ErrorEventArgs` class
- `FileLoadExceptions` class
- `FileSystemEventArgs` class
- `FileSystemWatcher` class
- `InternalBufferOverflowException` class
- `IODescriptionAttribute` class
- `RenamedEventArgs` class
- `WaitforChangedResult` structure

Creating a File Directory Browser

Unlike the Windows Mobile 5.0 Pocket PC SDK, the Smartphone SDK does not support the File Open or Directory Open dialog functions, nor is a file explorer shipped to any mobile devices running Microsoft Smartphone. This is probably because Smartphone devices have limited computing power and a fairly small amount of physical memory.

This section describes how to write a file and directory browser application that works more like a file explorer in a desktop PC.

Start a new Windows Mobile Smartphone device application and name the project `dirBrowse`. The UI design is relatively straightforward. First, rename the form to `BrowseFm` and change the caption of the form to **View Directory**. Then drag and drop a TreeView control to the form and make it occupy the entire client area. Then add a menu item Quit to the left soft key, as illustrated in Figure 5-3.

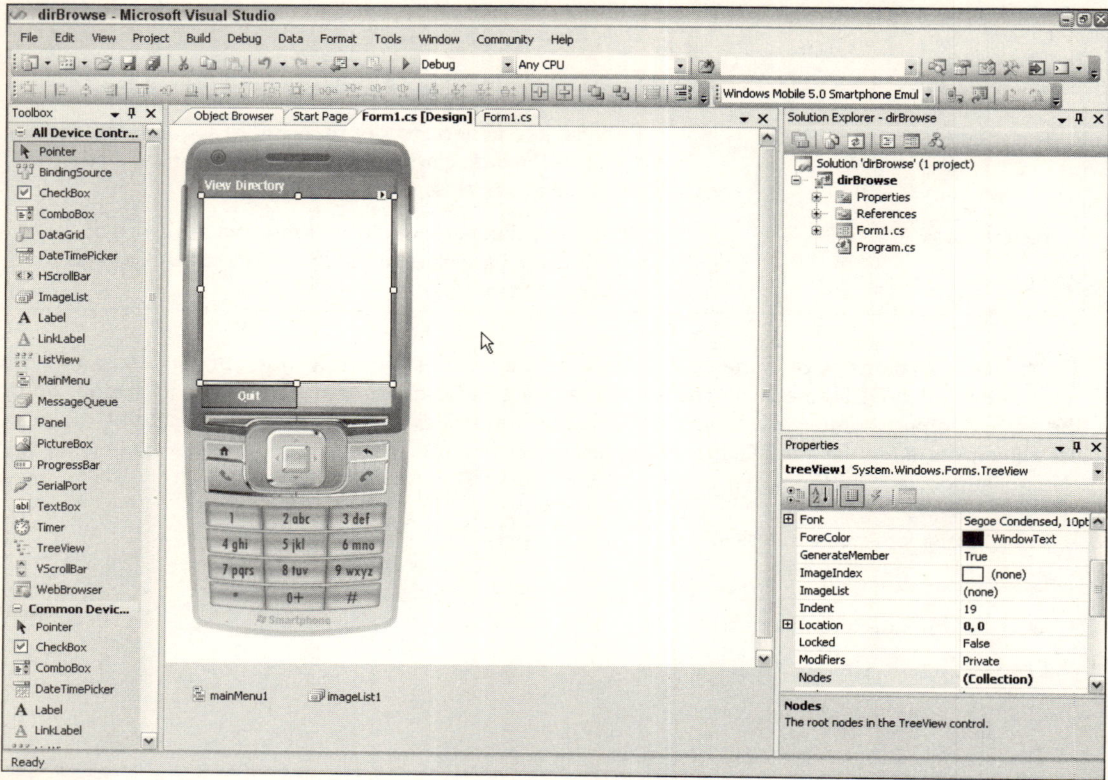

Figure 5-3

To make it more user-friendly, you can add an ImageList control that contains icons to represent different items, such as opened folders, closed folders, and so on. An ImageList control can contain images in different formats, including bitmaps, cursors, icons, JPEGs, and GIFs. The ImageList control provides a single repository for other windows controls — specifically, the ListView and TreeView controls on Windows Mobile Smartphone.

On the Form design window, drag and drop an ImageList control from the Toolbox to the form and use the default name `imageList1`. From the Properties Windows, select `imageList1` and click the Images button, as shown in Figure 5-4. The ImageCollection Collection Editor window appears. Click the Add button and add the icon images you would like to use in your project. In the example application, six images are added to represent the mobile device, the storage card, a closed folder, an open folder, an executable file, and a text file. Figure 5-5 illustrates the screen after six icon images are added. Click the OK button and close ImageCollection Collection Editor.

Figure 5-4

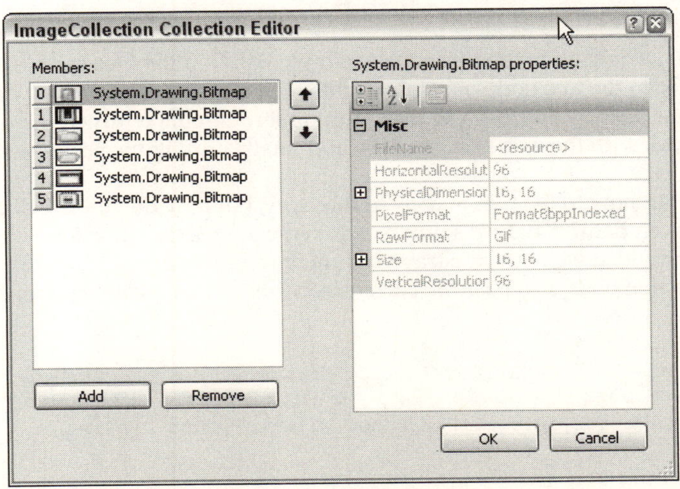

Figure 5-5

To traverse all the directories and files in a Smartphone system, it is better to write a recursive method to list all the directories and files for a given directory. To begin, the following is a method that lists all the files for a given directory:

```
private void GetAllFiles(DirectoryInfo curDir, TreeNode curNodes)
{
    //Get all the files in current directory
    try
    {
        foreach (FileInfo fi in curDir.GetFiles())
        {
            TreeNode tnFile = new TreeNode(fi.Name);

            //Set icons for executable files
            if (fi.Name.EndsWith("exe") || fi.Name.EndsWith("dll"))
            {
                tnFile.ImageIndex = 4;
                tnFile.SelectedImageIndex = 4;
            }
            //Set icons for non-executable files
            else
            {
                tnFile.ImageIndex = 5;
                tnFile.SelectedImageIndex = 5;
            }
            //Add the file to current treenode
            curNodes.Nodes.Add(tnFile);
        }
    }
    catch
    {
        //Do nothing if there are no files
        return;
    }

}
```

The `GetAllFiles()` method in the preceding example will find the files in a given directory `curDir` and add all the files to the tree node `curNodes`. This can be achieved by simply calling the `GetFiles()` method of the `DirectoryInfo` class and then looping through the returned `directoryInfo` array, creating new tree nodes with the names of the files, and finally adding the nodes to the given tree node.

It's easy to figure out how to list files in a given directory, but how do you list all the directories and subdirectories for a given path? A typical solution is to write a recursive function, which is a method that calls itself. For example, if we name the traverse function `GetAllDirs()`, it will call the `GetAllDirs()` method in the function body and pass the current subdirectory as the new parameter, as shown in the following code:

```
private void GetAllDirs(string DirName, TreeNode curNodes)
{
    DirectoryInfo curDir = new DirectoryInfo(DirName);

    try
    {
        foreach (DirectoryInfo di in curDir.GetDirectories())
```

```
            {
                TreeNode tnDir = new TreeNode(di.Name);

                  ...

                //Add current directory to
                curNodes.Nodes.Add(tnDir);

                //Recursively retrieving directories
                GetAllDirs(di.FullName, tnDir);
            }
            //Get files in current directories
            GetAllFiles(curDir, curNodes);
        }
        catch
        {
            //Do nothing if there are no more directories
            return;
        }
    }
```

As shown in the preceding example, for a given directory `DirName`, a new instance of `DirectoryInfo` `CurDir` is first created. Then all the subdirectories of `CurDir` can be obtained by calling the `GetDirectories()` method. Each subdirectory is added to the TreeView at the appropriate place and then a new search for the directories of each individual subdirectory is started by recursively calling the `GetAllDirs()` method.

The Windows Smartphone adopts the unified filesystem and eliminates the drive letter. This feature enables you to easily list all the directories and files in your Smartphone system, whether in RAM, ROM, or persistent storage, by passing the root of the filesystem \ to the `GetAllDirs()` method. However, how do you know whether a document is saved on a storage card, rather than the built-in ROM?

One way to determine whether a directory is on a storage card is to examine the `Directory` attribute bit and the `Temporary` attribute bit of a given file. For example, you can first define a storage card attribute `attrStorageCard` as follows:

```
static FileAttributes attrStorageCard =
    FileAttributes.Directory | FileAttributes.Temporary;
```

For any given directory, you can get its attributes by calling the `Attributes` function of the `DirectoryInfo` class. If both attribute bits are set, it will be a directory on a storage card; otherwise, it is not. The following code uses the bitwise `and` operation to determine whether a directory is stored on a storage card:

```
If ( (di.Attributes & attrStorageCard ) == attrSotrageCard )
    MessageBox.Show("This folder is saved on a storage card");
```

The following is the complete code of the file directory browser example:

```
using System;
using System.Collections.Generic;
using System.ComponentModel;
using System.Data;
```

```
using System.Drawing;
using System.Text;
using System.Windows.Forms;
using System.IO;

namespace dirBrowse
{
    public partial class BrowseFm : Form
    {

        static FileAttributes attrStorageCard =
            FileAttributes.Directory | FileAttributes.Temporary;

        public BrowseFm()
        {
            InitializeComponent();
        }

        private void BrowseFm_Load(object sender, EventArgs e)
        {
            //Create the root node of the TreeView
            treeView1.ImageList = this.imageList1;
            TreeNode rootNode = new TreeNode();

            rootNode.ImageIndex = 0;
            rootNode.SelectedImageIndex = 0;
            treeView1.Nodes.Add(rootNode);

            //Retrieve all the directories and files
            GetAllDirs(@"\", treeView1.Nodes[0]);
            treeView1.Nodes[0].Expand(); //Expand first layer
        }


        private void GetAllFiles(DirectoryInfo curDir, TreeNode curNodes)
        {
            //Get all the files in current directory
            try
            {
                foreach (FileInfo fi in curDir.GetFiles())
                {
                    TreeNode tnFile = new TreeNode(fi.Name);

                    //Set icons for executable files
                    if (fi.Name.EndsWith("exe") || fi.Name.EndsWith("dll"))
                    {
                        tnFile.ImageIndex = 4;
                        tnFile.SelectedImageIndex = 4;
                    }
                    //Set icons for non-executable files
                    else
                    {
                        tnFile.ImageIndex = 5;
                        tnFile.SelectedImageIndex = 5;
                    }
                    //Add the file to current treenode
                    curNodes.Nodes.Add(tnFile);
```

```
                }
            }
            catch
            {
                //Do nothing if there are no files
                return;
            }

        }

        private void GetAllDirs(string DirName, TreeNode curNodes)
        {
            DirectoryInfo curDir = new DirectoryInfo(DirName);

            try
            {
                foreach (DirectoryInfo di in curDir.GetDirectories())
                {
                    TreeNode tnDir = new TreeNode(di.Name);

                    //Not a storage card
                    if ( (di.Attributes & attrStorageCard) != attrStorageCard )
                    {
                        tnDir.ImageIndex = 2;
                        tnDir.SelectedImageIndex = 3;
                    }
                    else  //Storage Card
                    {
                        tnDir.ImageIndex = 1;
                        tnDir.SelectedImageIndex = 1;
                    }
                    //Add current directory to
                    curNodes.Nodes.Add(tnDir);

                    //Recursively retrieving directories
                    //String fullPath = di.Parent + @"\" + di.Name;
                    //GetAllDirs(fullPath, tnDir);
                    GetAllDirs(di.FullName, tnDir);
                }
                //Get files in current directories
                GetAllFiles(curDir, curNodes);
            }
            catch
            {
                //Do nothing if there are no more directories
                return;
            }
        }
    }
        //Close the application
        private void menuItem1_Click(object sender, EventArgs e)
        {
            this.Close();
        }

}
```

In the form initialized, a new `TreeNode` object, `rootNode`, is created. The root node will use the first image in `imageList1` for its icon, which represents the root a Smartphone. The `rootNode` object is then added to the control `treeView1`. Then the root of the filesystem and the head node of the `treeView1` control are passed to the `GetAllDirs()` methods, which recursively list all the directories and files and add them to the corresponding tree nodes. Figure 5-6 demonstrates the runtime result of the sample program.

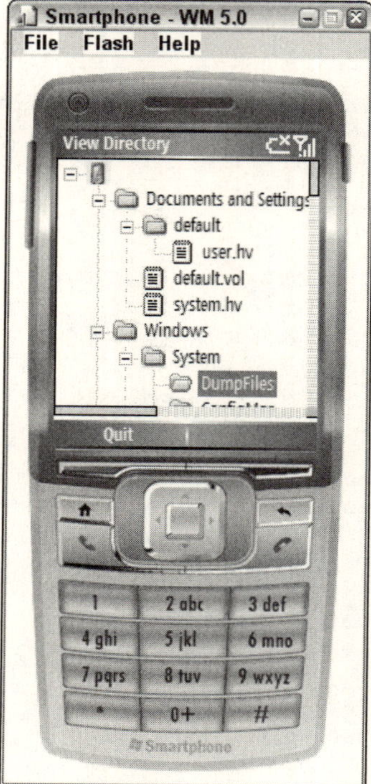

Figure 5-6

Implementing a Memo Application

The directory browsing example in the previous section illustrated how to use the `DirectoryInfo` and `FileInfo` classes to list all the directories and files, respectively. In this section, you learn how to read and write data from and to local files, as well as how to load and unload a control dynamically at runtime.

For this example, consider implementing a memo application that enables you to save memos into local files. The memo application should also give users the option to create, edit, and delete a memo. You can save all the memos to a single file with certain delimiters to separate each memo, or you can simply use one single text file for one memo.

For the user interface, there is not much to design because you have no clue how many memos are currently saved at design time. You should, however, provide menu items to enable users to perform operations such as creating, deleting, saving, and quitting memos. At runtime, you will need to read files to controls such as Textbox and display the information to users.

From Visual Studio 2005, create a new Windows Mobile Smartphone Device application. Name the project Memo and rename the name of the form from the default form1 to memoForm. Add a menuItem object to the left soft key and change the text to **Save and Quit**. Add another menuItem object Options to the right left key, and then add three submenu items to the Options menu item. Figure 5-7 shows the UI at design time.

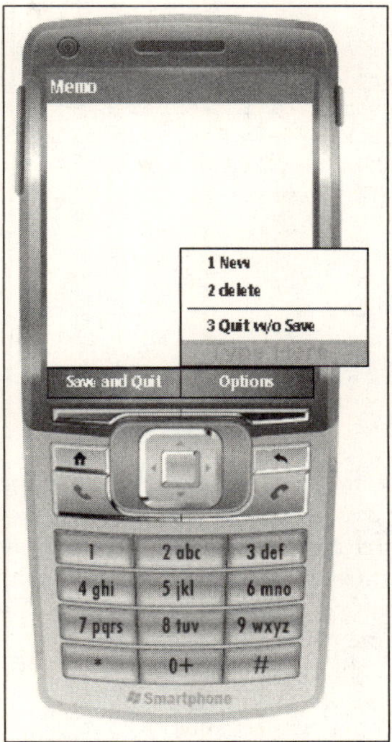

Figure 5-7

You may notice that there is no Edit menu item in the design. There is a reason for that. If you are going to use a TextBox control with the multiline property enabled to present the memo information, a built-in edit window enables you to view and edit the text of a memo in a full window.

After finishing the design of the user interface, you need to determine how to organize the data, how to access files dynamically, and how to respond to operations that a user will perform.

To make things easier, you can save the memo files to a fixed directory, such as \My Documents\Memos, and name the files numerically, such as 1.txt, 2.txt, and so on. How do you dynamically maintain the memo files? A simple solution could be to maintain an array of TextBox objects, with each TextBox object corresponding to a memo file. This way, it is also very easy to respond to user operations. When a memo is created, changed, or deleted, you do not have to save the changes right away. Instead, updated information is saved in RAM, and you can wait until users select the Save and Quit button to write data back to the files.

To dynamically display the TextBox objects in the self-defined ArrayList object, you first go through the ArrayList with the help of the IEumerator object. You then add each TextBox object to the Controls list of the form. After that, you can mark the Location and Size properties of each control so that it can be correctly positioned on the screen. The last thing, however, is to call the Show() method of the form to bring up all the controls on the window:

```
            int i = 0;
              IEnumerator tbEnum = tbList.GetEnumerator();

              while (tbEnum.MoveNext())
              {
                  this.aTxtbox = (TextBox)tbEnum.Current;

                  //Dynamically create a textbox and display it
                  //Height 1/10 client height, full width
                  this.Controls.Add(aTxtbox);
                  this.aTxtbox.Location = new System.Drawing.Point(0, 22 * i);
                  this.aTxtbox.Size = new System.Drawing.Size(ClientSize.Width, 22);
                  this.aTxtbox.Multiline = true;
                  i++;
              }

              this.aTxtbox.Focus();
              //Refresh display
              this.BackColor = System.Drawing.Color.White;
              this.ResumeLayout(true);
              this.Show();
              this.Refresh();
```

Another thing worth noting is how to delete a memo. When users select from the Delete menu, they expect the currently focused `TextBox` control to be deleted. You can achieve this by looping through the controls in the current form to search for the `TextBox` control that is currently focused. Once the focused control is found, you can remove it from both the `Controls` list and the `ArrayList` object, as follows:

```
        //Find the focused textbox
        foreach (Control c in this.Controls)
        {
            if (c.Focused && (c is TextBox))
            {
              string content = "Are you sure you want to delete memo content: \n" +
   c.Text;
              string title = "Delete Confirmation";

        DialogResult dr = MessageBox.Show(content, title, MessageBoxButtons.OKCancel,
                          MessageBoxIcon.Question, MessageBoxDefaultButton.Button2);

                //To delete control c in this.forms and from tblist
                if (dr == DialogResult.OK)
                {
                   this.tbList.Remove((TextBox)c);
                   this.Controls.Remove((TextBox)c);
                  ...
                 }
        }
```

Figures 5-8 through 5-10 illustrate some of the resulting screenshots when we run the application. There were three text files — namely, `1.txt`, `2.txt`, and `3.txt` — in the `\My Document\Memos` directory. The contents of each file were New Year Resolution, IP Phone, and XBOX 360, respectively, as indicated in Figure 5-8. Moving the arrow can change which `TextBox` object has the focus. When the first control has focus, pressing the Okay key provides access to the Edit window of this `TextBox` object, as shown in Figure 5-9. Figure 5-10 shows the confirmation dialog that appears when a user tries to delete the XBOX 360 memo, which happened to be the focused control when the Delete menu item was selected.

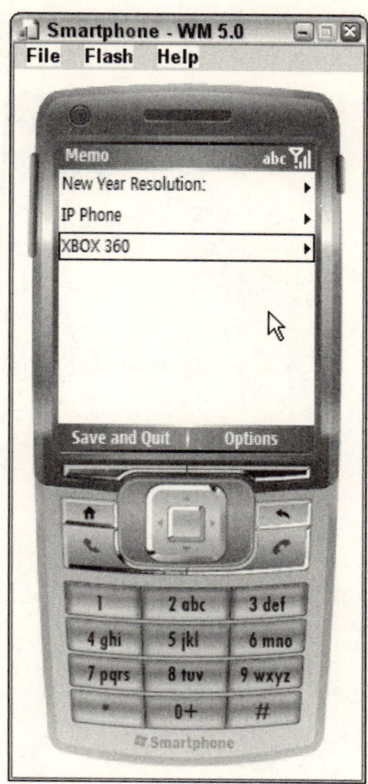

Figure 5-8

Figure 5-9

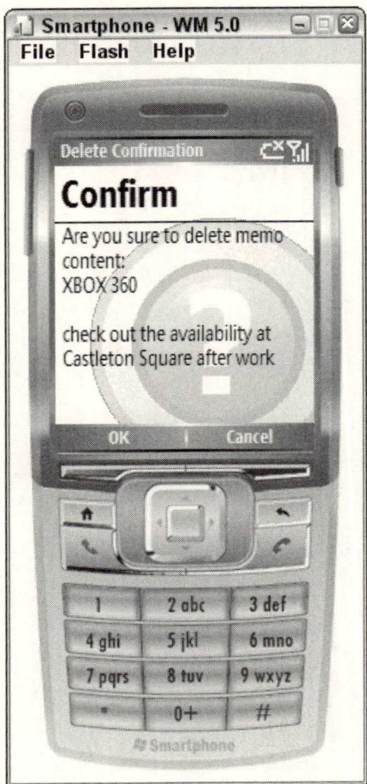

Figure 5-10

The following is the complete code of the memo example:

```
// Memos will be stored at "\My Documents\Memos"
// Files are named numerically

using System;
using System.IO;
using System.Collections;
using System.Collections.Generic;
using System.ComponentModel;
using System.Data;
using System.Drawing;
using System.Text;
using System.Windows.Forms;

namespace Memo
{

    public partial class memoForm : Form
    {
        static private string pathName = @"\My Documents\Memos";
        static private string txtExtension = ".txt";
        static private string extFiler = "*.txt";
        private System.Windows.Forms.TextBox aTxtbox;
```

```
private ArrayList tbList;

public memoForm()
{
    InitializeComponent();
    tbList = new ArrayList();

    //Create the memo directory if it does not exist
    CheckDirectory();

    //Read files to textbox array list
    readFile2TbList();

    //Read files to textboxes and add them to the ListView
    DisplayMemo();
}

private void CheckDirectory()
{
    //Check if the directory exists
    try
    {
        if (Directory.Exists(pathName)) return;
        Directory.CreateDirectory(pathName);
    }
    catch (Exception e)
    {
        MessageBox.Show( e.ToString() );
    }
}

//Read the textfile to tbList
//Return false if no files are found or there are errors
private bool readFile2TbList()
{
    try
    {
        DirectoryInfo di = new DirectoryInfo(pathName);
        FileInfo[] fis = di.GetFiles(extFiler);

        if (fis.Length == 0) return false;

        foreach (FileInfo fi in fis)
        {
            this.aTxtbox = new TextBox();
            this.aTxtbox.Name = fi.Name;

            //Open, read, and close the file
            StreamReader sReader =
                new StreamReader(
                    new FileStream(fi.FullName, FileMode.Open));
            this.aTxtbox.Text = sReader.ReadToEnd();
            sReader.Close();

            //Add aTxtbox to tbList
```

```
                this.tbList.Add((TextBox)aTxtbox);
            }
            return true;
        }
        catch (Exception e)
        {
            MessageBox.Show(e.ToString());
            return false;
        }
    }

    //Display items in textbox array list in the form
    private void DisplayMemo()
    {
        //Display no memo message when no items are in tblist
        if ( this.tbList.Count == 0 )
        {
            MessageBox.Show("There are no memos");
            //Disable "Delete" menu
            this.muItemDel.Enabled = false;
            return;
        }

        int i = 0;
        IEnumerator tbEnum = tbList.GetEnumerator();

        while (tbEnum.MoveNext())
        {
            this.aTxtbox = (TextBox)tbEnum.Current;

            //Dynamically create a textbox and display it
            //Height 1/10 client height, full width
            this.Controls.Add(aTxtbox);
            this.aTxtbox.Location = new System.Drawing.Point(0, 22 * i);
            this.aTxtbox.Size = new System.Drawing.Size(ClientSize.Width, 22);
            this.aTxtbox.Multiline = true;
            i++;
        }

        this.aTxtbox.Focus();
        //Refresh display
        this.BackColor = System.Drawing.Color.White;
        this.ResumeLayout(true);
        this.Show();
        this.Refresh();

    }

    //Save and close
    private void muItemSaveEx_Click(object sender, EventArgs e)
    {
        try
        {
            //Delete all previous files first
            DirectoryInfo di = new DirectoryInfo(pathName);
            foreach (FileInfo fi in di.GetFiles(extFiler) ) {
```

```
                fi.Delete();
            }

            int i = 0;
            IEnumerator tbEnum = tbList.GetEnumerator();

            while ( tbEnum.MoveNext() )  {
                string fileName = pathName + @"\" + i.ToString() +
txtExtension;
                i++;

                //Open, write, and close the file
                StreamWriter sWriter =
                    new StreamWriter(
                        new FileStream(fileName, FileMode.OpenOrCreate));

                        this.aTxtbox = (TextBox)tbEnum.Current;
                        sWriter.Write(aTxtbox.Text);
                        sWriter.Close();
            }

        }
        catch (Exception eQuit)
        {
            MessageBox.Show(eQuit.ToString());
        }
        this.Close();
    }

    //Add new files
    private void muItemNew_Click(object sender, EventArgs e)
    {
        //Create a new textbox and add it to the listview
        this.aTxtbox = new System.Windows.Forms.TextBox();
        int num = tbList.Count;
        this.aTxtbox.Name = num.ToString()+txtExtension;

        this.tbList.Add( (TextBox) aTxtbox);
        this.Controls.Add(aTxtbox);

        this.aTxtbox.Location = new System.Drawing.Point(0,22*num);
        this.aTxtbox.Size = new System.Drawing.Size(ClientSize.Width,20);
        this.aTxtbox.Multiline = true;

        this.aTxtbox.Focus();
        this.muItemDel.Enabled = true;
    }

    //Exit without saving
    private void muItemExt_Click(object sender, EventArgs e)
    {
        this.Close();
    }


    //Delete the textbox with the current focus
```

```
        private void muItemDel_Click(object sender, EventArgs e)
        {
            //Find the focused textbox
            foreach (Control c in this.Controls)
            {
                if (c.Focused && (c is TextBox))
                {

                    string content = "Are you sure you want to delete memo content:
\n" + c.Text;
                    string title = "Delete Confirmation";

                    DialogResult dr = MessageBox.Show(content, title,
MessageBoxButtons.OKCancel,
                        MessageBoxIcon.Question, MessageBoxDefaultButton.Button2);

                    //To delete control c in this.forms and from tblist
                    if (dr == DialogResult.OK)
                    {
                        this.tbList.Remove((TextBox)c);
                        this.Controls.Remove((TextBox)c);

                        //Display it again
                        DisplayMemo();
                        return;
                    }

                    //Not confirmed, return
                    return;
                }
            }
            //No focused textbox found
            MessageBox.Show("No memo is selected to delete\n Please try again");
            return;
        }
    }
}
```

During the initialization phase of the application, a new `ArrayList` object `tbList` is created that serves as a container for all the memos currently stored on the device. This application assumes that all memos are saved in the directory `\My Documents\Memos\`. If this directory does not exist, simply create it. A new `TextBox` control will be created for each text file in the memo directory, and then this newly created `TextBox` object is added to the `tbList`. Then the `DisplayMemo()` method is called to show all the memos. The content of each memo is copied to the `text` property of a `TextBox` control.

The `muItemNew_Click()` method handles the event when the menu item `New` is selected. This can easily be done by adding a new `TextBox` control to the `ArrayList` object `tbList` and carefully positioning the `TextBox` control on the screen. The `muItemDel_Click()` method removes the currently focused `TextBox` object from the `tbList`.

When users select the `Save and Quit` menu item, the application first deletes all the text files in the memo directory and the content of each `TextBox` control is written back sequentially as a text file. The application simply does nothing when users select the `Quit w/t Save` menu item.

Summary

In this chapter, you have learned the data storage structure of Window Mobile 5.0, especially the Smartphone edition. Persistent storage increases the battery life but may reduce overall system performance dramatically. Smartphone developers should always follow the best-practice guidelines to minimize the negative impact that this kind of storage has on mobile devices.

Accessing data locally basically involves using the classes provided in the `System.IO` namespace. `FileInfo` and `DirectoryInfo` are two classes you can use to perform file-related operations, such as copying, moving, creating, and deleting. Alternatively, you can use the static `File` and `Directory` classes to manipulate files and directories.

To read and write files in the basic byte format, use the `FileStream` class; otherwise, use `StreamReader` and `StreamWriter`. The `BinaryReader` and `BinaryWriter` are often used to access data in a specific encoding.

Dealing with file and directory operations is error prone. It is suggested that you catch exceptions with the `try...catch` statement. Even though the .NET Compact Framework supports a majority of the classes defined in the full .NET Framework, readers should be aware of unsupported classes and functions. This chapter emphasizes local file access; you are going to learn how to access data from a database in the next chapter.

6

Data Access with SQL Server Mobile

In the previous chapter you learned how to access data locally through the filesystem. This chapter deals with accessing relational data, whether it is saved locally or remotely on a database server. Knowing how to access relational data from a Smartphone is one of the fundamental skills that a programmer must grasp in today's data-driven world.

This chapter describes the architecture of SQL Server Mobile and provides step-by-step directions for installing and configuring the development, client, and server environments. Because SQL Server Mobile is installed locally on a smart device, it is ideal for Smartphone devices that are not always connected to the database. Two examples at the end of this chapter demonstrate how to programmatically access data, update data, and synchronize the data between the Microsoft SQL Server Mobile device and the Microsoft SQL Server via web services.

A good understanding of this chapter will also help you understand XML. In Chapters 13 and 14, you will learn how to enhance security when retrieving data from the database.

This chapter discusses the following topics:

- An overview of ADO.NET
- SQL Server Mobile architecture
- Installing SQL Server Mobile
- Creating databases and tables
- Creating publications
- Creating subscriptions
- A sample application with the DataGrid control

ADO.NET Overview

Simply put, ADO.NET is a set of managed classes in the `System.Data` namespace that are used to manipulate relational data. In the architecture of ADO.NET, a layer that can talk to various database systems is termed the *data provider*. The .NET Framework provides a variety of data providers, each handling one specific database. For instance, the .NET Framework data provider for SQL Server provides a means to easily connect and access data in a SQL Server 7.0 database, whereas the .NET Framework data provider for OLE DB supports accessing data in a SQL Server 6.5 or earlier database. Note that some data providers in the full ADO.NET, such as for Oracle and MySQL, are not supported in the .NET Compact Framework. The database systems that currently have their corresponding data providers implemented in the .NET Compact Framework are as follows:

- Microsoft SQL Server (version 7.0, 2000 and later)
- Microsoft SQL Server Mobile edition and its predecessor, Microsoft SQL CE
- IBM DB2 and DB Everywhere
- SyBase SQL
- Pocket Access

The ADO.NET classes provide two distinctive ways to access a database: connected mode and disconnected mode. In connected mode, applications access and update data directly via an open connection to the database. In disconnected mode, ADO.NET provides a `DataSet` class to cache data in memory. Applications can then access and manipulate data in memory without the open connections to the database. Because `DataSet` objects are independent of data sources, you can use `DataSet` objects to retrieve and access the data set from multiple data sources.

If you are not sure whether to use connected mode or disconnected mode to access the database, the following guidelines will help you make your decision:

- Use connected mode if
  - Your application requires forward-only and read-only data access
  - You want to free up more memory to other applications
- Use disconnected mode if
  - Your application interacts with remote data from XML web services
  - Your application interacts with multiple data sources
  - You want to free up the database connections to other applications

To connect and retrieve data in connected mode, first create a connection to the data source and then use a SQL command to directly retrieve the data. The following code snippet illustrates how to access the `students` table from a local SQL Server 2000 database `MyCourses`:

```
using System;
using System.Data;
using System.Data.SqlClient;
...

    SqlConnection dbConn = new SqlConnection("Data Source=localhost;
```

```
            Integrated Security = SSPI; Inital Catalog=myCourses");
        dbConn.Open();

        SqlCommand sqlCmd = dbConn.CreateCommand();
        sqlCmd.CommandText = "SELECT studentID, lastName FROM students";

        sqlDataReader stuReader = sqlCmd.ExecuteReader();

        while (stuReader.Read())
        {
            Console.WriteLine("{0}\t{1}", stuReader.GetInt32(0),
    stuReader.GetString(1));
        }

        stuReader.Close();
        dbConn.Close();
    ...
```

In the preceding code, a SQL connection is first created by specifying the connection string and invoking the `Open()` method of the `SqlConnection` class. Then a `SqlCommand` object is instantiated to retrieve records in the `studentID` field and the `lastName` field from the table `students`. The results of the query are executed by the `ExecuteReader()` method of the `SqlCommand` class and are passed to a `sqlDataReader` object. By iterating the `sqlDataReader` object with the `Read()` method, data can be retrieved row by row.

To retrieve the data in disconnected mode, a typical approach is to use a `DataAdapater` object to retrieve data from a data source and then populate the data to a `DataSet` object, which can hold multiple data tables in memory. The following code snippet demonstrates how to perform the same function in disconnected mode:

```
using System;
using System.Data;
using System.Data.SqlClient;
...
    SqlConnection dbConn = new SqlConnection("Data Source=localhost; Integrated
Security = SSPI; Inital Catalog=myCourses");

    string qStr = "SELECT studentID, lastName FROM students";
    SqlDataAdapter stuAdapter = new SqlDataAdapter(qStr, dbConn);

    DataSet stuSet = new DataSet();
    stuAdapter.Fill(stuSet, "students");

    for (int i=0; i<=stuSet.Tables["students"].Rows.Count-1; i++)
    {
        int ID = System.Convert.ToInt32(
stuSet.Tables["students"].Rows[i]["studentID"]);
        string name = stuSet.Tables["students"].Rows[i]["lastName"].ToString();
        System.Console.WriteLine("{0}\t{1}", ID, name);
    }
```

In the preceding example, a `SqlConnection` object is first created. However, you do not need to maintain an open connection to the database by calling the `open()` method. Rather, a `sqlDataApapter` object is created with the same SQL query command string and the same database connection string. When the `SqlAdapater.Fill()` method is executed, an implicit connection to the database is established and the data in the table is populated to the `DataSet` object. Once the `Fill()` operation is finished, it closes the connection to the database. You can then access each column of the data from the in-memory data set by stating the name of the table, the index of the row, and the name of the column.

This section only briefly introduces ADO.NET programming. For mobile devices, accessing a conventional database using either connected mode or disconnected mode may not be applicable because mobile devices do not have sufficient system memory and are not connected to the database servers all the time. In the following sections, you will learn how to access and manipulate data in a database that is designed for Windows Mobile devices.

Microsoft SQL Server 2005 Mobile Edition

The Microsoft SQL Server 2005 Mobile Edition (or, simply, SQL Server Mobile) is a lightweight database designed specifically for smart devices such as Smartphones, Pocket PCs, and Tablet PCs. Like its predecessor SQL Server CE, SQL Server Mobile is a trimmed-down version of Microsoft's desktop database. It allows faster and easier data access while disconnected, and synchronizes the data between mobile devices and desktop SQL servers while connected. In addition to enhanced reliability and performance, SQL Server Mobile adds a number of notable features, as summarized in Table 6-1.

Table 6-1 New Features of SQL Server Mobile

Feature	SQL Server CE	SQL Server Mobile
Multi-user support	No (single user only)	Yes
Multi-subscription support	No. You need to create a separate subscriptiondatabase for each subscription.	Yes
Column-level tracking	No. The minimum synchronization unit is a single row.	Yes. With both column- and row-level tracking, the minimum synchronization unit is a cell.
Auto reuse empty pages	No	Yes. The auto-shrink feature will reuse the empty pages, thereby saving storage space.
Integration with Visual Studio 2005 and SQL Server 2005	No	Yes

Note that SQL Server Mobile changed the database file format, so if you have a database file created with SQL Server CE, you will need to upgrade the database file from the command-line utility that SQL Server Mobile provides. By default, this upgrade utility is located at `C:\Program Files\Microsoft Visual Studio 8\SmartDevices\SDK\SQL Server\Mobie\v3.0\wce500\[processor]\upgrade.exe`.

Assume the old database is `oldDB.sdf` located in the `oldDir` folder with a password of `oldPass`. To upgrade it to `newDB.sdf` in the `newDir` folder with the password `newPass`, use the following command:

```
Upgrade.exe /s "oldDir\oldDB" /sp "oldPass" /d "newDir\newDB.sdf" /dp "newPass"
```

SQL Server Mobile Architecture

In a nutshell, SQL Server Mobile is a relational database that operates on a tiny runtime. As shown in Figure 6-1, the SQL Server Mobile architecture includes a development environment, a client environment, and a server environment.

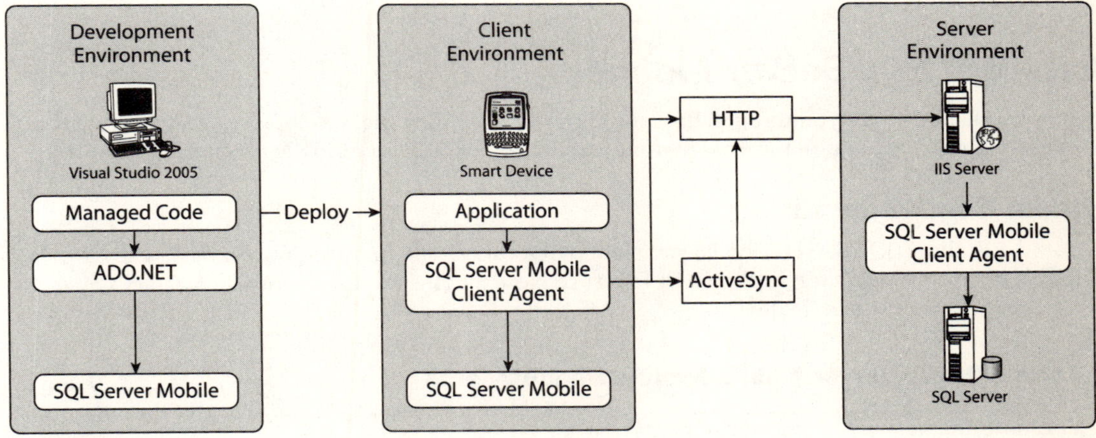

Figure 6-1

The *development environment*, of course, is where you develop your applications. For SQL Server Mobile applications, only Visual Studio 2005 is supported as the development environment. The managed code you develop relies on the ADO.NET layer in the .NET Compact Framework to call into SQL Server Mobile. If you have to develop your application using unmanaged code, you can get access to SQL Server through an OLE database provider.

The *client environment* is the smart device. Typically, SQL Server Mobile is preinstalled on a Windows Mobile 5 device and emulator. When you deploy your application to the device, the corresponding SQL Server Mobile databases are also copied to the mobile device as files in `.sdf` format. The Smartphone user can then simply access the local copy of the database for regular data manipulations. A connection to the servers is required when users need to synchronize their local databases with the copies stored on the database servers.

The *server environment* includes a Microsoft Internet Information Services (IIS) server, SQL Server Mobile Server Agent, and SQL Server. An IIS server is required in the server environment because the SQL Server Mobile Server Agent listens to the requests from the SQL Server Mobile Client Agent via HTTP requests. The SQL Server Mobile Client Agent can make such HTTP requests through either WiFi networks or by the ActiveSync connections. The desktop version of SQL Server in the server environment enables users to replicate and synchronize data between SQL Server Mobile databases and SQL Server databases, which in turn offers full functionality to manage and analyze data.

At the time of writing, Microsoft has announced SQL Server Everywhere Edition, which is very similar to SQL Server Mobile but it can be deployed not only to smart devices and Tablet PCs, but also to desktop computers and servers.

Installing SQL Server Mobile

This subsection walks you through the procedures for installing the development environment, the client environment, and the server environment to make SQL Server Mobile functioning correctly.

System Requirements

Before installing SQL Server Mobile, you should determine whether your system meets the hardware and software prerequisites, as shown in Table 6-2. Note that the required components vary depending on your SQL Server Mobile device.

Table 6-2 SQL Server Mobile System Requirements

Environment	Requirements
Development environment	Microsoft Visual Studio 2005
	One of the following operating systems:
	Microsoft Windows Server 2003, Windows XP Media Center Edition, Windows XP Professional, Windows XP Tablet PC Edition, Windows 2000 Professional SP4 or later versions, or Windows 2000 Server SP4 or later versions
	Microsoft Internet Explorer 6.0 or later to access SQL Server Mobile Books Online
	Microsoft ActiveSync 4.0 or later to debug and deploy applications

Environment	Requirements
Client environment	Any device that runs Microsoft Windows CE 5.0, Microsoft Windows XP Tablet PC Edition, Windows Mobile 2003 Software for Pocket PC, or Windows Mobile 5.0 2 to 3MB storage space
Server environment	Microsoft SQL Server 2000 SP3a or later SQL Server 2005: Intel or compatible Pentium 600 megahertz (MHz) or greater processor required (1 GHz or greater recommended), 256MB RAM minimum (512MB RAM or more recommended), 250MB hard disk space. IIS 5.0 or later versions: Supported on 32-bit Windows Server 2003, 32-bit Windows XP, and Windows 2000 SP4; 120MB of available disk space on the server. Microsoft ActiveSync 4.0 or later versions are required to use Management Studio to manage SQL Server Mobile databases on connected devices. Microsoft Internet Explorer 6.0 or later Microsoft Outlook 98 or later is required for synchronization of e-mail, calendar, contacts, tasks, and notes to the desktop or portable computer (Outlook 2003 recommended).

Installing the Development Environment

If you have already installed SQL Server 2005, you don't have to do anything. SQL Server Mobile files are already installed on the development computer. A number of classes in the `System.Data.SqlServerCe` namespace are available in the development environment to enable you to create databases and tables and to manipulate data in the databases.

If the developing tool is Visual C++ for Devices or Embedded C++, you need to include the `SsceOleDB.h`, `ca_mergex30.h`, and `Ssceerr30.h` files in your projects. These header files and libraries provide APIs to access SQL Server Mobile through OLE DB connections.

Installing the Server Environment

A typical setting for the server environment enables data exchange between SQL Server and SQL Server Mobile. You need to install IIS, SQL Server 2005/2000 with Replication Components, and SQL Server Mobile Server Tools. If the IIS server and SQL Server are on different computers, make sure the SQL Server Mobile Server Tools are installed on the one that runs IIS.

Installing the Client Environment

As mentioned earlier, SQL Server Mobile is well integrated with Visual Studio 2005. When you deploy a managed application that interacts with SQL Server Mobile, Visual Studio will determine whether the software and components needed to run applications are installed on the device. If not, those components will be installed automatically to the smart device. By default, the installation path is `\Windows` on the device. You can use SQL Server Management Studio in SQL Server to manage the SQL Server Mobile databases.

Things are little different if you need to deploy a native application. You need to manually copy SQL Server Mobile to the device. You may also want to copy the SQL Server Mobile Query Analyzer, which is a graphical management tool to manage SQL Server Mobile databases.

Setting Up the SQL Server Mobile Server Environment

In this section you will learn how to create and synchronize a SQL Server Mobile database. The process describes the most common scenario whereby a desktop version of SQL Server is running in the background and SQL Server Mobile is running on mobile devices. The synchronization between the two is made available through a web interface.

To begin, you need to install SQL Server Mobile Server Tools, which include the SQL Server Mobile Server Agent, the SQL Server Mobile Replication Provider, and the Configure Web Synchronization Wizard. This section assumes that you have already installed SQL Server 2005.

Installing SQL Server Mobile Tools

By default, Microsoft SQL Server Mobile Server Tools is not installed in SQL Server 2005. Navigate to `C:\Program Files\Microsoft SQL Server\90\Tools\Binn\VSShell\Common7\IDE` and find the setup program `sqlce30setupen.msi`.

Double-click the program to open the SQL Server 2005 Mobile Server Tools Setup Wizard. You can easily finish installing the tool; the settings are self-explanatory. The only thing to note here is to make sure that the Synchronize with SQL Server 2005 option is checked.

Creating a Database and Tables from SQL Server 2005

In the next step, you are going to create a database and a table from SQL Server.

First, from the machine on which SQL Server is installed, click Start⇨Program⇨Microsoft SQL Server 2005 and open SQL Server Management Studio. The Connect to Server Wizard appears, as indicated in Figure 6-2. Choose Database Engine as the server type and fill in the appropriate server name and authentication method of the server. When you are done, click the Connect button to connect to the SQL Server.

Figure 6-2

Once the SQL Server is connected, you can create a new database. Right-click the Databases icon from Object Explorer in SQL Server Management Studio and choose New Database (see Figure 6-3). In the resulting New Database window, shown in Figure 6-4, name the database **MyDB1**.

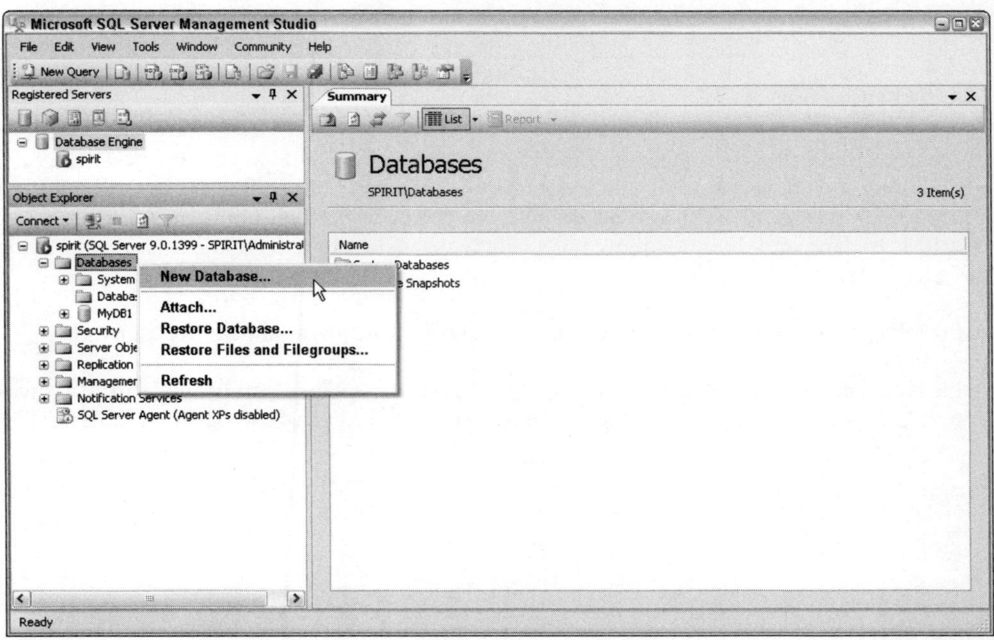

Figure 6-3

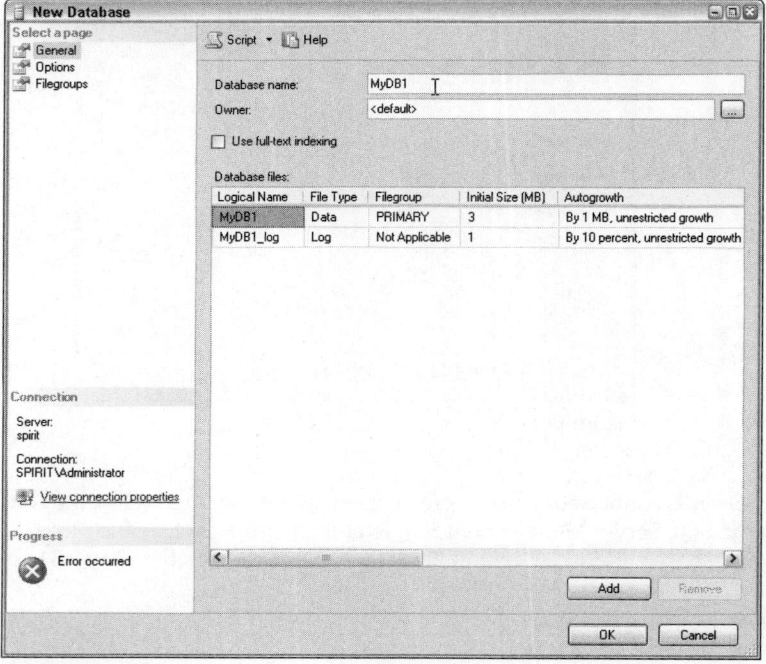

Figure 6-4

Next, create a simple StuGrades table in the MyDB1 database by right-clicking the MyDB1 icon from Object Explorer in SQL Server Management Studio and choosing New Table. In the Table designer of SQL Server Management Studio, add **StudentID** and **Grade** as the Column names. The data type of StudentID is `bigint` and does not allow nulls, whereas the data type of Grade is `smallint` and can take nulls, as indicated in Figure 6-5.

Once the table is created, it appears in the Object Explorer window, as shown in Figure 6-6.

The table is created, but it is currently empty. In SQL Server Management Studio, you can add rows to a table in a SQL Server database. Right-click the table you want to edit and choose Open Table (see Figure 6-7).

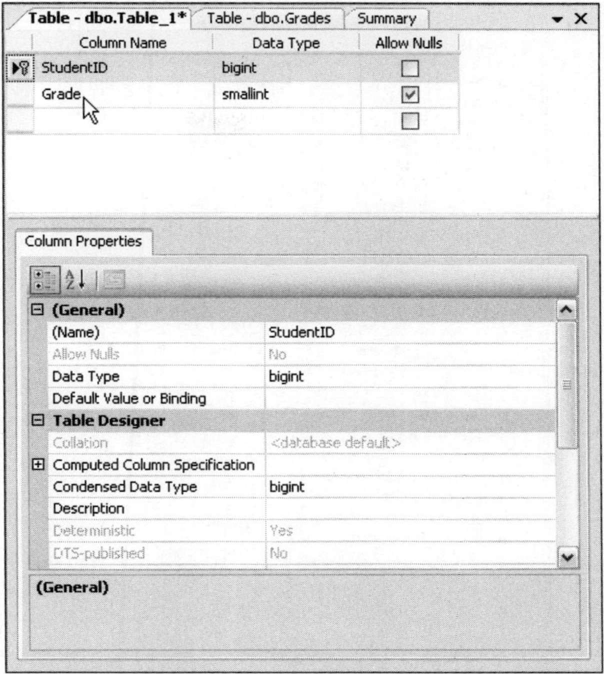

Figure 6-5

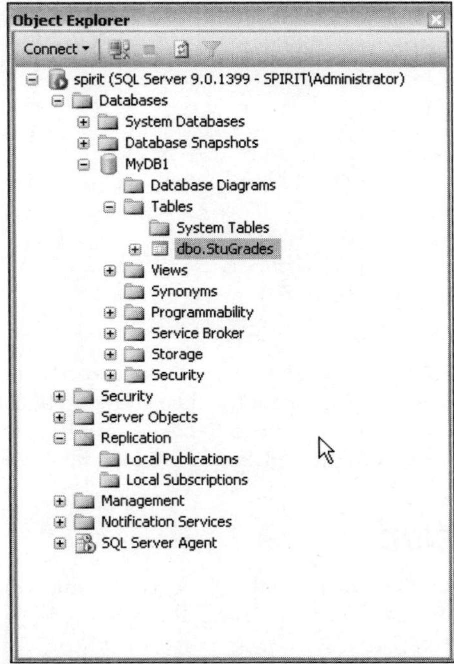

Figure 6-6

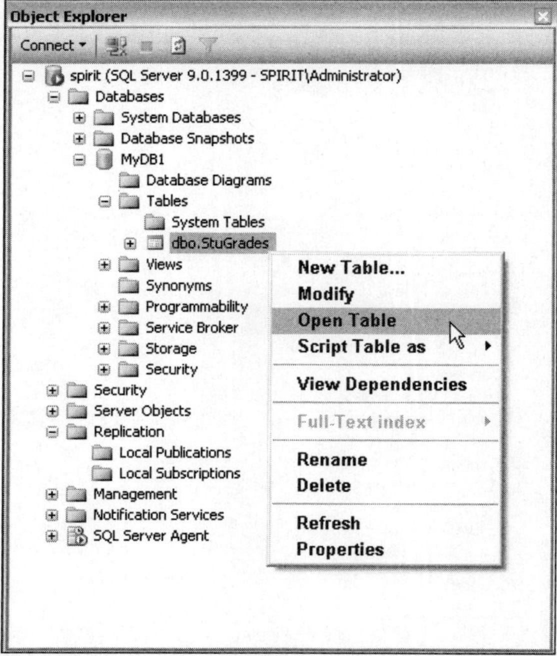

Figure 6-7

Randomly add entries to the table, as shown in Figure 6-8.

	StudentID	Grade
	1	98
	2	76
	3	52
	4	86
	5	79
	6	91
*	NULL	NULL

Figure 6-8

Now you have a simple database that contains a simple table, ready for action. In reality, the SQL Server database could be huge, and you certainly want the data stored in SQL Server to be easily replicated to a SQL Server Mobile database. To do that, you need to create a publication in a SQL Server database and then subscribe to this publication from the SQL Server Mobile database.

Creating a Publication

Before you create a publication, first ensure that SQL Server Agent, which appears in the bottom of the Object Explorer (see Figure 6-9), is currently running. If not, start it.

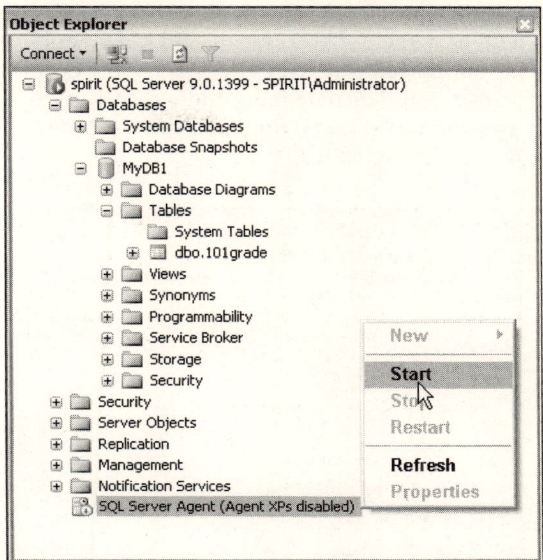

Figure 6-9

Expand the Replication folder from the Object Explorer and then launch the New Publication Wizard by right-clicking the Local Publication icon and choosing New Publication, as shown in Figure 6-10.

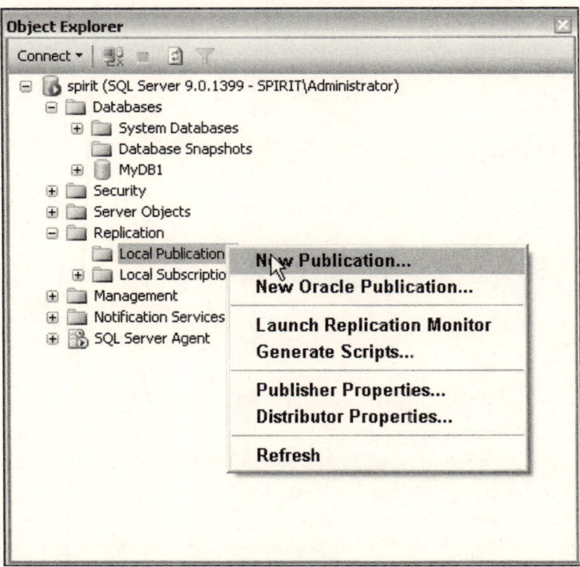

Figure 6-10

Four types of publications are available in SQL Server: Snapshot Publication, Transactional publication, Transactional publication with updatable subscriptions, and Merge publication. For SQL Server Mobile, Merge publication is the only supported way to synchronize with SQL Server. In Merge publication, both publisher and subscribers can update data independently. Changes are merged periodically. Choose Merge publication (see Figure 6-11) and don't forget to include SQL Server 2005 Mobile Edition in the Subscriber Types window (see Figure 6-12).

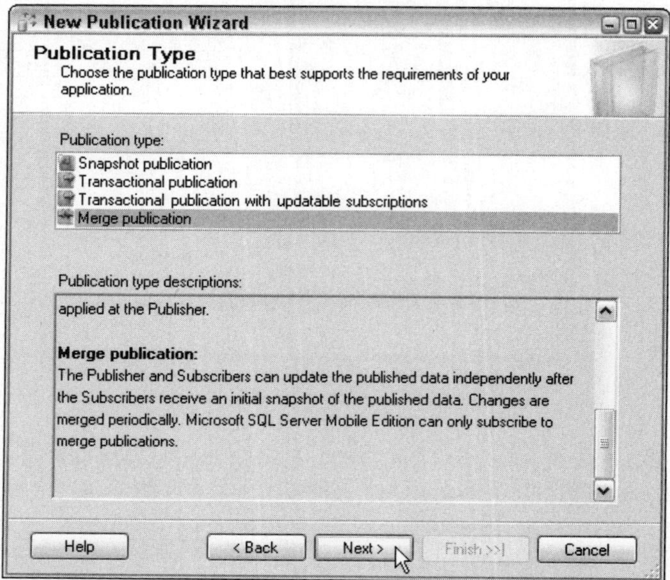

Figure 6-11

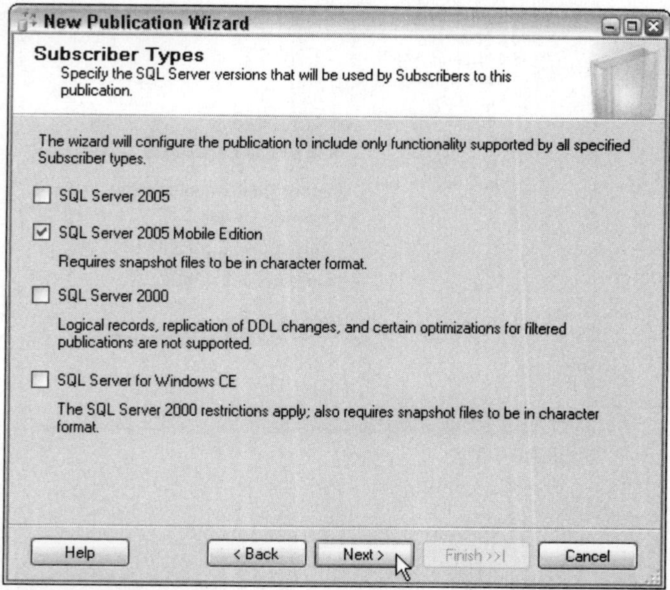

Figure 6-12

Next you will be prompted to choose which tables to publish. Check StuGrades from the Tables object, as shown in Figure 6-13.

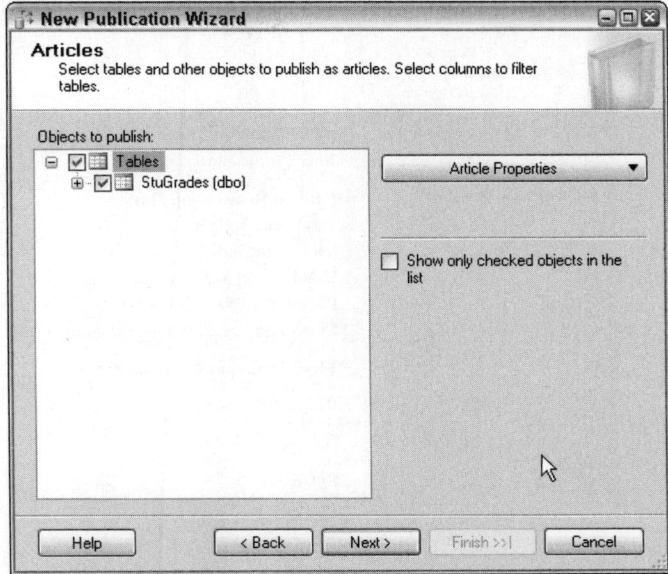

Figure 6-13

Configuring Web Synchronization

As you can see from the SQL Server Mobile architecture, synchronization between a smart device and a SQL server is handled by HTTP. That means you need to configure the web synchronization components of a published article. To do so, right-click the MyDB1 publication in Object Explorer and choose Configure Web Synchronization, as shown in Figure 6-14.

In the Synchronization wizard, add a new virtual directory to the default website, as shown in Figure 6-15.

SQL Server Mobile requires snapshot files to create a merge publication. By default, these snapshot files are located in the folder `C:\Program Files\Microsoft SQL Server\MSSQL.1\MSSQL\repldata`. You should share this folder and specify its UNC path name in the Snapshot Share Access window. The syntax for UNC naming is `\\ComputerName\ShareName`. In this example, the folder has a shared name of MyDB1 and a machine name of Spirit; therefore, the UNC path is `\\spirit\MyDB1`, as shown in Figure 6-16.

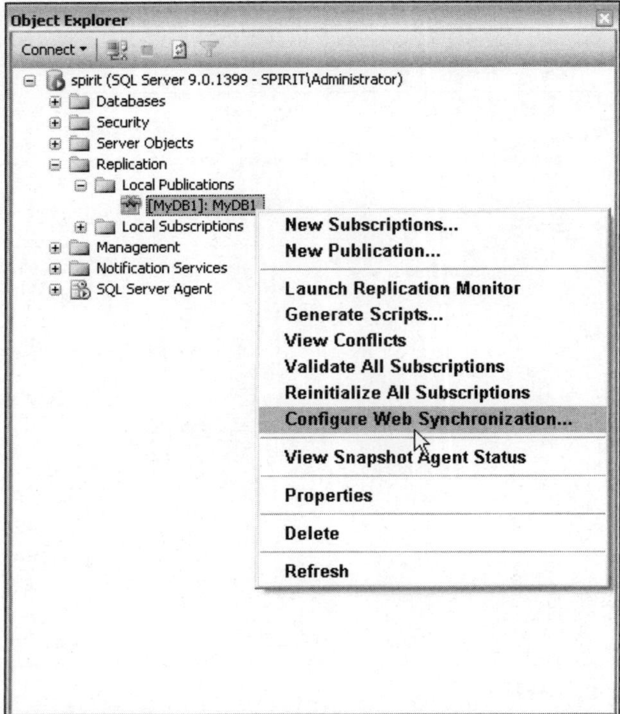

Figure 6-14

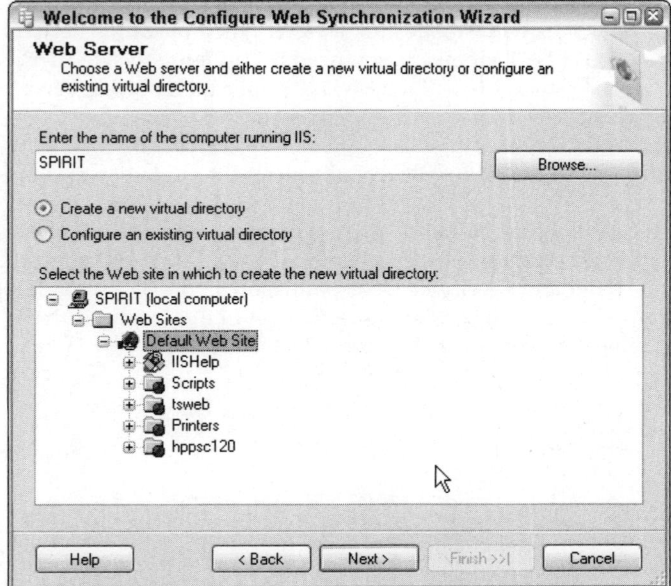

Figure 6-15

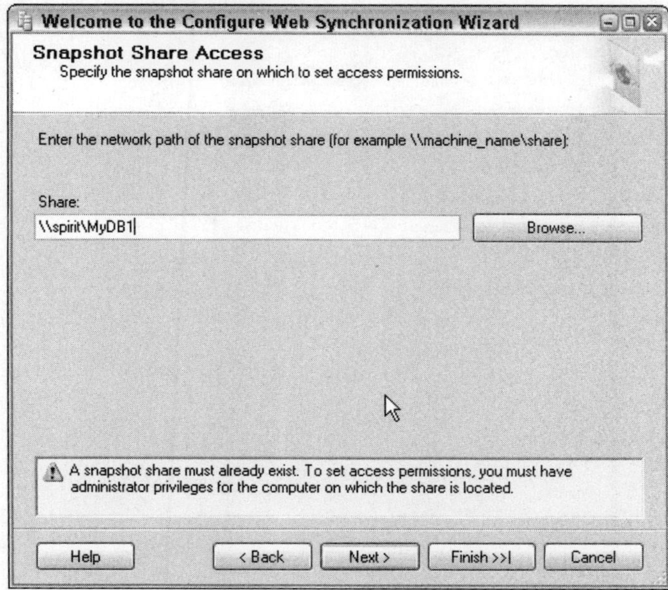

Figure 6-16

Creating a SQL Server Mobile Database

Now that the SQL server is set up and correctly configured, it is time to create and set up the SQL Server Mobile database.

Creating a SQL Server Mobile database is fairly simple using SQL Server Management Studio. From the Object Explorer, click Connect and choose SQL Server Mobile, as illustrated in Figure 6-17. When the Connect to Server window appears, click the drop-down list of Database files and choose <New Database>. You can then specify the name and location of the SQL Server Mobile database in the Create New SQL Server 2005 Mobile Edition Database dialog, as shown in Figure 6-18.

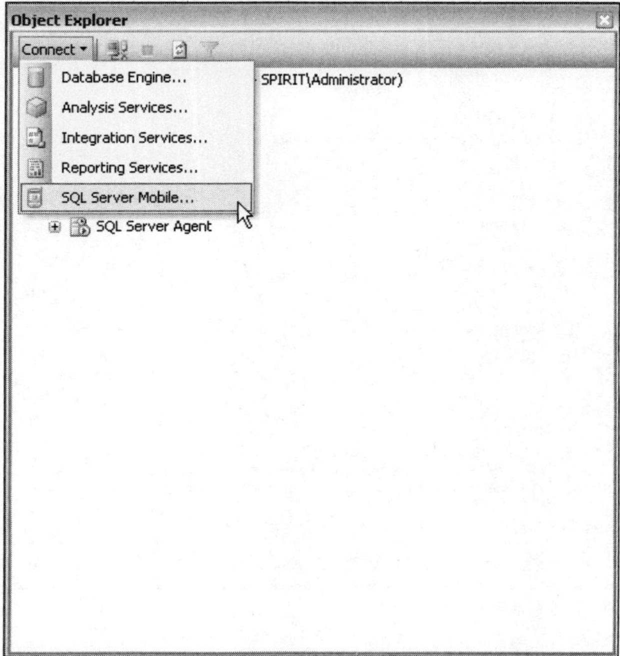

Figure 6-17

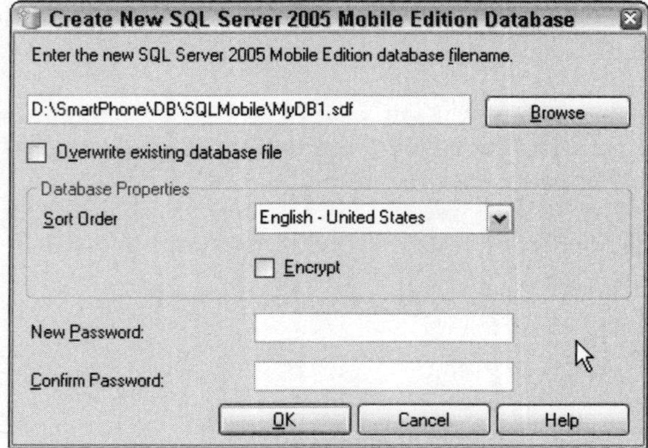

Figure 6-18

Creating Subscriptions in SQL Server Mobile

The last step to set up the server environment is to enable the SQL Server Mobile database to subscribe to the publications from SQL Server. To begin, run the New Subscriptions Wizard and select Find SQL Server Publisher, as shown in Figure 6-19 and Figure 6-20, respectively.

Object Explorer
Connect
spirit (SQL Server 9.0.1399 - SPIRIT\Administrator)
Databases
Security
Server Objects
Replication
Management
Notification Services
SQL Server Agent
SQL Server Mobile [My Computer\...\MyDB1]
Tables
Views
Programmability
Replication
Subscriptions
New Subscriptions...
Refresh

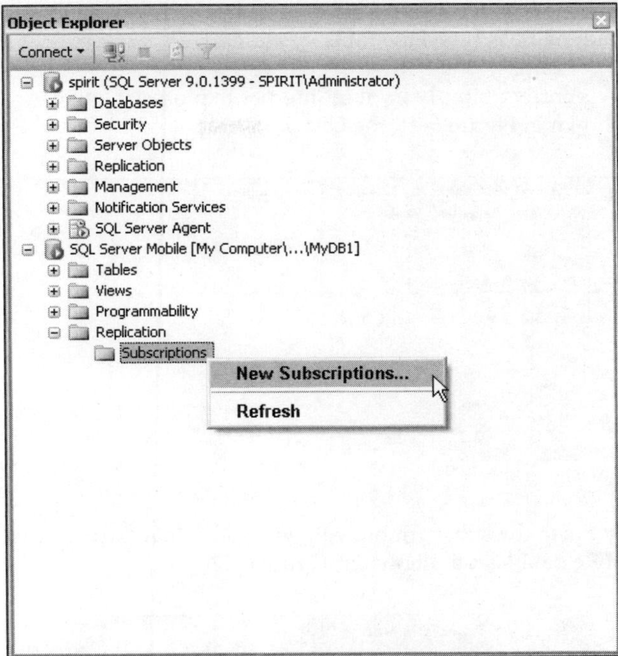

Figure 6-19

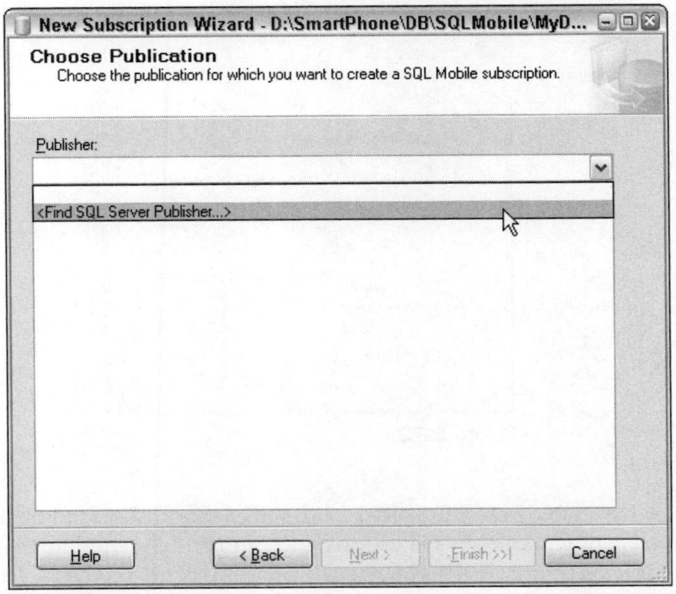

New Subscription Wizard - D:\SmartPhone\DB\SQLMobile\MyD...
Choose Publication
Choose the publication for which you want to create a SQL Mobile subscription.
Publisher:
<Find SQL Server Publisher...>
Help
< Back
Next >
Finish >>|
Cancel

Figure 6-20

During this process, you need to provide a URL to the SQL Server Mobile virtual directory located on the web server. The path name is typically `http://webserver_name/virutal_directoryname/sqlcesa30.dll`. In our example, the URL is `http://spirit/MyDB1/sqlcesa30.dll`. If you are not sure whether the URL is correct, simply try it in Internet Explorer. If you can read SQL Server Mobile Server Agent 3.0, as shown in Figure 6-21, the URL is correct.

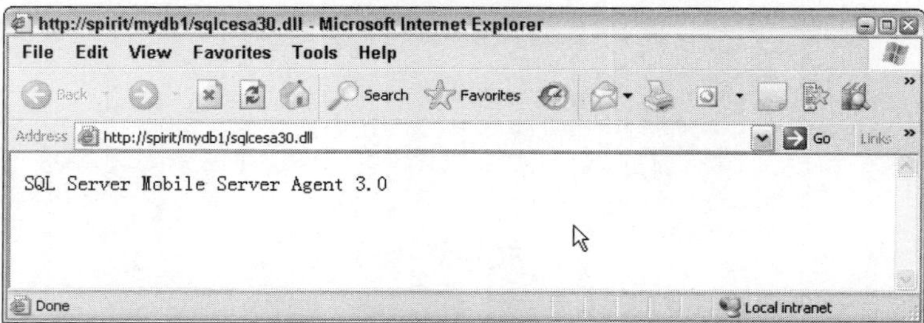

Figure 6-21

After you successfully finish the subscription, you will notice that the StuGrades table is now available in the SQL Server Mobile database, as shown in Figure 6-22.

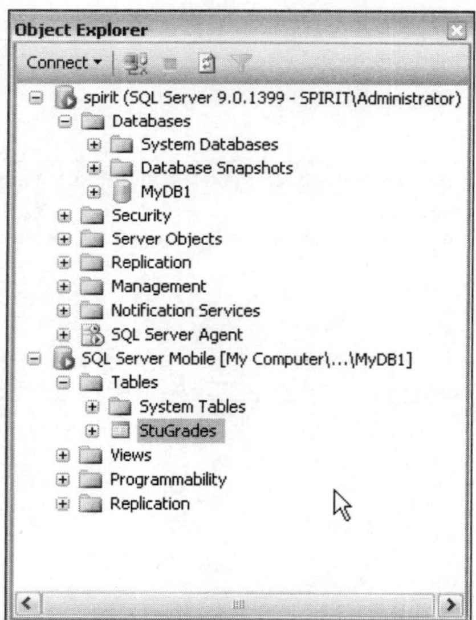

Figure 6-22

No graphical tools are available in Management Studio to enable you to read the rows in the SQL Server Mobile StuGrade table, but you can always use an SQL command to populate the data from the table. For example, the simple `Select` SQL statement enables you to compare the replicated table with the one you input manually. And, yes, they are the same, as you can tell from Figure 6-23.

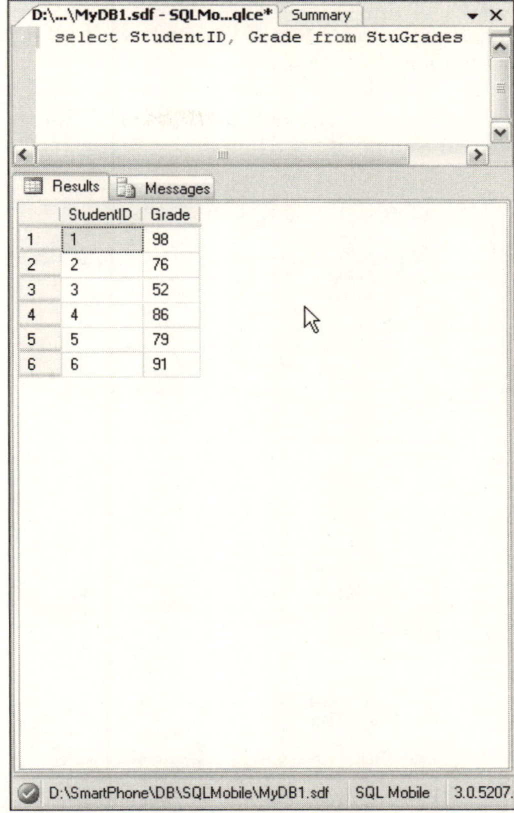

Figure 6-23

Writing SQL Server Mobile Applications

Programming SQL Server Mobile with Visual Studio 2005 is a relatively simple task. Indeed, you do not even have to write a single line of code to be able to retrieve data from a table. The following sections describe how to do just that.

A Simple Application with the DataGrid Control

A straightforward way to develop a SQL Server Mobile application is to add connections to the database and make those tables the data source for controls such as the DataGrid.

In the following example, you will be able to navigate the MyDB1 table that is replicated in the previous section.

First, create a new connection to the database. In Visual Studio 2005, click View➪Server Explorer to bring up the Server Explorer window. In the Server Explorer window, right-click Data Connections and choose Add Connection, as shown in Figure 6-24.

When the Add Connection dialog window appears (as shown in Figure 6-25), change the data source to SQL Server Mobile and specify the name of the database.

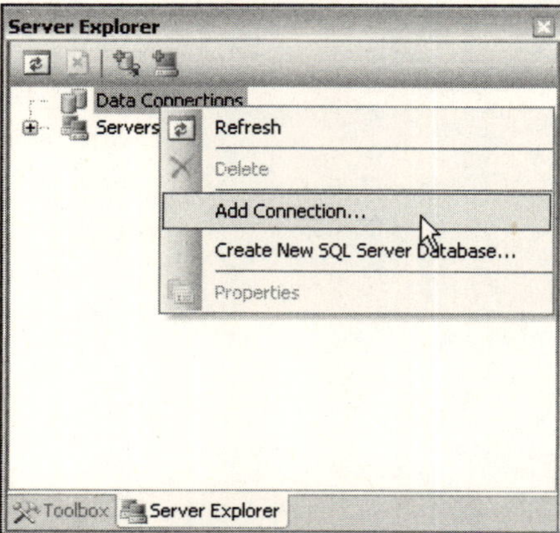

Figure 6-24

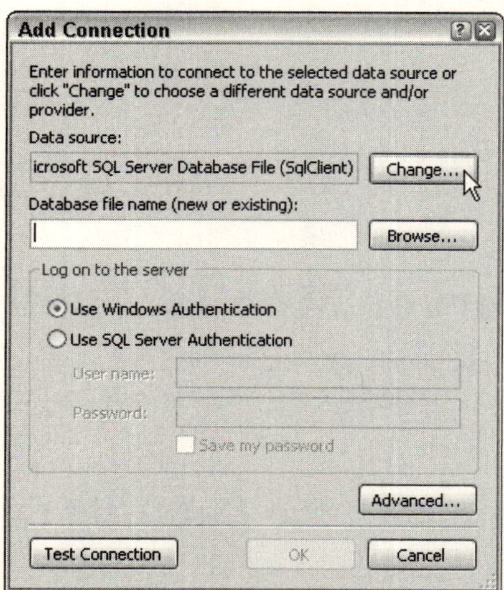

Figure 6-25

Now you will be able to see the table from the Server Explorer (see Figure 6-26). You can also use those tables or databases as the data source in your application.

The next step is to create a new Smartphone project. You can simply drag the DataGrid control to the form and specify the data source of the DataGrid control. As shown in Figure 6-27, from the Properties window of dataGrid1, click the DataSource drop-down list and then click Add Project Data Source to bring up the Data Source Configuration Wizard. Choose Database, as shown in Figure 6-28, and then specify the data connection that the application uses. In the example, the data connection is a SQL Server Mobile database MyDB1, shown in Figure 6-29.

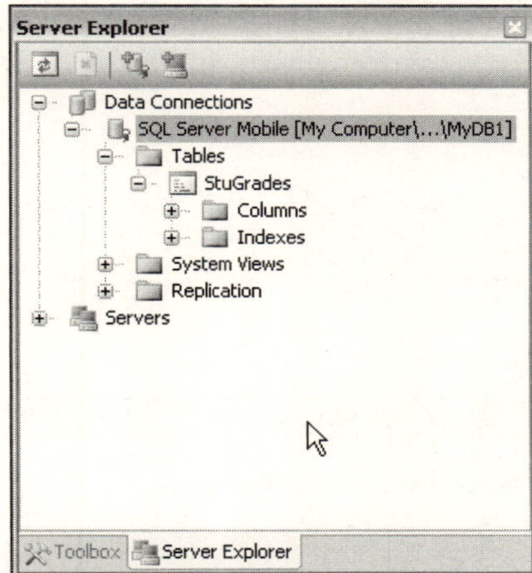

Figure 6-26

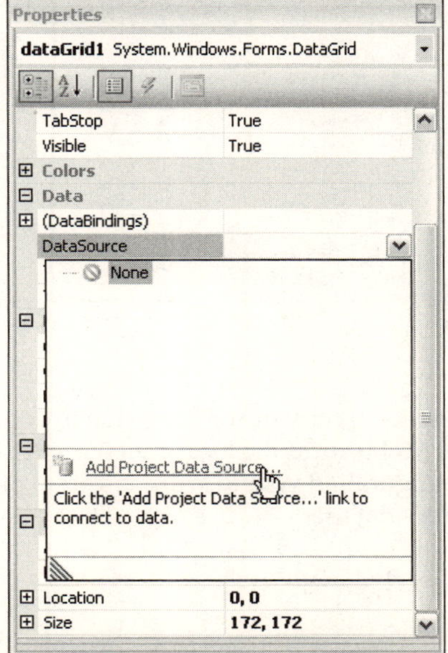

Figure 6-27

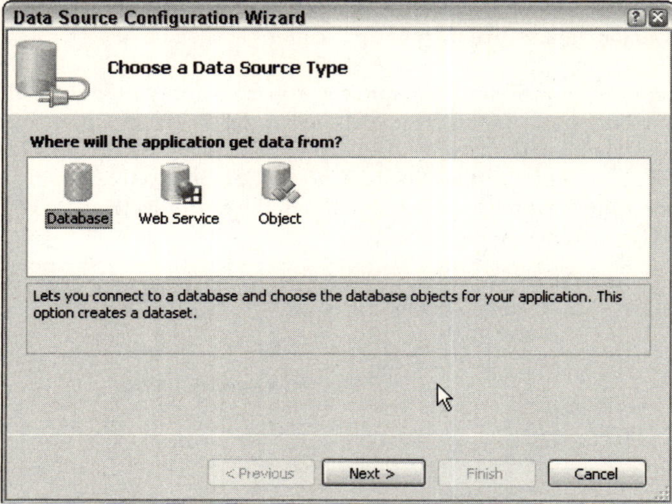

Figure 6-28

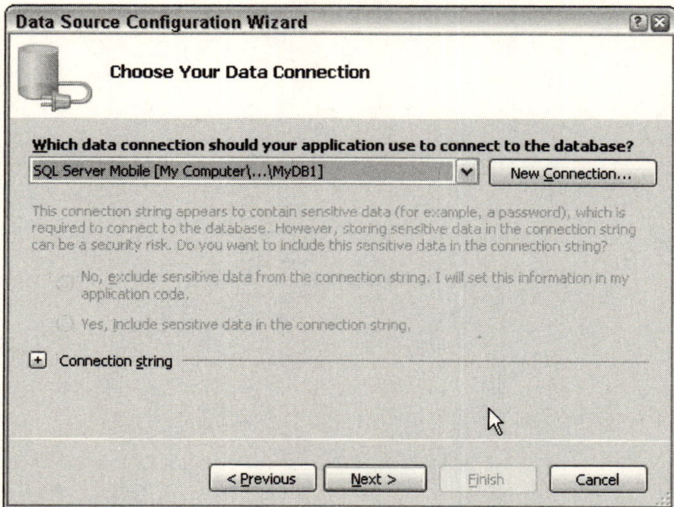

Figure 6-29

You are then asked if you would like to copy the database file to your Smartphone project, as shown in Figure 6-30. Because SQL Server Mobile uses the local data file, click Yes in the dialog box. As shown in Figure 6-31, the Data Source Configuration Wizard will ask you to pick the database objects that you wanted included in the in-memory data set.

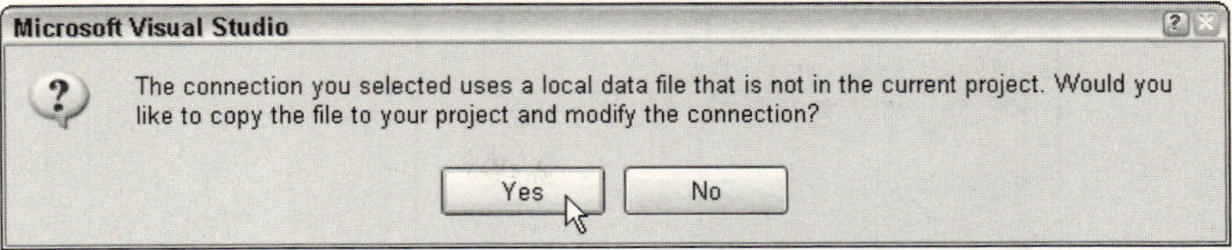

Figure 6-30

Figure 6-31

After data sources are added to the DataGrid control, the column names show up in the Form Designer. The simple UI of this example is shown in Figure 6-32.

Now just simply compile and deploy the application. Figure 6-33 shows that all the rows in table stuGrade are populated to the Data Grid control, and users can navigate the data with the navigation key from the Smartphone. Although it might not be what you expected to see, the DataGrid control works just fine even though you didn't write a single line of code.

In the next two examples, you will learn about some classes in the `SqlServerCe` namespace that enable you to create a database and tables and synchronize data.

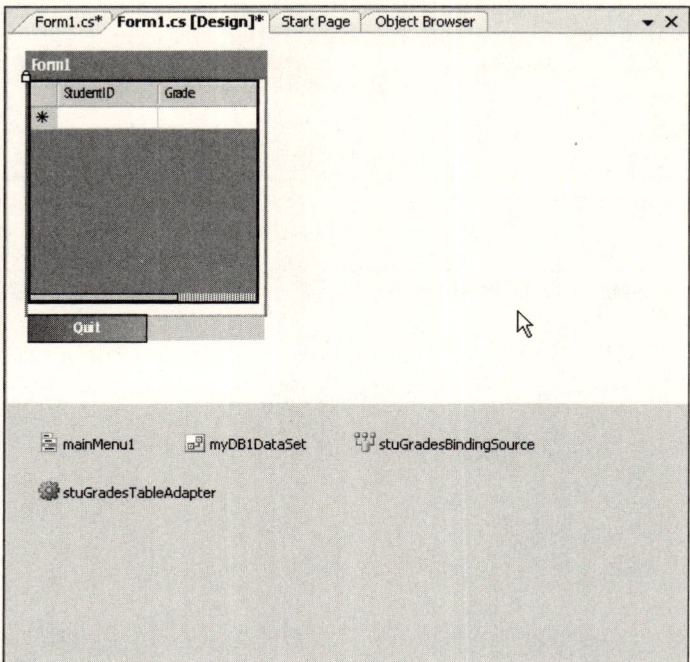

Figure 6-32

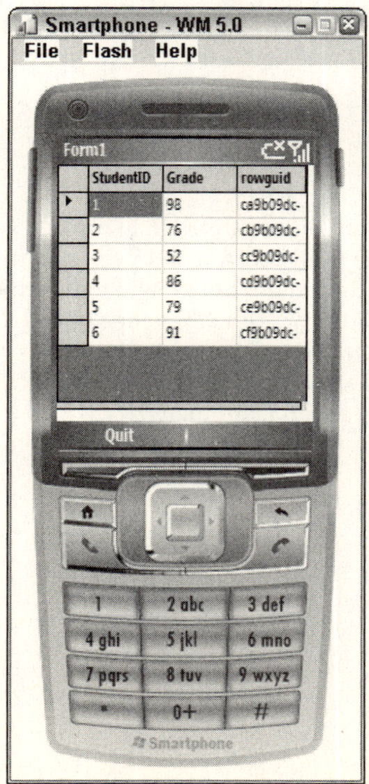

Figure 6-33

The SqlServerCe Namespace

The `SqlServerCe` namespace enables you to use SQL commands to manipulate data and manage the SQL Server Mobile database. To use APIs in this namespace, don't forget to add a reference. In Visual studio 2005, click Project➪Add reference and choose System.Data.SqlServerCe, as shown in Figure 6-34.

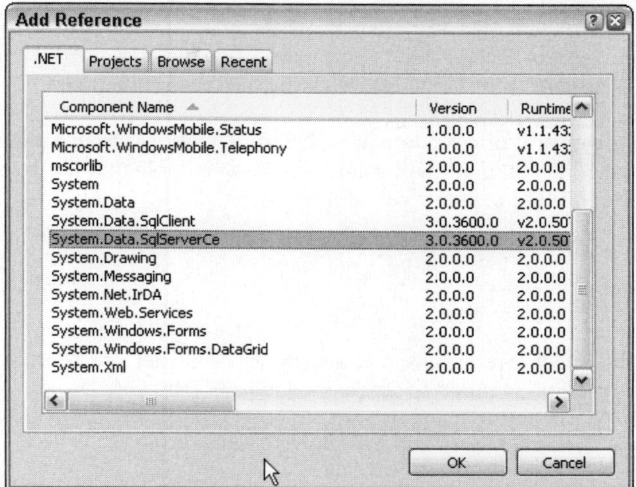

Figure 6-34

Recall from our earlier discussion that GUI tools are used to create a database and tables and to add rows. These steps can also be performed with the help of the `SqlServerCe` namespace.

To create a SQL Server Mobile database, use the `CreateDatabase()` method of the `SqlCeEngine` class. This operation requires a connection string to indicate the location of the database. The following code snippet demonstrates how to create a `MyDB2.sdf` database in the `My Documents` folder:

```
private string StrConn = @"Data Source=My Documents\MyDB2.sdf";
        private void addDB()
        {
            SqlCeEngine ceEngine = new SqlCeEngine();
            ceEngine.LocalConnectionString = StrConn;
            try
            {
                ceEngine.CreateDatabase();
            }
            catch (SqlCeException sqlEx)
            {
                MessageBox.Show(sqlEx.ToString());
            }
        }
```

Accessing relational data may cause exceptions to be raised. The best practice is to always put the database operation in a `try...catch` block.

The following example shows how to programmatically create a connection to a SQL Server Mobile database:

```
private string StrConn = @"Data Source=My Documents\MyDB2.sdf";
SqlCeConnection ceConn = new SqlCeConnection();
ceConn.ConnectionString = StrConn;
ceConn.Open();
```

A `SqlCeConnection` object is first created with a connection string pointing to the location of the database. The `open()` method is called to establish the database connection.

A simple solution to access and update data in SQL Server Mobile is to use the `SqlCeCommand` class. For example, the following SQL statement will create a stuGrades table that has two columns, studentID and Grade, where `studentID` is the primary key of the table:

```
CREATE TABLE StuGrades
   ( StudentID bigint not null CONSTRAINT PKStuGrades PRIMARY KEY ,
     Grade smallint )
```

The preceding SQL statement can be passed as a string to construct a `SqlCeCommand` object. Then the `ExecuteNonQuery()` method is called to create the table, as follows:

```
SqlCeCommand ceCmd = new SqlCeCommand();
ceCmd.Connection = ceConn;

string sqlCmd = new string (
    "CREATE TABLE StuGrades" +
    "( StudentID bigint not null" +
    "  CONSTRAINT PKStuGrades PRIMARY KEY , " +
    "  Grade smallint )"
    );
ceCmd.CommandText = sqlCmd;
ceCmd.ExecuteNonQuery();
```

Add data rows to a table using a similar approach. The following code snippet will add a grade of 98 to student ID 1:

```
SqlCeConnection ceConn = new SqlCeConnection();
SqlCeCommand ceCmd = new SqlCeCommand();
ceConn.ConnectionString = StrConn;

ceConn.Open();
ceCmd.Connection = ceConn;

ceCmd.CommandText = "INSERT StuGrades" +
                    " (StudentID, Grade)" +
                    " VALUES(1,98)";
ceCmd.ExecuteNonQuery();
```

The following code shows the full listing of the code that simulated what you practiced in the previous section:

```
using System;
using System.Collections.Generic;
using System.ComponentModel;
using System.Data;
using System.Drawing;
using System.Text;
using System.Windows.Forms;
using System.Data.SqlServerCe;

namespace DBapp2
{

    public partial class Form1 : Form
    {
        private string StrPath = @"My Documents\MyDB2.sdf";
        private string StrConn = @"Data Source=My Documents\MyDB2.sdf";

        public Form1()
        {
            InitializeComponent();
            addDB();
            addTable();
            addRows();
        }

        private void addDB()
        {
            SqlCeEngine ceEngine = new SqlCeEngine();
            ceEngine.LocalConnectionString = StrConn;
            try
            {
                ceEngine.CreateDatabase();
            }
            catch (SqlCeException sqlEx)
            {
                MessageBox.Show(sqlEx.ToString());
            }
        }

        private void addTable()
        {
            SqlCeConnection ceConn = new SqlCeConnection();
            SqlCeCommand ceCmd = new SqlCeCommand();
            ceConn.ConnectionString = StrConn;

            ceConn.Open();
            ceCmd.Connection = ceConn;

            string sqlCmd = new string (
                "CREATE TABLE StuGrades" +
                "( StudentID bigint not null" +
                "  CONSTRAINT PKStuGrades PRIMARY KEY , " +
                "  Grade smallint )"
                );
            ceCmd.CommandText = sqlCmd;
```

```
            ceCmd.ExecuteNonQuery();

            ceConn.Close();
        }

        private void addRows()
        {
            SqlCeConnection ceConn = new SqlCeConnection();
            SqlCeCommand ceCmd = new SqlCeCommand();
            ceConn.ConnectionString = StrConn;

            ceConn.Open();
            ceCmd.Connection = ceConn;

            ceCmd.CommandText = "INSERT StuGrades" +
                                " (StudentID, Grade)" +
                                " VALUES(1,98)";
            ceCmd.ExecuteNonQuery();

            ceCmd.CommandText = "INSERT StuGrades" +
                    " (StudentID, Grade)" +
                    " VALUES(2,76)";
            ceCmd.ExecuteNonQuery();

            ceCmd.CommandText = "INSERT StuGrades" +
                                " (StudentID, Grade)" +
                                " VALUES(3,52)";
            ceCmd.ExecuteNonQuery();

            ceCmd.CommandText = "INSERT StuGrades" +
                                " (StudentID, Grade)" +
                                " VALUES(4,86)";
            ceCmd.ExecuteNonQuery();

            ceCmd.CommandText = "INSERT StuGrades" +
                                " (StudentID, Grade)" +
                                " VALUES(5,79)";
            ceCmd.ExecuteNonQuery();

            ceCmd.CommandText = "INSERT StuGrades" +
                                " (StudentID, Grade)" +
                                " VALUES(6,91)";
            ceCmd.ExecuteNonQuery();

            ceConn.Close();
        }


    }
}
```

Everything in the example is straightforward except the path of the SQL Server Mobile database file. You should use the local path of the smart devices, rather than the development computer or the SQL servers.

Similarly, you can perform web synchronization by specifying the parameters of the publishers and subscribers, as follows:

```
using System;
using System.Collections.Generic;
using System.ComponentModel;
using System.Data;
using System.Drawing;
using System.Text;
using System.Windows.Forms;
using System.Data.SqlServerCe;

namespace DBapp3
{
    public partial class Form1 : Form
    {
        private string strPath = @"\Program Files\VS_SQLMobile\MyDB1.sdf";
        private string strConn = @"Data Source=\Program Files\VS_SQLMobile\MyDB1.sdf";

        public Form1()
        {
            InitializeComponent();
            updateArow();
            webSync();
        }

        private void updateArow()
        {
            SqlCeConnection ceConn = new SqlCeConnection();
            SqlCeCommand ceCmd = new SqlCeCommand();
            ceConn.Open();
            ceCmd.Connection = ceConn;

            ceCmd.CommandText = "UPDATE StuGrades" +
                                "SET grade = 84" +
                                "WHERE StudentID = 4";
            ceCmd.ExecuteNonQuery;
            ceConn.Close();
        }

        private void webSync()
        {
            SqlCeReplication repl = new SqlCeReplication();

            repl.InternetUrl = @"http://192.168.0.88/MyDB1/sqlcesa30.dll";
            repl.Publisher = @"spirit";
            repl.PublisherDatabase = @"MyDB1";
            repl.PublisherSecurityMode = SecurityType.NTAuthentication;
            repl.PublisherLogin = @"administrator";
            repl.PublisherPassword = @"password";
            repl.Publication = @"MyDB1";
            repl.Subscriber = @"MyDB1";

            repl.SubscriberConnectionString = "Data Source=" + strPath;

            try
```

```
            {
                repl.Synchronize();
            }
            catch (SqlCeException sqlEx)
                {
                        MessageBox.Show(sqlEx.ToString());
                }
        }

    }
```

The `updateArow()` function in the preceding example changes the grade to 84 for the student who has an ID of 4. This is accomplished by first establishing a `SqlCeConnection` and then executing a `SqlCeCommand` object, as follows:

```
private void updateArow()
{
    SqlCeConnection ceConn = new SqlCeConnection();
    SqlCeCommand ceCmd = new SqlCeCommand();
    ceConn.Open();
    ceCmd.Connection = ceConn;

    ceCmd.CommandText = "UPDATE StuGrades" +
                        "SET grade = 84" +
                        "WHERE StudentID = 4";
    ceCmd.ExecuteNonQuery;
    ceConn.Close();
}
```

To synchronize data through the web, you must instantiate a `SqlCeReplication` object. In addition, you need to correctly set many properties of the `SqlCeReplication` object. In the following `WebSync()` function, the URL of the web service is set to `http://192.168.0.88/MyDB1/sqlcesa30.dll`, and the `Publisher`, `Subscriber`, `PublisherLogin`, `PublisherPassword`, and `subscriberConnectionString` are all needed to perform a web synchronization:

```
private void webSync()
{
    SqlCeReplication repl = new SqlCeReplication();

    repl.InternetUrl = @"http://192.168.0.88/MyDB1/sqlcesa30.dll";
    repl.Publisher = @"spirit";
    repl.PublisherDatabase = @"MyDB1";
    repl.PublisherSecurityMode = SecurityType.NTAuthentication;
    repl.PublisherLogin = @"administrator";
    repl.PublisherPassword = @"password";
    repl.Publication = @"MyDB1";
    repl.Subscriber = @"MyDB1";

    repl.SubscriberConnectionString = "Data Source=" + strPath;

    try
        {
```

```
        repl.Synchronize();
    }
    catch (SqlCeException sqlEx)
        {
                MessageBox.Show(sqlEx.ToString());
        }
```

An IP address is used in the example instead of the computer name of the IIS server. This is because the Smartphone emulator may have trouble resolving computer names and therefore may not be able to connect the IIS server even though everything else is correct.

Summary

This chapter introduced two different ways to access a database in ADO.NET. The connected mode requires open connections all the time, whereas the disconnected mode provides an in-memory data set to cache the data from the database; therefore, it is not relying on an open connection. Usually, the `DataReader` class is used in connected mode and the `DataSet` class is instantiated in disconnected mode.

A Smartphone is characterized by a relatively slow processor and limited RAM, and there is no guarantee it is connected to the server or the Internet at all times. It therefore makes sense to have a slim version of the database stored on the smart device and a full-powered version of the database server running on a desktop. Replication is needed to ensure that both databases are on the same page.

To set up a programming environment for SQL Server Mobile–related applications, you should install and set up the client environment, the server environment, and the development environment. Programmatically accessing SQL Server Mobile data is pretty straightforward; it usually requires a database connection and a valid SQL statement.

In the next chapter, you will learn how to access data or an application through a variety of networks.

7

Networking

This chapter introduces you to one of the most exciting parts of Smartphone software development: networking! We start with an overview of the networking support in the .NET Compact Framework for Smartphone and then introduce topics ranging from application layer web access to TCP clients and sockets.

Be aware that due to the resource-constrained nature of Smartphones, networking applications should be carefully designed so that they don't consume too much memory and CPU cycles, especially when multiple threads and sockets are used in the application. You should not use low-level networking classes such as sockets unless you absolutely need them.

The following topics are discussed in this chapter:

- HTTP-based web access, including synchronous and asynchronous access
- TCP servers and TCP clients, and related IP endpoint classes
- TCP sockets and UDP sockets

An Overview of Smartphone Networking

Depending on your choice of application development environment, you will likely be able to use a largely different set of classes. For C# or Visual Basic applications targeting the Microsoft .NET Compact Framework, the exposed managed classes in the .NET Compact Framework library essentially enable fast and type-safe application development. For C++ applications targeting the underlying operating system, more direct control over the networking layers and the wireless connections are available in the form of the C++ API. Table 7-1 lists networking-related classes in the .Net Compact Framework.

Table 7-1 .NET Compact Framework Networking Classes

Namespace	Classes	Description

`System.Net`	`Dns`
	`IPEndPoint`
	`IPAddress`
	`IPHostEntry`
	`IrDAEndPoint`
		For identifying network endpoints

`System.Net`	`HttpWebRequest` `HttpWebResponse`	For building and processing HTTP-related requests and responses
`System.Net`	`WebRequest` `WebResponse`	For building web-based pluggable protocol implementations

`System.Net.Sockets`	`IrDAClient`
	`IrDAListener`
	`Socket`
	`TcpClient`
	`TcpListener`
	`UdpClient`
		For traditional stream- and datagram-based communication

The `Socket` class in the `System.Net.Sockets` namespace allows transport-layer and IP-layer communication with respect to different types of sockets. `TcpListener`, `TcpClient`, and `Udpclient` each implement a specific type of transport-layer socket.

Web and HTTP classes are for the application layer on top of the transport layer, TCP. The Dns class implements basic domain name resolution functionality; IPAddress and IPEndPoint represent the IP address and IP endpoint (IP address and a port number), respectively. `IPHostEntry` is a helper class for the `Dns` class, providing domain name-to-IP address mapping. `IrDA` classes, including `IrDAEndpoint`, `IrDAClient` , and `IrDAClient`, implement `IrDA` connection functionality.

As part of the networking capability, XML web services are so important that a separate chapter is devoted to them. Please refer to Chapter 9 for more details.

The .NET Compact Framework, at the time of writing, does not support Bluetooth networking. However, as more and more smart phones are equipped with Bluetooth and possibly other wireless technologies, it is fairly reasonable to expect support for those technologies in .NET Compact Framework for Smartphone application development.

Win32 APIs to control Bluetooth are available in the Smartphone SDK.

Emulator Networking

The Smartphone emulator implements the complete network stack of a Smartphone operating system. Hence, it can access the network through its hosting desktop computer via ActiveSync if the desktop computer is connected to an Ethernet or wireless LAN. In this case, the hosting computer works as a transparent network proxy for the emulator. No configuration is needed on the emulator. Essentially, the emulator comes with an emulated NE2000 PCMCIA network card that can be bound to a physical network card on the development machine. If you want to have a network totally contained within the desktop computer, you choose Hosts-only in the emulator configuration. If the development computer has no network connection or does not have a physical network card, you must install a Microsoft Loopback Adaptor via Add/Remove Hardware in the Control Panel of the desktop computer for the hosts-only network.

It your development computer has been configured to use a web proxy, you need to set the same proxy on the emulator. To do this, select Settings⇨Connections⇨Proxy, and then add the `proxy:port` *to the list.*

To check whether the Smartphone can access the network, open Visual Studio 2005 and run the emulator manager and cradle it. When the emulator appears, select the icon for Internet Explorer Mobile from the top menu and visit one of the default Favorite websites.

Figure 7-1 shows the Smartphone emulator running Internet Explorer Mobile accessing the MSN Mobile website. The network connection is made through the hosting desktop computer.

Figure 7-1

Once you have confirmed that the Smartphone emulator is able to connect to the network, you are ready to explore the networking facilities provided by the .NET Compact Framework.

Web Access

The `WebRequest` and `WebResponse` classes are abstract classes that implement basic web access functionality, including HTTP web requests and responses, and local filesystem access using Uniform Resource Identifiers (`file://`). The classes that implement these two types of web access are as follows:

- **HttpWebRequest (derived from WebRequest) and HttpWebResponse (derived from WebResponse)** — For typical HTTP web access to a website
- **FileWebRequest (derived from WebRequest) and FileWebResponse (derived from WebResponse)** — For local filesystem access via a web interface

The two web request types support both synchronous access using `GetResponse()` and `GetRequestStream()` and asynchronous access using `BeginGetResponse()/EndGetResponseStream()` and `BeginGetRequestStream()/EndGetRequestStream()`.

The HttpWebRequest and HttpWebResponse Classes

The `HttpWebRequest` class enables you to specify security credentials and a web proxy, and a timeout value for a request. This class is useful when you need to specify some settings for web access. Table 7-2 describes the class's properties.

Table 7-2 HttpWebRequest Properties

Properties	Description	Example
`Credentials`	Authentication information of the requester; can be either a `NetworkCredential` instance or a `Credential Cache` instance	`NetworkCredential myCredentials = new Network Credential(username,passwd);` `myCredentials.domain = domain;` `MyWebRequest.Credential = myCredentials;`
`ContentType`	HTTP header `Content-Type`	`MyRequest.ContentType = "text/ html";`
`ContentLength`	HTTP header `Content-Length`; `-1` if the request does not upload data	`string data = "data to be uploaded";` `ASCIIEncoding ascii = new ASCIIEncoding();` `Byte[] encodedBytes = ascii .GetBytes(data);` `MyWebRequest.ContentLength = encodedBytes.Length;`
`Method`	The method of the HTTP request; the default is `Get`	`MyWebRequest.Method = "POST";`
`Proxy`	The web proxy to be used for the request; the default is `Global ProxySelection.Select();`	`WebProxy myProxy = new WebProxy("http://myproxy: 8000");` `MyWebRequest.Proxy = myProxy;`
`Timeout`	The timeout value of the request, in milliseconds; the default is 100 seconds	`MyWebRequest.Timeout = 5000;`
`ProtocolVersion`	The HTTP version of the request; only 1.0 and 1.1 are allowed	`myWebRequest.ProtocolVersion= HttpVersion.Version10;`

In addition, the `HttpWebRequest` and `HttpWebResponse` classes have a `Headers` property of a `WebHeaderCollection` type that, along with the properties in Table 7-2, make up the headers for the request. A `WebHeaderCollection` type is a list of name-value pairs of the protocol headers.

`HttpWebResponse` also has a `ContentLength` property indicating the total length of the response data, but not all web servers support this HTTP header while generating a response.

After you consume the response data, be sure to release the resource of `StreamReader` and the `WebResponse` by calling their `Close()` methods.

Creating HTTP Request

To create a specific web request, you can call the static method `Create()` of the `WebRequest` class with a URI (universal resource identifier), such as an HTTP address in the form of `http://company.com/page.html` or a file address in the form of `file://computer_name/folder/filename`. The `Create()` method will return an instance of the `HTTPWebRequest` class or a `FileWebRequest` class. Then you can use either synchronous or asynchronous methods to send the request. Once you have the `WebResponse` instance, you can obtain the stream of the `WebResponse` instance and read and write the stream using a `StreamReader` object. The following sections provide examples of both synchronous access and asynchronous access.

Synchronous Access

Synchronous access means that when the application sends the request to a remote server, it blocks until a response is received or a timeout occurs. Synchronous access is generally simpler to program than asynchronous access. To retrieve a web page from a web server synchronously, you create a `WebRequest` object with a certain URL, and then call the `GetResponse()` method of the object. This method returns a `WebResponse` object. Then, using a `StreamReader` object, you can read the response stream by calling the `GetResponseStream()` method of the `WebResponse` object. The following example demonstrates this procedure:

```
//Create an HTTP web request with an HTTP URL. Note that the HttpWebRequest
instance is a WebRequest class instance because HttpWebRequest is a descendant
class of WebRequest
WebRequest myWebRequest = WebRequest.Create(http://www.msn.com);

//Send the request; here we use the default timeout value of the system
WebResponse myWebResponse = MyWebRequest.GetResponse();

//Create a StreamReader to access the data. The second parameter is the encoding of
the response stream. For Unicode text, use System.Text.UnicodeEncoding.
StreamReader myStreamReader = new
StreamReader(myWebResponse.GetResponseStream(),Encoding.ASCII);

//Now we can read the stream
Char[] buffer = new Char[1024];
Int count = myStreamReader.Read(buffer, 0, 1024);
While(count > )
```

```
  count = myStreamReader.Read(buffer, 0, 1024);

...//Use the data here
myStreamReader.Close();
myWebResponse.Close();
......
```

The call to `GetResponse()` will block the calling thread until the response is received or a timeout occurs. After a `WebResponse` object is instantiated, you can use a `StreamReader` object to read the data in the response according to an encoding scheme such as ASCII or Unicode.

Asynchronous Access

In the case of asynchronous access, a call to send requests does not block the calling thread. This means the calling thread can continue to do something else. Once a response is ready to be picked up, a callback method is invoked to process the response. Asynchronous access essentially enables applications to perform tasks concurrently while waiting for a response. Therefore, it should be used for better performance if responses are often slow. Asynchronous web access is slightly more complicated than synchronous access. You need to specify a callback method as one of the parameters in the `BeginGetResponse()` method. The call to `BeginGetResponse()` will return immediately. The .NET runtime creates a separate thread that executes the callback method. The callback method should call `EndGetResponse()`, which will block the calling thread until a response arrives. Then a `WebResponse` object associated with this asynchronous call operation is returned. Another object needed for asynchronous access is a user-defined state object that distinguishes the request from other requests.

In the following example, as in the synchronous access case, a `WebRequest` object is created with a specified URL. Note that a `RequestState` object is used to pass the `WebRequest` object and `WebResponse` object to the callback method. Here you have the freedom to define whatever class will do the job. Then the `BeginGetResponse()` method of the `WebRequest` object is called with two parameters: a delegate that points to the callback method (listed below) and the `RequestState` object:

```
WebRequest myWebRequest = WebRequest.Create("http://google.com");
myRequestState = new RequestState(); //Assuming we have defined a RequestState
class that contains a WebRequest property and a WebResponse property
myRequestState.request = myWebRequest;
//Get request asynchronous result
//The first parameter is an AsyncCallBack delegate to be called when this operation
is over
//The second parameter is user-defined object that distinguishes this request
IAsyncResult asyncResult = (IAsyncResult)myWebRequest.BeginGetResponse(
                    new AsyncCallback(MyCallback), myRequestState);
....
```

The following snippet is the corresponding callback method:

```
private void MyCallback(IAsyncResult ar)
{
  myRequestState = (RequestState)ar.AsyncState;  //Get the state object
  WebRequest webRequest = myRequestState.request; //Get the WebRequest object of
this call

  myRequestState.response = webRequest.EndGetResponse(ar); //Complete the
asynchronous call, return the response
```

```
  Stream responseStream = myRequestState.response.GetResponseStream(); //Ready to
read
...
}
```

The advantage of the asynchronous access method is that your main thread does not need to wait until the HTTP response arrives, which may take from hundreds of milliseconds to even a few seconds, depending on many factors on the Internet. You can perform a GUI update or other tasks while waiting for the response. Note that you can also read the stream asynchronously.

While using the `WebRequest` and `WebResponse classes`, you might want to check whether a `WebException` or `NotSupportedException` exception has been raised while accessing object properties and methods. You need to add a `try...catch` block for the `GetResponse()` and `BeginGetResponse()` methods to catch those exceptions.

An Example of Web Access

The form class (shown in Figure 7-2) demonstrates both synchronous and asynchronous access to a web page. The form has a left softkey menu bar (Connect!) and right softkey menu bar (Exit). In addition, a combo box right above the left softkey menu bar enables the user to select Sync or Async for the desired web access method. The web response is displayed in a text box. When a user selects a method on the combo box and selects the Connect menu bar, the corresponding operation will be performed and raw page text will be displayed on the screen. A progress bar next to the combo box is used to indicate that the operation is in progress (although it does not strictly follow the exact progress).

The following code shows the method of the Connect menu bar. Depending on the value of the combo box `cbxConnectMethod`, either an asynchronous or a synchronous connection will be established:

```
        private void menuItem1_Click(object sender, EventArgs e)
        {
            progressBar1.Value = 0;
            if (cbxConnectMethod.SelectedIndex == 1) AsyncGet();
            else SyncGet();
            return;
        }
```

The following example code shows the asynchronous connection method. A callback method named `MyCallback()` will be called (in a thread other than the main UI thread) when a response is received. Access to the data stream of the established connection is also done asynchronously — a `MyReadCallback()` callback method is called when specified amount of data is read into the buffer:

```
        private void AsyncGet()
        {

            try
            {
                WebRequest myWebRequest =
WebRequest.Create("http://www.yahoo.com");

                myWebRequest.Proxy = GlobalProxySelection.GetEmptyWebProxy();

                myRequestState = new RequestState();
```

```
                myRequestState.request = myWebRequest;

                IAsyncResult asyncResult =
(IAsyncResult)myWebRequest.BeginGetResponse(
                    new AsyncCallback(MyCallback), myRequestState);

            }
            catch (WebException e)
            {
                MessageBox.Show(e.ToString());
            }
            catch (Exception e)
            {
                MessageBox.Show(e.ToString());
            }

        }
```

The following are two callback methods used for asynchronously processing the web response and reading the response stream. The response stream is displayed in the TextBox `txtContent`.

```
        private void MyCallback(IAsyncResult ar)
        {
            try
            {
                myRequestState = (RequestState)ar.AsyncState;
                WebRequest webRequest = myRequestState.request;

                myRequestState.response = webRequest.EndGetResponse(ar); //To check
the operation status, get the response object

                Stream responseStream =
myRequestState.response.GetResponseStream();

                myRequestState.responseStream = responseStream;

                //Start to read the stream asynchronously into the buffer
                IAsyncResult asynchronousResultRead =
responseStream.BeginRead(myRequestState.bufferRead,
                    0, RequestState.BUFFER_SIZE, new AsyncCallback(MyReadCallBack),
myRequestState);

            }
            catch (WebException e)
            {
                MessageBox.Show(e.ToString());
            }
            catch (Exception e)
            {
                MessageBox.Show(e.ToString());

            }
        }
        private void MyReadCallBack(IAsyncResult ar)
        {
```

```
            try
            {

                myRequestState = (RequestState)ar.AsyncState;
                Stream responseStream = myRequestState.responseStream;
                int read = responseStream.EndRead(ar);
                //Read the contents of the HTML page and then print to the console
                if (read > 0)
                {

                         myRequestState.requestData.Append(
Encoding.ASCII.GetString(myRequestState.bufferRead, 0, read));


                    txtContent.Invoke(new EventHandler(this.UpdateUI));
                    IAsyncResult asynchronousResult = responseStream.BeginRead(
                        myRequestState.bufferRead, 0, RequestState.BUFFER_SIZE, new
AsyncCallback(MyReadCallBack),
                        myRequestState);
                }
                else
                {

                   //Close everything
                    myRequestState.bFinished = true;
                    txtContent.Invoke(new EventHandler(this.UpdateUI));

                    responseStream.Close();
                    myRequestState.response.Close();

                }
            }
            catch (WebException e)
            {
                MessageBox.Show(e.ToString());
            }
            catch (Exception e)
            {
                MessageBox.Show(e.ToString());
            }

        }
```

The progress bar and the text box on the form need to be updated every time the asynchronous `MyReadCallback()` method is called. The `UpdateUI()` method performs this task and will be called by the UI control's `Invoke()` method via an `EventHandler` delegate:

```
        public void UpdateUI(object sender, EventArgs e)
        {
            if (myRequestState.bFinished)
            {
                progressBar1.Value = 100;
                return;
            }

            txtContent.Text = myRequestState.requestData.ToString();
```

```
            //Here we do a trick: we don't update the progress bar according to the
real progress
            if (progressBar1.Value < 80) progressBar1.Value += 20;
            else progressBar1.Value = 100;
            //long uiFinished =  myRequestState.requestData.Length * 100 /
myRequestState.lTotalResponseLength;
            //Why the content-length  is -1?
            //progressBar1.Value = (int)uiFinished; //Set the progress bar
        }
```

The synchronous connection method will be blocked at the `GetResponse)` call until a response is received. Access to the stream is also done synchronously:

```
        private void SyncGet()
        {

            WebResponse webResponse = null;
            Stream dataStream = null;
            StreamReader dataReader = null;
            try
            {
                WebRequest req = WebRequest.Create("http://www.google.com");

                progressBar1.Value = 10;  //Do the trick

                //Because the URI indicates the HTTP protocol, the WebResponse
object is actually an HTTPWebResponse object
                webResponse = req.GetResponse();

                //Get the result from the response using a Stream object
                dataStream = webResponse.GetResponseStream();
                dataReader = new StreamReader(dataStream);
                string str = dataReader.ReadToEnd();

                progressBar1.Value = 100;


                //Set the text to the response string
                txtContent.Text = str;
            }
            catch (WebException webEx)
            {
                //If somehow a connection cannot be made, show
                //the exception string
                MessageBox.Show(webEx.ToString());
            }
            finally
            {
                //Always close the objects
                if (webResponse != null)
                {
                    dataReader.Close();
                    dataStream.Close();
                    webResponse.Close();
```

```
            }
        }
    }
```

A data object is needed to manage an asynchronous I/O state. The object is needed when `WebRequest.BeginGetResponse()` and `Stream.BeginRead()` are called. In this example, we define a `RequestState` class (shown below) to manage parameters and track request and response objects:

```
public class RequestState
{
    //This class stores the state of the request
    public const int BUFFER_SIZE = 1024;

    public long lTotalResponseLength;
    public bool bFinished = false;

    public StringBuilder requestData; //Web response
    public byte[] bufferRead;  //The buffer
    public WebRequest request; //Associating WebRequest
    public WebResponse response; //Associating WebResponse
    public Stream responseStream;

    public RequestState()
    {
        bufferRead = new byte[BUFFER_SIZE];
        requestData = new StringBuilder("");
        request = null;
        responseStream = null;
        lTotalResponseLength = 0;
    }
}
```

Figure 7-2 shows the display of both access methods.

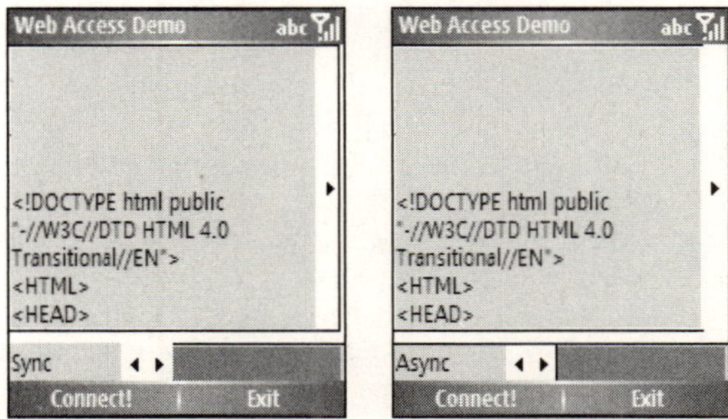

Figure 7-2

The `WebClient` class that encapsulates web-based common file download and upload operations in the .NET Framework is not available in the .NET Compact Framework.

TCP Servers and Clients

The TcpListener class in the System.Net.Sockets namespace provides a basic implementation of a TCP server that works in blocking synchronous mode, meaning the calling thread will be blocked while waiting for incoming TCP client requests. Once a TcpListener is created and started, you can check whether there is any incoming TCP client request. Conversely, if you want to implement a TCP client, you can use the TcpClient class or the Socket class (explained below).

The IPEndPoint Class

For TCP and UDP connections working on the transport layer of the TCP/IP stack, the two sides are identified by an *IP endpoint,* a composite data structure consisting of an IP address and a port number. For example, an HTTP web server usually runs an IP address of the hosting machine at port 80. The IPEndPoint class implements an IP endpoint.

The IPAddress and IPHostEntry Classes

The .NET Compact Framework uses an IPEndPoint class to encapsulate properties and methods of an IP endpoint. To construct an IPEndPoint object, you need an IPAddress object (which can be obtained from an IPHostEntry object) and a port number. The procedure is as follows:

1. Use Dns.GetHostEntry() with a domain name to obtain an IPHostEntry object that contains IP addresses of the given domain name.
2. Use an IPAddress array to get all the IPAddress objects from the IPHostEntry object.
3. Use one of the IPAddress objects (there may be just one) obtained from Step 2 and a port number to build the IPEndPoint object.

You can also create an IPEndPoint from a SocketAddress object (explained below). The following code snippet shows a number of ways:

```
IPAddress ipadr = IPAddress.Parse("66.102.7.147"); //Get the IP address from a
dotted-quad IP addresses
IPEndPoint ipe = new IPEndPoint (ipadr, 80); //Create an IP endpoint

IPHostEntry addresses = Dns.GetHostEntry("www.msn.com");
IPAddress[] ipadrs = addresses.AddressList;
IPAddress ipadr = ipadrs[0]; //Get the first element of the address array
IPEndPoint ipe = IPEndPoint (ipadr, 80);

IPEndPoint clonedIPEndPoint = (IPEndPoint) ipe.Create(socketAddress); //Assuming
socketAddress is a SocketAddress object
```

The IPAddress object provides several public static fields for well-known IP addresses for both Ipv4 and Ipv6:

- IPAddress.Any — 0.0.0.0
- IPAddress.Broadcast — 255.255.255.255

- `IPAddress.Loopback` — 127.0.0.1
- `IPAddress.None` — 255.255.255.255
- `IPAddress.IPv6Any` — 0:0:0:0:0:0:0:0 (::)
- `IPAddress.IPv6Loopback` — 0:0:0:0:0:0:0:1 (::1)
- `IPAddress.IPv6None` — 0:0:0:0:0:0:0:0 (::0)

You can convert an `IPAddress` object into a byte array, representing the 4 bytes of an IPv4 address or the 6 bytes of an IPv6 address, as follows:

```
Byte[] ipbytes = IPAddress.GetAddressBytes();
//To obtain a dotted-quad value of an IPAddress object
ipr.ToString();
```

Network and Host Byte Order Conversion

A common task in network application development is converting data into network byte order before sending it to the network, as computers on the network may have a different architecture: either big-endian such as Sun Sparc processors on which high-order bytes are stored at low-memory addresses, or little-endian such as Intel x86 processors on which high-order bytes are stored at high-memory addresses. For example, an integer of 4 bytes — byte 3, byte 2, byte 1, byte 0 — is stored in memory as follows:

- Big-endian:
  - Base address — Byte 3
  - Base address + 1 — Byte 2
  - Base address +2 — Byte 1
  - Base address +3 — Byte 0
- Little-endian:
  - Base address — Byte 0
  - Base address + 1 — Byte 1
  - Base address +2 — Byte 2
  - Base address +3 — Byte 3

Computer networks are historically big-endian, so little-endian computers, including most PCs, must perform the byte order conversion for network communication. Both IP header data and payload data must be converted before a packet is generated. You probably have used Windows socket functions such as `htons()`, `htonl()`, `ntohs()`, and `ntohl()` to convert short and long values between the host byte order, which is little-endian, and the network byte order, which is big-endian. If the computer is big-endian, these functions will simply return the parameters. In the .NET Compact Framework, these can be done using `IPAddress.HostToNetworkOrder()` and `IPAddress.NetworkToHostOrder()`. These static methods can perform conversion for short (Int16), int (Int32), and long (Int64) values:

```
short x = 256; //x is 0x0100 in hexadecimal on an Intel computer
short networked_x = IPAddress.HostToNetworkOrder(x); //network_x is 0x0001
short y = IPAddress.NetworkToHostOrder(networked_x); //y is 0x0100, the same as x
```

The TcpListener and TcpClient Classes

The `TcpListener` and `TcpClient` classes can be used to build a simple TCP-based network application with a pluggable protocol that does not require fine-grain controls, such as online chat room and instant messaging applications. Smartphone applications are likely to use `TcpClient` rather than `TcpListener` because in most cases the Smartphone application acts as a client to access services on a remote server, not the other way around. However, if you are building a peer-to-peer application that acts as both a client and a server to communicate with applications on other devices, then a TcpListener is needed to wait, accept, and process incoming requests.

There are two ways to create a `TcpListener` object. One method is to use an `IPAddress` object and a port number; the other method is to use an `IPEndPoint` object as the only parameter. The following shows an example of using `IPAddress` to create a `TcpListener object:`

```
IPAddress localAddress = IPAddress.Parse("127.0.0.1");  //IP address
Int port = 4400;  //Port number
TCPListener tcpServer = new TCPListener(localAddress, port); //The TCP server will
listen at port 4400
```

The following is an example of using `IPEndPoint` to create a `TcpListener` object:

```
//Another way to create a TcpListener object
IPEndPoint ipe = new IPEndPoint(localAddress, port);
TCPListener tcpServer = new TCPListener(ipe);
```

Once the `TcpListener` is started, it is ready to accept incoming TCP connection requests, which will be queued while waiting for the server to process. You can use the `AcceptTcpClient()` method or the `AcceptSocket()` method to accept an incoming request by dequeueing a connection request from the queue. Note that both of these methods will block until a pending connection request is present.

The `AcceptTcpClient()` method will return a `TcpClient` object, whereas the `AcceptSocket()` method will return a `Socket` object. Generally, you would use a `Socket` object if you want to have fine-grain control over the TCP connection. Conversely, if you just want to send and receive data over the connection and don't want to bother with the low-level details of the socket, use the `TcpClient` object instead.

Then a `NetworkStream` object can be obtained for reading and writing. The `Read()` method of the `NetworkStream` class will block until data is available:

```
Public int Read(byte[] buffer, int offset, int size);
```

Then the read operation will read as much data as is available, up to the number of bytes specified by the `size` parameter. The `Read()` method returns the number of bytes being read. If the other side closes the connection and all data has been read, the method returns `0` immediately. You can use the `DataAvailable` property of the `NetworkStream` class to avoid the blocking; before a read operation, check to see whether `DataAvailable` is `true` or `false`.

The following code shows how a connection is handled. The `AcceptTcpClient()` call blocks until a connection request arrives. You can use the `Pending()` method to determine whether there are connection requests:

```
tcpServer.Start();
TcpClient tcpClient = tcpServer.AcceptTcpClient(); //The call blocks until a
connection request arrives
NetworkStream ns = tcpClient.GetStream();

//Now we can read and write the network stream
//Write an encoded data to the network stream
string hello = "Hello From Server";
byte[] data = System.Text.Encoding.ASCII.GetBytes(hello);
ns.Write(data, 0, data.Length);

//Read from the network stream
byte[] buffer = new byte[1024];
//Read all data until the other side closes the connection
while((size = ns.Read(buffer, 0, buffer.Length)!=0)  {
  //Use the received data
  string message = System.Text.Encoding.ASCII.GetString(buffer, 0, size);
  ...
}
ns.Close();
tcpClient.Close();
...
tcpServer.stop(); //Stop the server when needed
```

As with many other network classes, the `TcpListener` and `TcpClient` methods throw a `System.Net.Sockets.SocketException`.

To avoid the blocking calls, you can use a method called `Pending()` of the `TcpListener` class to check whether there are queued incoming connection requests. If so, the connection request can be processed; otherwise, the server code can do something else or wait for some specified time and poll again:

```
if(tcpServer.isPending())
{
  TcpClient tcpClient = tcpServer.AcceptTcpClient();
  NetworkStream ns = tcpClient.GetStream();
  ...
}
else
{
  //Do something else
}
```

A frequently used pattern for network servers is to create a thread for each incoming connection request; thus, the main thread of the server code does not need to block while processing a connection request. The following example shows a portion of the TCP server code. Each incoming connection request will be processed in a `WorkerThreadProc()` method of a user-defined `ThreadWithState` object named `worker`:

```
try
{
  //Create the TcpListener object at port 4400
  IPAddress ipaddr = IPAddress.Parse("127.0.0.1");
  IPEndPoint ipe = new IPEndPoint(ipaddr, 4400);
  tcpServer = new TcpListener(ipe);
  tcpServer.Start();
  while (true)
  {
    //Once we see an incoming message, process it in another thread
    //leaving the current server thread for other incoming messages
    if (tcpServer.Pending())
    {
      ThreadWithState worker = new ThreadWithState(myForm);
      Thread t = new Thread(new ThreadStart(worker.WorkerThreadProc));
      t.Start();
    }
    Thread.Sleep(200);
  }
}
catch (SocketException ex)
{
  MessageBox.Show(ex.ToString());
}
```

An Example of TcpListener and TcpClient

This section presents a simple ping-pong application using `TcpListener` and `TcpClient`. This section shows only part of the code; to read all the code for this example, download the sample project from the book's website (the TCPDemo project under Chapter 7). The protocol works as follows:

1. The server starts and waits at port 4400.
2. A client starts and connects to port 4400 of the given server address.
3. The server accepts the client and sends the message "Ping From Server."
4. The client receives the ping message and replies with the pong message "Pong From Client."
5. The client closes the connection to the server.

The TCP Client

As shown in Figure 7-3, the UI of the client is quite simple. A label is used to display the server's message. Two soft key menu bars, Start Client and Exit, are used to initiate a `TcpClient` and connect to the server, and to exit the application, respectively. All the major work is done when the Start Client menu bar is clicked. Once it starts to receive messages from the server, the client always checks whether the ping message is present, as it is an indication the client can close the connection and start to send a pong message.

Figure 7-3

The TCP client code uses a hard-coded IP address, `192.168.0.188`, to specify the TCP server IP address. When you run the code, change this to the IP address of the machine that you will use to run the TCP server application, which is a desktop application:

```
using System;
using System.Collections.Generic;
using System.ComponentModel;
using System.Data;
using System.Drawing;
using System.Text;
using System.Windows.Forms;
using System.IO;

using System.Net;
using System.Net.Sockets;

namespace TCPDemo
{
    public partial class Form1 : Form
    {
        public Form1()
        {
            InitializeComponent();
        }

        private void menuItem1_Click(object sender, EventArgs e)
        {
            TcpClient myClient = null;
            NetworkStream ns = null;
            try
            {
                int buffer_len = 1024;
                int size = 0;
                byte[] buffer = new byte[buffer_len];
                byte[] res = System.Text.Encoding.ASCII.GetBytes("Pong From
Client\r\n");
                int res_len = res.Length;

                //Get an IPEndpoint object of the remote TcpListener server
```

```
                //Assuming we know the IP address and port number of the server
                IPAddress ipaddr = IPAddress.Parse("192.168.0.188");

                //Alternatively, you can use Dns.Resolve() if you know the domain
name of the server
                //IPAddress ipaddr = Dns.Resolve("www.contoso.com").AddressList[0];
                IPEndPoint ipe = new IPEndPoint(ipaddr, 4400);

                myClient = new TcpClient();
                myClient.Connect(ipe);
                ns = myClient.GetStream();
                //Read server's "Ping" message
                while ((size = ns.Read(buffer, 0, buffer_len)) > 0)
                {
                    string response = Encoding.ASCII.GetString(buffer, 0, size);
                    string newText = lblContent.Text + response;
                    lblContent.Text = newText;
                    //Once receiving "Ping From Server", break the read and send a
reply
                    if (newText.IndexOf("Ping From Server") >= 0) break;
                }
                ns.Write(res, 0, res_len);
            }
            catch (SocketException ex)
            {
                MessageBox.Show(ex.ToString());
            }
            catch (IOException ex)
            {
                MessageBox.Show(ex.ToString());
            }
            finally
            {
                ns.Close();
                myClient.Close();
            }
        }

        private void menuItem2_Click(object sender, EventArgs e)
        {
            Application.Exit();

        }
    }
}
```

The TCP Server

The TCP server is a desktop application running on the .NET Framework. It also has a UI to display the messages from all clients. Two buttons are added: Start/Stop Server and Exit. The first enables the user to start and stop the TCP server. Once the server is started, you will see only Stop Server on the button. Clicking the Exit button will terminate the application. The main thread is the UI thread handling the form. The TCP server's operations, including creating the `TcpListener` object, starting the server, and listening for a connection request, are performed in another thread. Once a connection request arrives, a third thread (called a *worker* thread) is created to handle this specific request. Many worker threads can be serving connection requests simultaneously. Figure 7-4 depicts the three types of threads used in the TCP server.

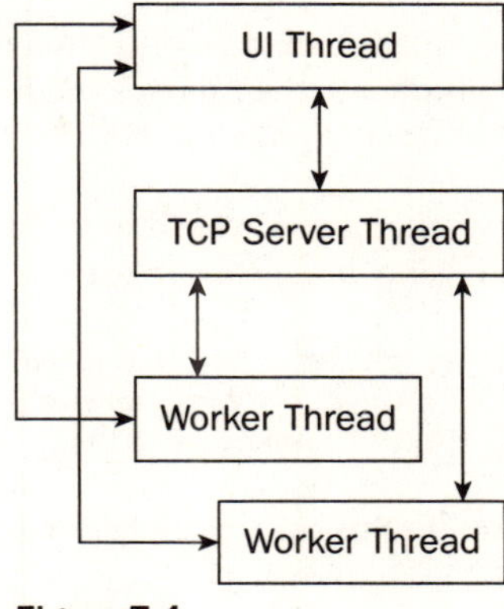

Figure 7-4

The basic construction of a thread is to create a `ThreadStart` delegate that refers to a method. Then the method will be executed in the thread when the `Invoke()` method of the `Thread` class is called. The following code shows an example:

```
using System.Threading;
...
class Worker
{
  ThreadProc()
  {
  ...
  }
}
ThreadStart myDelegate = new ThreadStart(WorkerClass.ThreadProc);
Thread myThread = Thread(myDelegate);
MyThread.Start();  //The new thread starts to run the ThreadProc() method
//Continue to do something in the main thread
```

This scheme enables you to create threads for different tasks and to control each thread's priority. However, you have to manage the thread state, and handle wait/sleep events for each of them. A simple way to create multiple threads that are mainly in wait state and do work briefly is to use `ThreadPool`. A `ThreadPool` manages a pool of worker threads that have been optimized for system performance. This enables developers to focus on application tasks rather than thread scheduling. One of the advantages of using `ThreadPool` is that a `ThreadPool` object takes into account the systemwide states in addition to the process that owns the `ThreadPool` object. One process can have only one `ThreadPool`, which provides 25 threads by default. Because the `ThreadPool` class is an abstract class, you cannot create an instance of it directly. To request a specific method to be handled by a thread in the `ThreadPool`, use the `ThreadPool.QueueUserWorkItem()` method. The parameter to this method is a `WaitCallBack` delegate pointing to the callback method you want to run:

```
using System.Threading;
//Assuming a ThreadProc method is created in the class
//A WaitCallback delegate is needed
WaitCallback waitDelegate = new WaitCallback(ThreadProc);
ThreadPool.QueueUserWorkItem(waitCallback);
```

The callback method `ThreadProc()` will be called when a thread pool thread is available. There is no need for a "`Start()`" method. To pass data to the callback method, you can create a data object (defined yourself) and pass it to the `QueueUserWorkItem()` method:

Note that the server thread and the worker threads need to update the UI. You can do this by calling the `Invoke()` method of the main form, using an `InvokeDelegate` object as the parameter:

```
//To update the UI in a worker thread, assuming an InvokeMethod is defined to
modify content of the controls on the form "myForm"
myForm.Invoke(new InvokeDelegate(InvokeMethod));
```

Figure 7-5 shows the TCP server form. In this case, the server has serviced two connection requests.

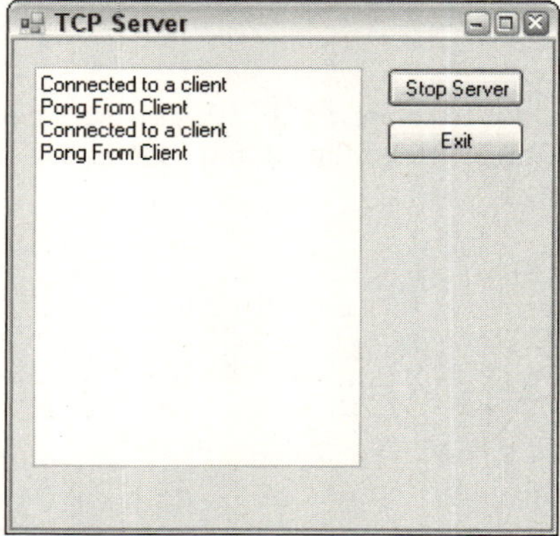

Figure 7-5

The Start/Stop Server routine in the form class starts or stops the `TcpListener`, depending on the current status of the server. This routine is executed in the main UI thread. Once connected to a `TcpClient` object, the server will send a `Ping From Server` message to the client; and the client will reply with a `Pong From Client` message. The server's worker thread will process the reply message:

```
        private void button1_Click(object sender, EventArgs e)
        {
            if (button1.Text == "Start Server") //Start the TcpListener server
            {
                button1.Text = "Stop Server";
                //Create the TcpListener object at port 4400
                IPAddress ipaddr = IPAddress.Parse("127.0.0.1");
                IPEndPoint ipe = new IPEndPoint(ipaddr, 4400);
                myServer = new TcpListener(ipe);
```

```
                myServer.Start();
                //Start a server thread to process incoming messages
                //It will block current UI thread
                ThreadWithState server = new ThreadWithState(this);
                Thread t = new Thread(new ThreadStart(server.ThreadProc));
                t.Start();
            }
            else  //Stop the server
            {
                myServer.Stop();
                button1.Text = "Start Server";
            }
        }
```

The form class also has a public method, `UpdateLog()`, for UI updates. It will be called by the server's worker thread to update the text box on the UI:

```
        //This allows the worker thread to update the UI
        public void UpdateLog(string data)
        {
            txtLog.Text += data;
        }
```

The tasks of the server thread and worker thread are defined in the `ThreadWithState` class. It uses a private field, `myForm`, to refer to the main Windows form class in the underlying namespace of this project. Its `ThreadProc()` method is associated with the server thread that creates a worker thread that executes `WorkerThreadProc()` whenever a request arrives. The worker thread does the real job by obtaining the client stream, writing into it and reading from it. In addition, to enable the two threads to update the UI, a delegate named `InvokeDelegate` is used, which is mapped to the `InvokeMethod()` that will be run in the context of the main UI thread:

```
    public class ThreadWithState
    {
        private MyForm myForm;
        private string data = null;
        private delegate void InvokeDelegate(); //We can update the text log
  control using Control.Invoke()
        //The constructor obtains the state information
        public ThreadWithState(MyForm form)
        {
            myForm = form;
        }

        public void ThreadProc()
        {
            try
            {
                while (true)
                {
                    if (myForm.myServer.Pending())
                    {
                        ThreadWithState worker = new ThreadWithState(myForm);
```

```
                    Thread t = new Thread(new
ThreadStart(worker.WorkerThreadProc));
                    t.Start();
                }
                Thread.Sleep(200);
            }
        }
        catch (SocketException ex)
        {
            MessageBox.Show(ex.ToString());
        }
    }

    private void WorkerThreadProc()
    {
        TcpClient client = null;
        NetworkStream ns = null;
        try
        {
            int size = 0;
            int buffer_len = 1024;
            byte[] buffer = new byte[buffer_len];

            //Message to the client
            string hello = "Ping From Server\r\n";
            byte[] res = System.Text.Encoding.ASCII.GetBytes(hello);
            int res_len = res.Length;

            //The server blocks here, waiting for a client
            client = myForm.myServer.AcceptTcpClient();
            //Here comes a client
            ns = client.GetStream();
            //Send our "Ping" message
            ns.Write(res, 0, res_len);
            //Receive "Pong" message from the client
            while ((size = ns.Read(buffer, 0, buffer_len)) > 0)
            {
                data += System.Text.Encoding.ASCII.GetString(buffer, 0, size);
            }

            //Update UI using the delegate method
            myForm.Invoke(new InvokeDelegate(InvokeMethod));
        }
        catch (SocketException ex)
        {
            MessageBox.Show(ex.ToString());
        }
        catch (IOException ex)
        {
            MessageBox.Show(ex.ToString());
        }
        finally
        {
            ns.Close();
            client.Close();
```

```
                }
        }

        //This method will run in the context of the UI thread
        //Don't place network processing code here because it will block UI update
        private void InvokeMethod()
        {
                myForm.UpdateLog("Connected to a client\r\n");
                myForm.UpdateLog(data);
        }

    }
}
```

This example doesn't use an additional UI update thread in the `TcpClient` because its operations are not sophisticated; you don't need to update the UI while waiting for the server's response. If you have some more controls on the UI and would allow the user to access the controls while the `TcpClient` blocks for the server's response, you need to place the `TcpClient`-related tasks into a thread other than the main UI thread (the form thread).

Network Sockets

The `TcpListener` class does not provide all the socket-level functionality, such as socket options, socket polling, and selection. If a pluggable protocol requires more flexible control over network connections on the socket level, a `TcpListener` does not suffice. For this purpose, you can use the `System.Net.Sockets.Socket` class, which implements the Berkeley socket interface. `TcpListener` and `TcpClient` are actually wrappers of a `Socket` class.

Unlike the `TcpListener` and `TcpClient` classes, which support only TCP, a `Socket` class can be associated with a list of protocol types defined in the `Socket.ProtocolType` enumeration. Some of the protocols that may be used for Smartphone application development are TCP, UDP, IP, IPv6, ICMP, and Raw. The Raw protocol type enables you to implement your own network layer protocols.

The `Socket` class provides both synchronous and asynchronous data transfer between two communication endpoints. Synchronous methods such as `Send()`, `SendTo()`, `Receive()`, and `ReceiveFrom()` have their asynchronous counterparts: `BeginSend()` and `EndSend()`, `BeginSendTo()` and `EndSendTo()`, `BeginReceive()` and `EndReceive()`, and `BeginReceiveFrom()` and `EndReceiveFrom()`.

The constructor of a `Socket` object requires three parameters:

```
public Socket(
   AddressFamily addressFamily,
   SocketType socketType,
   ProtocolType protocolType);
```

`AddressFamily` indicates the address for the socket. Frequently used address families for Smartphone development are `AddressFamily.InterNetwork`, `AddressFamily.InterNetworkV6`, and `AddressFamily.Irda`.

SocketType could be one of the following widely used socket types: Dgram for UDP-based connectionless, unreliable communication; Stream for TCP-based, connection-oriented, reliable, and duplex communication; and Raw for protocol implementations that require the developer to handle IP headers. In addition, the .NET Compact Framework supports two extended socket types: Rdm for connectionless but reliable and in-order message delivery, and Seqpacket for connection-oriented and in-order message delivery. Note that Steam socket types do not guarantee in-order delivery of messages.

As you can see, some socket types implicitly specify a ProtocolType. For example, a SocketType of Stream indicates a ProtocolType of Tcp. Conversely, a SocketType of Dgram requires a ProtocolType of Udp. If you pass an incompatible SocketType and ProtocolType, a SocketException will be raised.

The following examples create a TCP socket, a UDP socket, and a Raw socket that is used to implement ICMP, respectively:

```
Socket serverSocket = new Socket(AddressFamily.InterNetwork, SocketType.Stream,
ProtocolType.Tcp);
Socket serverSocket = new Socket(AddressFamily.InterNetwork, SocketType.Dgram,
ProtocolType.Udp);
Socket serverSocket = new Socket(AddressFamily.InterNetwork, SocketType.Raw,
ProtocolType.Icmp);
```

TCP Sockets

Once created, a TCP server socket can be bound to a specific IP address and a port number. Then the server socket can start to listen at a local port using the Listen() method, and accept incoming connection requests using the Accept() method. The Accept() method returns a Socket object that you can further use to send or receive data. A TCP socket can directly connect to a remote TCP server that is listening.

TCP Server Sockets

The following example demonstrates the common procedure for creating a TCP server socket:

```
//Assuming an IPEndPoint object ipe is created
//Create the socket
Socket serverSocket = new Socket(AddressFamily.InterNetwork, SocketType.Stream,
ProtocolType.Tcp);
//Bind to a local endpoint
ServerSocket.Bind(ipe);
//Start to listen; this method does not block
ServerSocket.Listen(queueLength);  //queueLength specifies the number of
connections that can be queued
//The server socket blocks waiting for an incoming request
Socket oneConn = serverSocket.Accept();
```

The Accept() method will block the calling thread until a connection request is received. Then a second Socket object, oneConn in this example, is created. The initial socket object will continue to queue incoming connection requests. Sending and receiving data over the connected socket is very similar to common I/O operations, with one exception: a SocketFlags parameter can be used to control how the data is sent over the socket. The following are two examples of the overloaded methods:

```
public int Send(byte[] buffer, SocketFlags flags);
public int Receive(byte[] buffer, SocketFlags flags);
```

The `Send()` operation does not send data to the network; it actually copies data from your buffer to the system's protocol stack buffer. Similarly, the `Receive()` operation does not directly pick up data from the network interface; instead, it retrieves data from the protocol stack buffer to the user's buffer. In both cases, it is the underlying system's task to eventually send data to the network.

The `SocketFlags` parameter can be a combination of several items of the `SocketFlags` enumeration, including `SocketFlags.None`, `SocketFlags.DontRoute`, `SocketFlags.OutOfBand`, and so on.

The following sample code shows how to send and receive data synchronously over a connected socket `oneConn`. When the data transfer is over, you need to shut down the socket first, and then close it. The `Socket.Shutdown()` method will ensure that all data has been sent and received. The `Socket.Shutdown()` method needs a parameter to indicate whether to disable sending, receiving, or both. Then the `Socket.Close()` method can be called to release all resources used by the socket.

```
try{

Byte[] message = Encoding.ASCII.GetBytes("Hello!");
Byte[] buffer = new byte[1024];
OneConn.Send(message, SocketFlags.None);
OneConn.Receive(buffer, SocketFlags.None);
...
}
catch (SocketExcpetion ex)
{...
}
finally
{
  oneConn.Shutdown(SocketShutdown.Both);  //Shut down the connection first to
ensure all data has been sent
  oneConn.Close(); //After shutting down the socket, close it
}
```

The .NET Framework has an additional method, `Disconnect()`*, in the* `System.Net.Sockets.Socket` *class, which disconnects the socket and provides an option for the developer to specify whether the endpoint of the socket can be reused. The* `Socket` *class of the .NET Compact Framework does not have this method.*

In the previous section, you saw code for a desktop Windows TCP server application. Now let's examine a console-based TCP server also running on desktop Windows. This example is in the TCPConsoleServer project of Chapter 7. The underlying class `ServerSocket` defines a field of the server socket and three methods: `StartServer()`, `ProcessConnection()`, and `ShutdownServer()`. The `StartServer()` method creates a server socket object that binds to a local address and port 4400 (you can choose any port as long as it is not used by other programs), and then calls `Socket.Accept()` waiting for incoming connections. When a connection request arrives (from desktop TCP software such as telnet or a Smartphone TCP client, introduced below), a new thread will be created to handle this connection, denoted by a socket object returned by the `Accept()` method. The `ProcessConnection()` method first sends a ping message. Then, for each message it receives through the underlying socket, it replies with a pong message. This sequence will continue until the other side shuts down the connection. If the client forcibly closes the connection, a `SocketException` with an error code of 10054 (connection reset by peer) will be raised. The `ProcessConnection()` method does not handle this exception because it is considered a normal case; the socket will be shut down and closed without presenting any error message to the user. If, however, another `SocketException` occurs, the user will see a warning message.

```
using System;
using System.Collections.Generic;
using System.Text;

using System.Net.Sockets;
using System.Net;
using System.IO;
using System.Threading;

namespace TCPConsoleServer
{
    class ServerSocket
    {
        Socket serverSocket = null;
        private void StartServer()
        {
            Console.WriteLine("=============");
            Console.WriteLine("TCP server @ " + IPAddress.Any + ":port 4400");
            Console.WriteLine("=============");
            Socket serverSocket = new Socket(AddressFamily.InterNetwork,
            SocketType.Stream, ProtocolType.Tcp);
            serverSocket.Bind(new IPEndPoint(IPAddress.Any, 4400));
            serverSocket.Listen(30);
            while (true)
            {
                Socket oneSocket = serverSocket.Accept();
                ThreadPool.QueueUserWorkItem(new
WaitCallback(ProcessConnection),oneSocket);
            }

        }
        private void ProcessConnection(object o)
        {
            Socket oneSocket = (Socket)o;
            byte[] buf = new byte[1024];
            byte[] message = Encoding.ASCII.GetBytes("Ping From Server");
            try
            {
                String remoteEnd = oneSocket.RemoteEndPoint.ToString();
                Console.WriteLine("Got a Client from {0}", remoteEnd);
                /*
                Console.WriteLine("SO_REUSEADDRESS:" + oneSocket.GetSocketOption(
                    SocketOptionLevel.Socket, SocketOptionName.ReuseAddress));
                Console.WriteLine("This application will timeout if Send does not
return within " +
                    oneSocket.GetSocketOption(SocketOptionLevel.Socket,
SocketOptionName.SendTimeout));
                */
```

```
                oneSocket.Send(message);
                Console.WriteLine("Send Ping to {0}", remoteEnd);
                int size = 0;
                do
                {
                    size = oneSocket.Receive(buf, 1024, SocketFlags.None);
                    if (size > 0)
                    {
                        Console.WriteLine("\r\nReceived from {0}", remoteEnd);
                        String tmp = Encoding.ASCII.GetString(buf, 0, size);
                        Console.Write(tmp);
                        //Check to see if a line of message is received
                        if(tmp.IndexOf("\r\n") >= 0)
                            oneSocket.Send(message); //Send "Ping" for each
received message
                    }
                } while (size > 0);
                Console.WriteLine();
            }
            catch (SocketException sockEx)
            {
                //Skip the exception that was raised when a client forcibly closes
the socket
                if(sockEx.ErrorCode != 10054)
                Console.WriteLine("\r\nException Code: {0} : {1}",
sockEx.ErrorCode,
                    sockEx.Message);
            }
            catch (IOException ioEx)
            {
                Console.WriteLine(ioEx.Message);
            }
            finally
            {
                oneSocket.Shutdown(SocketShutdown.Both);
                oneSocket.Close();
            }

        }
        private void ShutdownServer()
        {
            if (serverSocket.Connected)
            {
                serverSocket.Shutdown(SocketShutdown.Both);
                serverSocket.Disconnect(true);  //Close the server socket and allow
reuse of the port
            }
        }
        static void Main(string[] args)
        {
            ServerSocket myServer = new ServerSocket();
            myServer.StartServer();
        }
    }
}
```

Figure 7-6 shows an example of the output of the TCP console server.

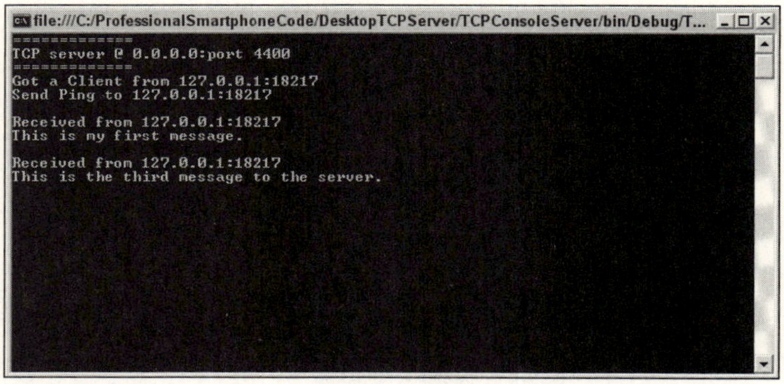

Figure 7-6

TCP Client Sockets

A TCP client socket does not listen at a port. Once created, you need to call the `Connect()` method to connect an endpoint. After that, you can use `Send()` and `Receive()` for data transfer, as shown in the TCP server socket. The following code snippet shows a TCP client that is connected to a remote endpoint and receives data:

```
//Assuming an IPEndPoint ipe has been created
mySock = new Socket(AddressFamily.InterNetwork,
                    SocketType.Stream, ProtocolType.Tcp);
mySock.Connect(ipe);
try
{
    byte[] buf = new byte[1024];
    int size = 0;
    do
    {
        size = mySock.Receive(buf, 1024, SocketFlags.None);
        if (size > 0)
        {
            //Consume received data
        }
    } while (size > 0);
}
catch (SocketException sockEx)
{
    ...}
```

For a Smartphone application with a GUI, it is common to place the data reception code of a socket into a separate thread other than the main UI thread. This can be done using a `ThreadPool`. Sending data from the socket is often triggered by some controls on the UI. For example, when a user selects some local data and clicks a button, the `click` event handler of the button will start the `Send` operation.

The following is an example of a TCP socket client that communicates with the aforementioned TCP console server. As shown in the first screen of Figure 7-7, the main UI features a windows form containing a text box showing the message log and two menu items, Send a Msg and Connection. The Connection menu item has two submenu items, Connect and Exit. Note that in the .NET Compact Framework, only the right menu of a windows form can have submenus. When the program starts, the Send a Msg menu item is temporarily disabled. Users must first select Connection⇨Connect to make a connection to a predefined server. Once connected, they can start to send a message by clicking Send a Msg. A new form will show on the screen (the second screen in Figure 7-7). Here users can enter a message and send it to the server by clicking Send! Then the main form will show up again, and the message log will be updated to show the user's sent messages and the messages from the server (the third screen in Figure 7-7).

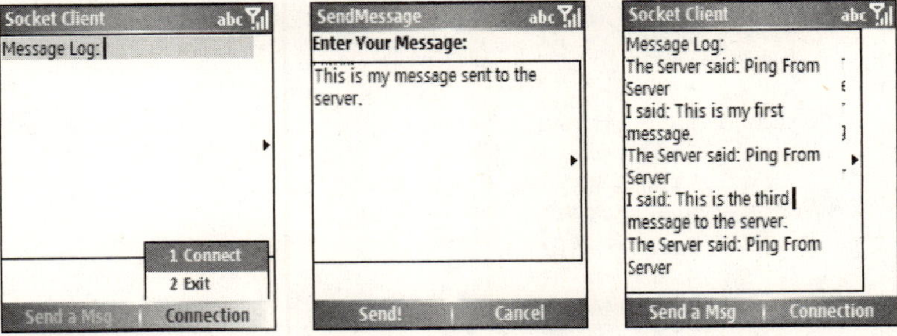

Figure 7-7

The fields and the constructor of the `SocketClient` class are shown in the following code. The call has a `Socket` member of `mySock` and two string members of `ReceivedMessage` and `SendMessage`. The delegate member of the class is used by a socket thread that runs `ReceiveProc()` to update the UI control in the context of the main UI thread through `Control.Invoke()`:

```
public partial class SocketClient : Form
    {
        Socket mySock = null;
        string ReceivedMessage = null;
        string SendMessage = null;
        private delegate void InvokeDelegate();

        public SocketClient()
        {
            InitializeComponent();
            mnuSendMessage.Enabled = false;
        }
...
```

The Connect menu item event handler and the `Connect()` routine together make a connection to the server. They are shown as follows. The `Connect()` method further uses a `ThreadPool` for the socket `ReceiveProc()` method:

```
        private void mnuConnect_Click(object sender, EventArgs e)
        {
            if (mySock!=null && mySock.Connected)
            {
                //Disconnect the socket
                mySock.Shutdown(SocketShutdown.Both);
                mySock.Close();
                mnuConnect.Text = "Connect";
            }
            else
            {
                mnuConnect.Text = "Disconnect";
                Connect();
            }
        }
      private void Connect()
        {
            IPEndPoint ipe = new IPEndPoint(IPAddress.Parse("192.168.0.188"),4400);
            mySock = new Socket(AddressFamily.InterNetwork,
                SocketType.Stream, ProtocolType.Tcp);
            mySock.Connect(ipe);
            if (mySock.Connected)
            {
                mnuSendMessage.Enabled = true;
                ThreadPool.QueueUserWorkItem(new WaitCallback(ReceiveProc));
            }
            else
            {
                MessageBox.Show("Cannot connect to server.");
                mnuConnect.Text = "Connect";
            }
        }
```

The `ReceiveProc()` method performs a blocking `Socket.Receive()` that continuously reads a fixed length of bytes into a buffer. As long as some data has been received in an iteration, the method will loop for the next `Socket.Receive()` call. If no data is available (in the protocol stack buffer), then the `Socket.Receive()` call will block. In some cases the other side closes the connection gracefully (i.e., outstanding data is sent successfully), and then `Socket.Receive()` returns `0`, which will terminate the loop:

```
        private void ReceiveProc(Object stateInfo)
        {
            try
            {
                byte[] buf = new byte[1024];
                int size = 0;
                do
                {
                    size = mySock.Receive(buf, 1024, SocketFlags.None);
                    if (size > 0)
                    {
```

```
                        ReceivedMessage = Encoding.ASCII.GetString(buf, 0, size);
                        //Update the UI
                        this.Invoke(new InvokeDelegate(InvokeMethod));
                    }

                } while (size > 0);
            }
            catch (SocketException sockEx)
            {
                //Skip the exception that was raised when we close the socket while
still receiving
                if(sockEx.ErrorCode != 10004)
                    MessageBox.Show(sockEx.ErrorCode.ToString() + ":" +
sockEx.Message);
            }
        }

        private void InvokeMethod()
        {
            txtLog.Text += "\r\nThe Server said: " + ReceivedMessage;
        }
```

The `SocketClient` form must provide a method for the `SendMessage` form to update the message log from within the `SendMessage` form. In addition, when a user clicks the Exit submenu, the ongoing socket will be shut down for both sending and receiving, and then closed:

```
        public void UpdateSendLog(String myMessage)
        {
            SendMessage = myMessage;
            txtLog.Text += "\r\nI said: " + SendMessage;
        }
        private void mnuExit_Click(object sender, EventArgs e)
        {
            try
            {
                if (mySock != null)
                {
                    if (mySock.Connected)
                    {
                        mySock.Shutdown(SocketShutdown.Both);
                        mySock.Close();
                    }
                }
            }
            catch (SocketException sockEx)
            {
                MessageBox.Show(sockEx.ErrorCode.ToString() + ":" +
sockEx.Message);
            }

            Application.Exit();
        }
```

The SendForm class implements a separate Windows form that enables the user to enter a message and send it out by selecting the Send! menu. Because when the SendForm is created it has been initialized with the already established socket, the SendForm class does not create any new Socket object. The Socket.Send() method blocks until all the data the user entered in the text box txtMessage has been sent:

```
namespace SocketClient
{
    public partial class SendForm : Form
    {
        Socket mySocket = null;
        Form returnForm = null;

        public SendForm(Socket mainSocket, Form caller)
        {
            InitializeComponent();
            mySocket = mainSocket;
            returnForm = caller;
        }

        private void menuItem1_Click(object sender, EventArgs e)
        {
            try
            {
                mySocket.Send(Encoding.ASCII.GetBytes(txtMessage.Text+"\r\n"));
                SocketClient parentForm = (SocketClient)returnForm;
                parentForm.UpdateSendLog(txtMessage.Text);
            }
            catch (SocketException sockEx)
            {
                MessageBox.Show("Cannot send message: ", sockEx.Message);
            }
            this.Close();
            returnForm.Show();
        }

        private void menuItem2_Click(object sender, EventArgs e)
        {
            this.Close();
            returnForm.Show();
        }
    }
}
```

You can test the TCP server and TCP client locally on the same machine. You need to run the server first in Visual Studio or from the command line, and then launch the client on the emulator or a Smartphone using Visual Studio.

UDP Sockets

So far this chapter has used TCP sockets in all the examples to show how to leverage the networking classes in the .NET Compact Framework to build Smartphone applications. Now let's look at the facility for another popular type of transport: UDP.

Because UDP is a connectionless protocol, you don't need to establish a connection before sending and receiving data. Thus, no `Socket.Connect()` call is needed, saving some round-trip time between two hosts. Unlike TCP, UDP does not provide a stream of unlimited length for endpoints. It is most often used to send small packets. UDP is generally not reliable because it does not provide any acknowledgment and retransmission mechanism; UDP packets may get lost silently without being noticed, which may require the application layer to deal with the problem.

To create a UDP socket, you need to specify `SocketType` as `Dgram` and `ProtocolType` as `Udp`. Then you can call `SendTo()` or `ReceiveFrom()` to send and receive data, respectively. You must pass the endpoint to `SendTo()` as the destination. For `ReceiveFrom()`, simply create an endpoint object, and pass it to the `ReceiveFrom()` method by reference (using `ref`). The `ReceiveFrom()` method will capture the remote endpoint into the object. This is useful when you don't know who will be sending data to your UDP endpoint. Both `SendTo()` and `ReceiveFrom()` will block (in the default blocking mode) until data is sent or data is available. Note that for `ReceiveFrom()` used in UDP, the first queued datagram received will be saved into the application buffer `buf`. If the datagram is too large to fit into the buffer, extra data will be lost.

The following is an example of a UDP client sending packets to a UDP server:

```
IPHostEntry hostEntry = Dns.Resolve(Dns.GetHostName("time.mycompany.com"));
IPEndPoint endPoint = new IPEndPoint(hostEntry.AddressList[0],123);

//The remoteEnd object is used to save the sender's end point
IPEndPoint sender = new IPEndPoint(IPAddress.Any, 0);
IPEndPoint remoteEnd = (IPEndPoint)sender;

Socket mySocket = new Socket(IPAddress.AddressFamily,
        SocketType.Dgram,
        ProtocolType.Udp);

byte[] msg = Encoding.ASCII.GetBytes("message to the server");
byte[] buf = new byte[1024];
mySocket.SendTo(msg, 0, msg.Length, SocketFlags.None, endPoint);  //Send data
mySocket.ReceiveFrom(buf, ref remoteEnd); //Receive data
s.Close();
```

A UDP server does not listen at a port for connection. However, it does need to bind to a local endpoint before calling the `ReceiveFrom()`:

```
IPHostEntry hostEntry = Dns.Resolve(Dns.GetHostName("time.mycompany.com"));
IPEndPoint endPoint = new IPEndPoint(hostEntry.AddressList[0],123);


Socket mySocket = new Socket(IPAddress.AddressFamily,
        SocketType.Dgram,
        ProtocolType.Udp);

//Create an end point object to store remote end point in the ReceiveFrom() call
IPEndPoint sender = new IPEndPoint(IPAddress.Any, 0);
```

```
IPEndPoint remoteEnd = (IPEndPoint)sender;

s.Bind(endPoint);  //Bind to local end point

byte[] buf = new Byte[1024];

s.ReceiveFrom(msg, ref remoteEnd);  //Receive from anyone
s.Close();
```

Nonblocking Mode and Asynchronous Methods

In the .NET Compact Framework, a socket can be in either blocking mode or nonblocking mode. The mode setting of a socket will affect the way a socket works. A socket is in blocking mode by default, meaning that `Connect()`, `Accept()`, `Send()`, `SendTo()`, `Receive()`, and `ReceiveFrom()` will block the calling thread until the call succeeds. Blocking mode has no effect on asynchronous operations such as `BeginSend()` and `EndSend()`, `BeginReceive()` and `EndReceive()`, and so on. Note that if no data is available for reading, then the `Receive()` and `ReceiveFrom()` methods will block until data is available or TCP times out. The `Send()` method and `SendTo()` method will block until all data in the buffer is sent.

Nonblocking mode will make synchronous methods such as `Send()` and `Receive()` return immediately, in which case either it completes successfully or an exception is raised. In nonblocking mode, to check whether a specific operation is finished, you have to poll the socket. Asynchronous socket methods such as `BeginConnect()/EndConnect()`, `BeginSend()/EndSend()`, and `BeginReceive()/EndReceive()` are intrinsically nonblocking calls and do not require socket polling. The "Begin" calls will return immediately, and the corresponding callback methods are executed in a separate thread and will block on the "End" calls. Under the hood, these operations are implemented using Win32 I/O completion ports, a system mechanism to use a pool of kernel threads to process asynchronous I/O requests. You can consider asynchronous socket operations as a combination of threading and synchronous operations.

The following code snippet is an example of an asynchronous send using `BeginSend()` and `EndSend()`. The `BeginSend()` call requires a state object as the last parameter, which is used in the `AsyncCallback` method `MyAsyncSend`. We obtain the underlying socket from the `IAsyncResult` object. Of course, you can define your own socket state class to carry more information.

```
//In the message send routine
byte[] toSend = Encoding.ASCII.GetBytes(txtMessage.Text+"\r\n");
mySocket.BeginSend(Encoding.ASCII.GetBytes(txtMessage.Text + "\r\n"),
                   0, toSend.Length, SocketFlags.None,
                   new AsyncCallback(MyAsyncSend), mySocket);
...
//The async send call back
private void MyAsyncSend(IAsyncResult ar)
{
  Socket theSocket = (Socket)ar.AsyncState;
  theSocket.EndSend(ar);
  ...
}
```

If you are in nonblocking mode and no data is available in the protocol stack buffer, the `Receive` method will complete immediately and throw a `SocketException`. You can use the `Available` property to determine whether data is available for reading. In nonblocking mode, the send operations may not send all the data in the buffer; thus, your application must check to see whether a resend is necessary.

Smartphones are usually used to access some backend services on the Internet or an enterprise network. Thus, a Smartphone network application is most likely a socket client instead of a server because it merely connects and consumes a service, rather than provide a service. For low-volume traffic applications, the default blocking mode and synchronous operations within a single thread seem a good choice. For high-volume traffic applications, asynchronous operations (thus nonblocking) in association with asynchronous callbacks will generally lead to better performance. In both cases, the rule of thumb is that the GUI thread should not block on socket calls.

Summary

This chapter described the networking support in .NET Compact Framework for Smartphone application development. Our focus is clearly on the HTTP networking facility (web access) and socket classes, as they are the most widely used networking software construct. In conjunction with the introduction to the `System.Net` and `System.Net.Sockets` namespaces, you have gained some experience in multithreading for supporting responsive GUI and asynchronous socket operations. The chapter also discusses general design guidelines of using sockets, TCP clients, and TCP servers for Smartphone networking applications.

The following chapter covers Smartphone messaging, including e-mail and SMS. You will also learn how to leverage the .NET Compact Framework for Smartphone to develop applications accessing personal information management (PIM) data.

8

E-mail, SMS, and PIM Data

The ability to manage your e-mail, calendar, and tasks on-the-go is a must in today's dynamic business environment. The Outlook Mobile application that ships with Windows Mobile Smartphone provides the applets that enable you to manage your e-mail, text messages, and personal information manager (PIM) data. In this chapter, you will learn how to interact with those components in your application.

The chapter begins with an introduction to the Pocket Outlook Object Model (POOM), which provides a number of native APIs to programmatically access Pocket Outlook objects with unmanaged code. The attention then turns to using the `Microsoft.WindowsMobile.PocketOutlook` namespace in C#, which can be conceived of as the managed code implementation of the POOM. Note that the `Microsoft.WindowsMobile.PocketOutlook` namespace is currently available only in Windows Mobile 5.0 and is not supported in Smartphone 2002 and Smartphone 2003.

This chapter covers the following topics:

- An introduction to POOM and using POOM with C++
- The `Microsoft.WindowsMobile.PocketOutlook` namespace
- Writing e-mail applications with managed APIs
- Managing PIM data with managed APIs
- Writing text messaging applications with managed APIs

Pocket Outlook Object Model (POOM)

Microsoft provides a series of APIs for developers to access Microsoft Outlook from Windows CE-based devices. The classes of those APIs are termed Pocket Outlook Object Model (POOM). In a nutshell, POOM is a subset of the Outlook object model available on the desktop.

The goal of POOM is to provide a means for developers to manipulate e-mail and PIM data, such as contacts, calendars, and tasks data.

Traditionally, POOM offers a family of COM-based APIs in native format. For C# or VB.NET developers, it means you will have to use P/Invoke, or even switch to C++ to be able to utilize those Windows native APIs.

The steps to access data in Outlook Mobile are as follows:

1. Establish a POOM session.
2. Create a reference to PIM item folders.
3. Retrieve the PIM item from the Pocket Outlook database.
4. Close the POOM session.

To establish a POOM session using C++, perform the following steps:

1. Declare an `IPOutlookApp` interface object by initializing an `IPOutlookApp` pointer. For instance, the following code creates an empty `IPOutlookApp` pointer `pMyPoomApp`:

```
IPOutlookApp *pMyPoomApp = NULL;
```

2. Initialize the COM using the `CoInitializeEx` API:

```
CoInitializeEx(Null,0);
```

3. Create an `Application` COM object with the `CoCreateInstance` API:

```
CoCreateInstance(CLSID_Application, NULL, CLSCTX_INPROC_SERVER,
                 IID_IUnknown, (void **) &pUnknown);
```

4. Link the `Application` COM object to the `IPOutlookApp` interface object with the `QueryInterface` method:

```
pUnknown->QueryInterface(IID_IPoutlookApp, (void **) &pMyPoomApp);
```

5. Log on to the Pocket Outlook COM server using the `Logon()` method:

```
pMyPoomApp->Logon(NULL);
```

After you complete these steps, an Outlook Mobile session `pMyPoomApp` is established. You can then create a PIM item and retrieve the data from the corresponding Outlook Mobile data store.

The PIM data is organized into different folders in POOM. The default folders are defined in the `OlDefaultFolders` enumeration, as follows:

```
enum OlDefaultFolders {
  olFolderCalendar = 9,       //Default Calendar folder
  olFolderContacts = 10,      //Default Contacts folder
  olFolderTasks    = 13,      //Default Tasks folder
  olFolderCities   = 101,     //Default Cities folder
  olFolderInfrared = 102,     //Default Infrared folder
};
```

To access the PIM information, simply get the folder that contains the information you want. For instance, if you are interested in finding out the items in the Tasks folder, you can use the `GetDefaultFolder()` method of the `IPOutlookApp` interface to retrieve the `olFolderTasks` folder, as follows:

```
IFolder *pFolder;
pMyPoomApp->GetDefaultFolder(olFolderTasks, &pFolder);
```

Now that you have the Tasks folder, you can retrieve the items in the default Tasks folder by calling the `get_Items()` method of the `IFolder` interface. First, however, you need to declare a generic PIM item collection, as shown in the following sample code:

```
IPOutlookItemCollection *pGenericItems;
pFolder->get_Items(&pGenericItems)
```

In POOM, a task item is defined as an `ITask` object and is stored in the default Tasks folder, which is exposed by the pGenericItems object. You can then manipulate task items by accessing the pGenericItems object. For example, the following code lists all the task items:

```
//Declare a reference to ITask object
ITask *pTask;

//Get the total items in the Tasks folder
int len;
pGenericItems->get_Count(&len);

//Go through the items in the Tasks folder with a for loop
for (int i =0 ; i < len; i++) {
    //Retrieve a Task item
    pGenericItems->Item(i, &pTask);
}
```

As you can see, accessing PIM data using native POOM APIs is not difficult, although the process is a little complicated. It also requires a developer to be familiar with C++ and Win32 API programming. The good news is that a number of managed POOM APIs are shipped with the Windows Mobile 5.0 SDK that greatly simplify your code and make your program easier to manage and understand.

This book focuses on managed code. You will learn how to use those new APIs in the following sections. Note, however, that the managed APIs that ship with Windows Mobile 5.0 do not apply to previous platforms and do not include all the functionality provided by POOM. For example, if you are writing a program for a device running Windows Mobile Smartphone 2003, or if your application needs access to the Infrared folder, which currently cannot be accessed through managed APIs, you will have to stick to the POOM native APIs.

The WindowsMobile.PocketOutlook Namespace

One of the greatest features introduced in Windows Mobile 5.0 is the `Microsoft.WindowsMobile.PocketOutlook` namespace, which provides similar functions to POOM but in managed code. This section summarizes what the namespace offers and how to use those managed APIs.

Note that the `Microsoft.WindowsMobile.PocketOutlook` namespace is not part of the .NET Compact Framework. Therefore, you first need to add the reference to your application. To add a reference in a solution or project in Visual Studio 2005, click Project⇨Add Reference. In the .NET tab of the Add Reference dialog box, choose Microsoft.WindowsMobile.PocketOutlook, as shown in Figure 8-1.

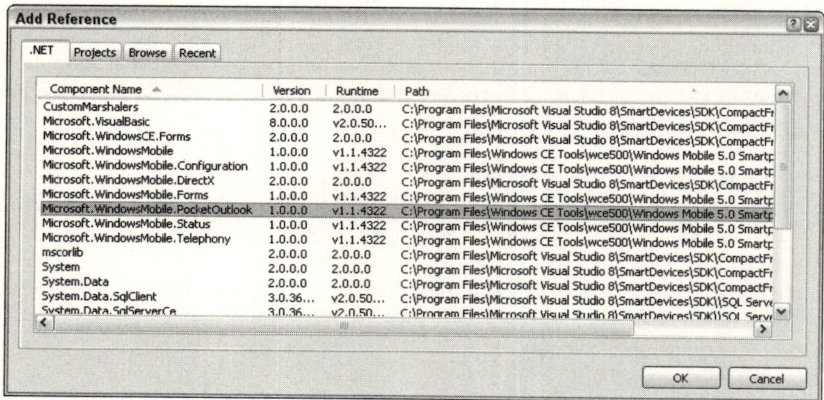

Figure 8-1

The process of using the managed APIs in `Microsoft.WindowsMobile.PocketOutlook` is much simpler than using the native POOM APIs, as illustrated here:

1. Create a new Outlook session:

```
OutlookSession aSession = new  OutlookSession();
```

2. Access the PIM object. Once an Outlook session is created, it exposes five properties to enable you to send e-mail, SMS messages, and PIM data:

    - `Appointments`
    - `Contacts`
    - `Tasks`
    - `EmailAccounts`
    - `SmsAccount`

    `EmailAccounts` and `SmsAccount` contain the accounts saved on the Outlook Mobile applet. You will need those accounts to send e-mail and to send/retrieve text messages. The `Appointments`, `Contacts`, and `Tasks` properties enable you to access those PIM folders. For example, the following code will list all the tasks in the Tasks folder of Outlook Mobile and print the task subject using the static `Show()` method of the `MessageBox` class:

```
TaskCollection taskItems = aSession.Tasks.Items;
foreach (Task t in taskItems) {
    Messagebox.Show(t.subject);
}
```

3. Release the Outlook session. Before closing your application, remember to release the resources by calling the `Dispose()` method. This is necessary because the `Microsoft.WindowsMobile.PocketOutlook` namespace is not part of the .NET Compact Framework; thus, the resources cannot be collected by the garbage collector.

```
aSession.Dispose();
```

The managed APIs in the `Microsoft.WindowsMobile.PocketOutlook` namespace are easy to use. You will learn the details for sending e-mail, accessing PIM data, and using text messaging in the next three sections.

Creating E-mail Applications with Managed APIs

The Outlook Mobile application that ships with Windows Mobile devices can manage both e-mail and text messages. From the home screen of a mobile device, users can click Start⇨Messaging⇨Outlook E-mail to enter the Outlook E-mail mobile application.

Users can create a new e-mail message by pressing the left soft key. If users are starting from scratch and no e-mail accounts are associated with this application, they will need to add e-mail accounts by clicking the right soft key and choosing Options⇨New Account, as indicated in Figure 8-2 and Figure 8-3.

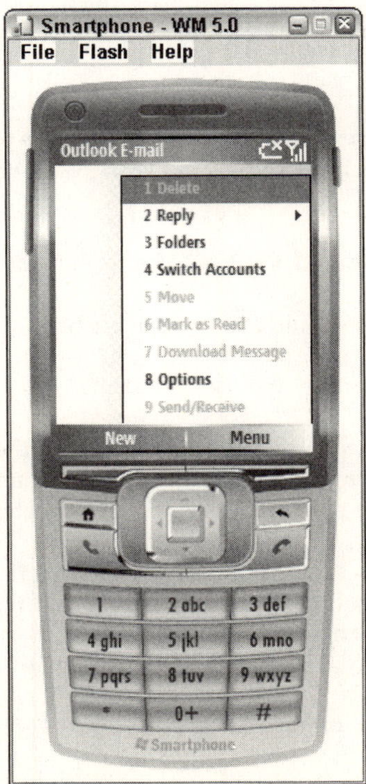

Figure 8-2

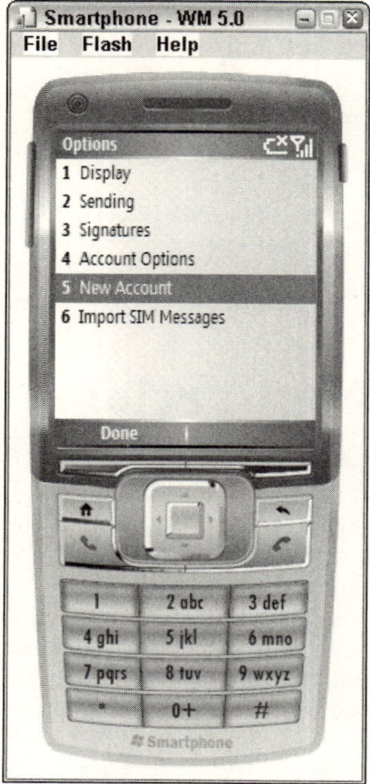

Figure 8-3

This will bring up the E-mail Setup wizard, which enables users to configure their name, e-mail address, e-mail server address, and other e-mail options. In Figure 8-4, an e-mail account IMAP4 is set up with the e-mail address byang@bsu.edu.

Once the e-mail accounts are configured, Outlook Mobile is ready to send and retrieve e-mail.

To write an e-mail application, you should start by establishing an Outlook Mobile session, as follows:

```
OutlookSession aSession = new  OutlookSession();
```

The three basic elements of an e-mail application — the sender, the recipient, and the message — are all defined in the `Microsoft.WindowsMobile.PocketOutlook` namespace. They are the `EmailAccount` class, the `Recipient` class, and the `EmailMessage` class, respectively.

The `EmailAccounts` property of an `OutlookSession` class is defined as a collection of the `EmailAccount` object. It contains all the e-mail accounts users have configured with the Outlook Mobile application. A particular e-mail account can be identified either by index or by its name. For example, to use the second e-mail account of the `EmailAccounts` property, you can declare an `EmailAccount` object as follows:

```
EmailAccount anEmailAcct = aSession.EmailAccounts[1];
```

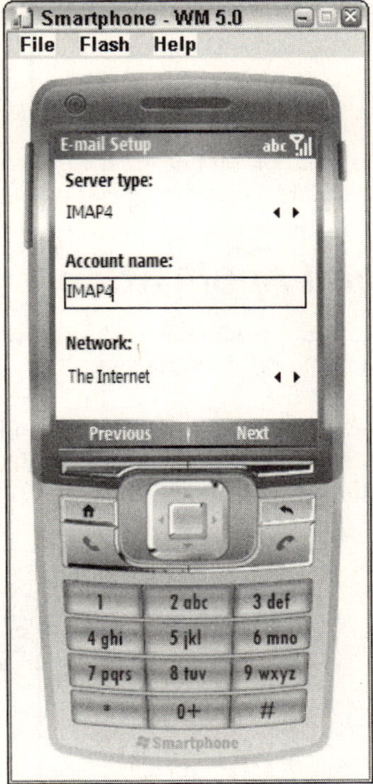

Figure 8-4

Alternatively, if you know the name of the e-mail account, you can refer to that account by name. The following code declares an `EmailAccount` object `anEmailAcct`, which refers to the account name IMAP4 in the Outlook Mobile application:

```
EmailAccount anEmailAcct = aSession.EmailAccounts["IMAP4"];
```

The `Recipient` class has two properties: `Address` and `Name`. You can create a new `recipient` object by passing the receiver's name and e-mail address to the `recipient` constructor, as follows:

```
Recipient recv = new Recipient("FirtName LastName", "FLastName@somewhere.com");
```

The `EmailMessage` class is the key managed API in an e-mail application. It exposes eight properties for constructing an e-mail message: `To`, `CC`, `Bcc`, `Attachments`, `Subject`, `Sensitivity`, `Importance`, and `BodyText`. Of course, you don't have to use all of them to compose a new message. Just as you would write an e-mail message in Outlook Mobile, you can simply fill those properties with the desired information. For example, the following code creates a simple `EmailMessage` object that says hello to our readers:

```
EmailMessage mesg = new EmailMessage();

mesg.To.Add(recv);
mesg.Subject = "Hello";
mesg.BodyText = "Dear readers, we hope you enjoy learning Smartphone programming";
```

The last action is to send the e-mail out by calling the `Send()` method of the `EmailAccount` class, as follows:

```
anEmailAcct.Send(mesg);
```

Depending on how the network and Outlook Mobile are configured on the device, the e-mail may be sent out directly or users may have to wait until the synchronization is completed when the device is cradled.

Creating a Simple E-mail Application

The following simple e-mail application includes all the aforementioned pieces. It will send a simple e-mail message to the editor of this book, Mr. John Sleeva.

Start a new Windows Mobile 5.0 device application in Visual Studio 2005, and name the project **email1**. Figure 8-5 illustrates the simple UI. It contains only two menu items: The left soft key triggers the quit event and the right soft key sends the e-mail.

Figure 8-5

Before you actually do the coding, don't forget to add a reference to the `Microsoft.WindowsMobile.PocketOutlook` namespace; otherwise, the following `using` statement will not even pass the compile:

```
using Microsoft.WindowsMobile.PocketOutlook;
```

Note also that you should always call the `Dispose()` method when you terminate the Outlook session. Failure to do so may cause errors when you launch the application again.

The following is the full listing of the sample code:

```
using System;
using System.Collections.Generic;
using System.ComponentModel;
using System.Data;
using System.Drawing;
using System.Text;
using System.Windows.Forms;
//Don't forget to add a reference to Microsoft.WindowsMobile.PocketOutlook
using Microsoft.WindowsMobile.PocketOutlook;

namespace email1
{
    public partial class Form1 : Form
    {
        //Declare an OutlookSession object aSession
        OutlookSession aSession;

        public Form1()
        {
            InitializeComponent();

            //Create a Pocket Outlook session
            aSession = new OutlookSession();
        }

        // When the left soft key is pressed, quit the application
private void menuItem1_Click(object sender, EventArgs e)
        {
            //Release the Pocket Outlook session
            aSession.Dispose();

            Application.Exit();
        }

        //When the right soft key is pressed, compose and send out email message

        private void menuItem2_Click(object sender, EventArgs e)
        {

            //Assumes an account called IMAP4 exists
            //Use email account IMAP4 as the sender
            EmailAccount myAcct = aSession.EmailAccounts["IMAP4"];

            //Create a new EmailMessage object
            EmailMessage mesg = new EmailMessage();

            //Compose the body text of this email message
            mesg.BodyText = "John, thanks a lot. We really appreciate it! ";

            //Create a new recipient with name and email address
            Recipient recv = new Recipient("John Sleeva", "JSleeva@wiley.com");

            //Add the recipient to the To field of the message
```

```
                mesg.To.Add(recv);

                //Send the message
                myAcct.Send(mesg);

                //Inform users the operation is finished
                MessageBox.Show("Done!", "Notice");

                //Disable the send menu;
                this.menuItem2.Enabled = false;

            }
        }
    }
```

In the preceding example, an OutlookSession object, aSession, is declared as a property of the Form1 class. aSession is instantiated when a Form1 object is constructed. The right soft key click handler menuItem2_click() creates the e-mail message and recipient and chooses the e-mail account IMAP4 as the sender. After the Send() method is called, a MessageBox.Show() is called to inform users that the action is performed. As mentioned earlier, this e-mail message may not be delivered directly. It really depends on how users have configured their Outlook Mobile applications. One thing you do not want users to do is to send the same e-mail message repeatedly by pressing the right soft key. Therefore, the last line of the menuItem2_click() function disables the Send menu item to avoid duplicated e-mail messages.

Creating an E-mail Application with Attachments

The sample application email1 illustrated how to send an e-mail message from an application, but everything is hard-coded and it is not user friendly: Users don't get to see the message and cannot even modify it. What if users want to send an e-mail attachment? Furthermore, what about displaying a graphic user interface that actually enables users to review and edit a message?

Adding an attachment to an e-mail message is not difficult to implement because the EmailMessage class has an Attachments property. For instance, to attach a file myFile1 to an EmailMessage object myMessage, first create a new Attachment object, attFile, from myFile1, and then insert the attachment to the myMessage.Attachments by calling its Add() method:

```
//Create a new attachment where myFile1 is the filename
Attachment attFile = new Attachment (myFile1);

//MyMessage is an EmailMessage object
myMessage.Attachments.Add (attFile);
```

Although adding an attachment to an e-mail message is rather straightforward, dealing with filenames can be tricky. To avoid potential human errors with the name of the file, it is suggested that you let users select the file from a folder, rather than key in the name of the file. For example, if the attachment is an image file, you can use a SelectPictureDialog object to pick the picture file.

SelectPictureDialog is a class in the new Microsoft.WindowsMobile.Forms namespace that ships with Windows Mobile 5.0. Like the Microsoft.WindowsMobile.PocketOutlook namespace, the Microsoft.WindowsMobile.Forms namespace is one of the new features in Windows Mobile 5.0 and is not part of the .NET Compact Framework. As a result, when using the classes in this namespace, you need to add the reference to your project or solution file.

The following code snippet demonstrates how to use a `SelectPictureDialog` object to retrieve the filename of a picture file:

```
using Microsoft.WindowsMobile.Forms;
...

SelectPictureDialog picDlg = new SelectPictureDialog();
picDlg.InitialDirectory = @"\Images";

DialogResult result = picDlg.ShowDialog();

if (result == DialogResult.OK)
{
    MessageBox.Show(picDlg.FileName,"Information");
}
```

In the preceding code, a `SelectPictureDialog` object, `picDlg`, is first created. Then the directory displayed in the Select a Picture dialog box is set to `\Images` by specifying the `InitialDirectory` property. The execution result of the Select a Picture dialog is passed to a `DialogResult` object. If the user selects a picture, the `DialogResult` is equal to `DialogResult.OK`. The filename of the selected the picture can be retrieved by getting the `FileName` property of the Select a Picture dialog box.

How do you construct a user-friendly e-mail application user interface? You can certainly design your own form, but a better solution would be to use the `MessagingApplication` class in the `Microsoft.WindowsMobile.PocketOutlook` namespace.

The `MessagingApplication` class provides automation of the messaging application's user interface. The key function of this class is the static `DisplayComposeForm()` method, which is overloaded and can be used to display an e-mail message form as well as an SMS message form. A typical way to display an e-mail compose form is to pass an `EmailAccount` object and an `EmailMessage` object to the `DisPlayComposeForm()` method, as follows:

```
//mesg is an instance of EmailMessage class
//myEmail is a reference to one of the email accounts
MessagingApplication.DisplayCompseForm (myEmail, mesg);
```

With the help of the Select a Picture dialog box and the `MessagingApplication` class, you can create an enhanced version of an e-mail application. This application can enable users to select an image as an e-mail attachment and will display the compose form on the screen.

Start a new Windows Mobile Smartphone application project and name it **email2**. Add references to both the `Microsoft.WindowsMobile.Forms` namespace and the `Microsoft.WindowsMobile.PocketOutlook` namespace. Then use the same code from application email1, except for the click event handler of menuItem2.

When users click the Send menu, you need to first create a new Select a Picture dialog box. After the user selects the picture, construct a new message and pass the filename of the picture to the e-mail message attachment and display the compose form.

The following example shows the complete code of application email2:

```
using System;
using System.Collections.Generic;
using System.ComponentModel;
using System.Data;
using System.Drawing;
using System.Text;
using System.Windows.Forms;

//Need to add references to the following two namespaces
using Microsoft.WindowsMobile.Forms;
using Microsoft.WindowsMobile.PocketOutlook;

namespace email2
{
    public partial class Form1 : Form
    {
        OutlookSession aSession;

        public Form1()
        {
            InitializeComponent();
            aSession = new OutlookSession();
        }

        // When the left soft key is pressed, quit the application
        private void menuItem1_Click(object sender, EventArgs e)
        {
            //Release the Mobile Outlook session
            aSession.Dispose();

            Application.Exit();
        }

        //When the right soft key is pressed, compose and send out email message
        private void menuItem2_Click(object sender, EventArgs e)
        {
            //Create a new Select Picture dialog box
            SelectPictureDialog picDlg = new SelectPictureDialog();
            picDlg.InitialDirectory = @"\Images";

            //Do not forward a Digital Rights Management protected file
            picDlg.ShowForwardLockedContent = false;

            //Get the dialog result
            DialogResult result = picDlg.ShowDialog();

            //After the user selects a picture
            if (result == DialogResult.OK)
            {
                //Create a new email message
                EmailMessage mesg = new EmailMessage();
                mesg.Subject = "Email with Picture Attachment";
```

```
                mesg.BodyText = "Open the attachment. It is not a virus. ";
                //Create and add a new recipient
                Recipient resv = new Recipient("John Doe","JDoe@somewhere.com");
                mesg.To.Add(resv);

                //Add the picture to the attachment
                Attachment picture = new Attachment(picDlg.FileName);
                mesg.Attachments.Add(picture);

                //Use the default email account
                EmailAccount myEmail = aSession.EmailAccounts[0];

                //Display the email compose form
                MessagingApplication.DisplayComposeForm(myEmail, mesg);
            }
        }
    }
}
```

After you compile and run the application, a Select a Picture dialog box appears after users click the right soft key, as shown in Figure 8-6. Assuming the `Waterfall.jpg` file is selected, the e-mail compose form then appears, as illustrated in Figure 8-7.

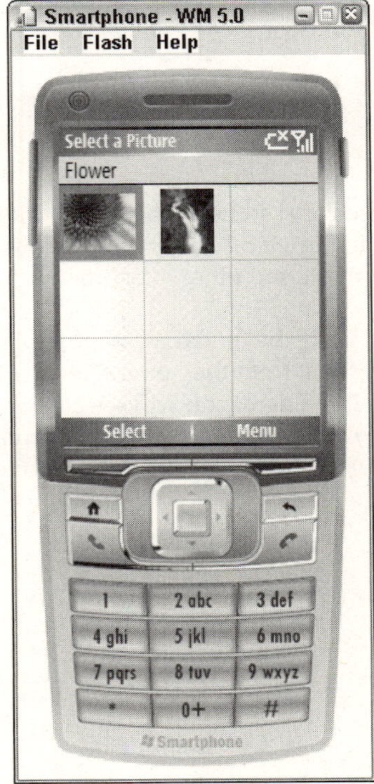

Figure 8-6

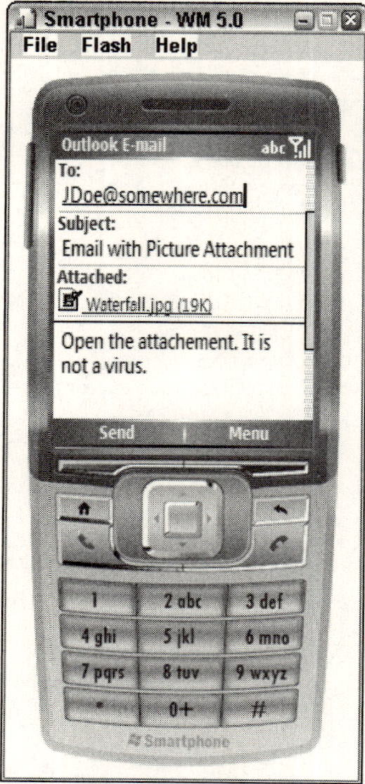

Figure 8-7

Note that this is a compose box. Users can still edit the distribution list, the topic of the message, and the body text. Moreover, as displayed in Figure 8-8, the menu options of the compose form give users additional options, such as adding recipients and other attachments.

When users choose the Send menu item, the e-mail message is placed in the outbox of the Outlook Mobile application. To verify the results, from the home screen of the Smartphone device, click Start⇨Messaging⇨Outlook Email⇨Menul⇨Folders, You will see a new message is added to the Outbox folder. Click the Outbox folder to further review the summarized information of the e-mail message. The screenshots of Outlook's E-mail Outbox folder are displayed in Figure 8-9 and Figure 8-10.

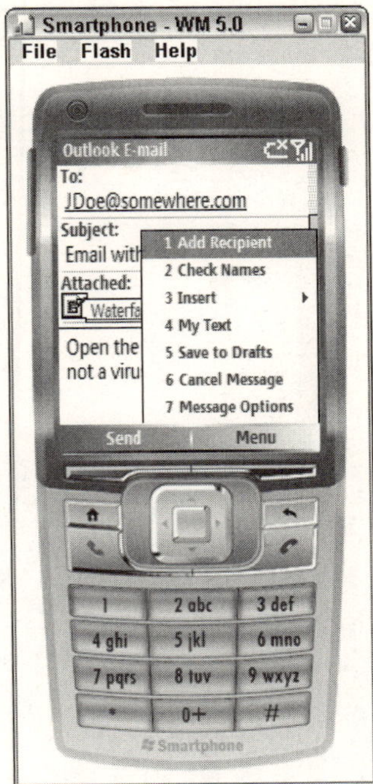

Figure 8-8

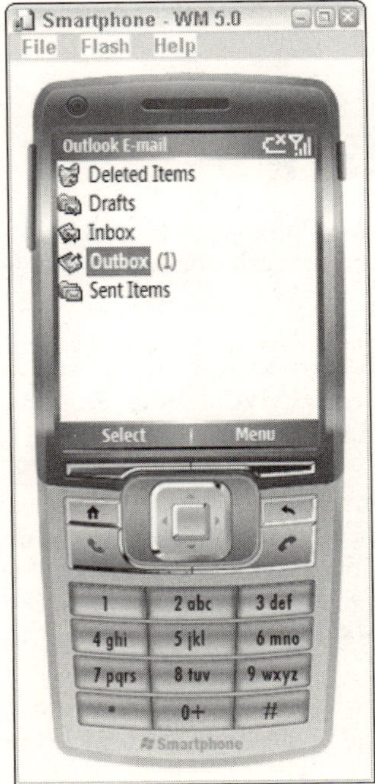

Figure 8-9

Figure 8-10

Accessing PIM Data

Users are interested primarily in three major types of PIM data: appointments, contacts, and tasks. This section describes a simple application that can populate a ListView control with all the appointments in the calendar application.

Create a new Windows Mobile 5.0 Smartphone application and name it **GetCalendar**. On the Form Designer, change the name of the form to **MyCalendar**. Then drag and drop a ListView control onto the MyCalendar form, and rename the ListView control **CalView**.

When using a ListView control in Smartphone, you should always keep in mind that the viewable area is fairly small, so it is not a good practice to add too many columns. In the example, two columns will be shown on the screen: date and subject. Because the default behavior of a ListView control displays only one column in large icons, you need to change the default `view` property of a ListView control to `Details`, as demonstrated in Figure 8-11.

Next, you need to edit each column's header. Click Columns in the property editor and add date and subject columns in the ColumnHeader Collection Editor window, as shown in Figure 8-12.

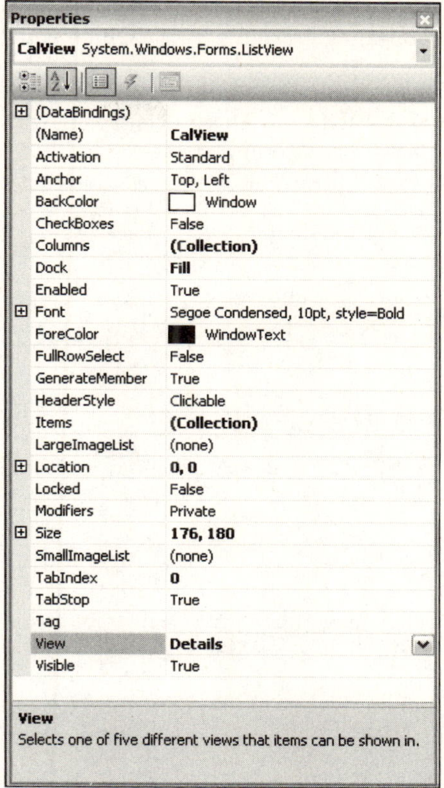

Figure 8-11

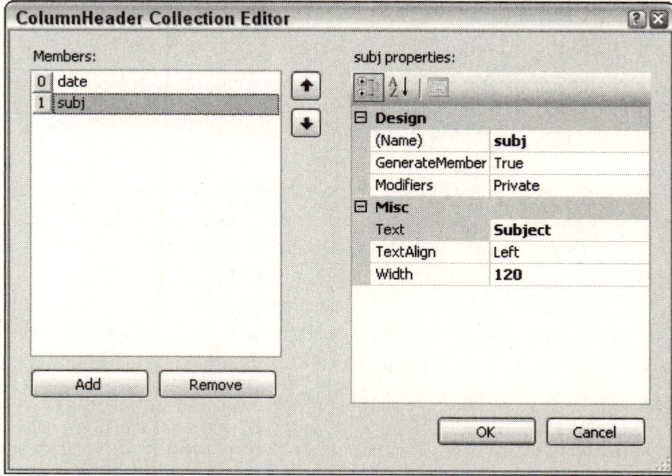

Figure 8-12

After finishing the design of the ListView control, you can make some changes to the user interface to make it friendlier, if not more appealing. In Figure 8-13, a Quit menu item is added on the left side and a Menu menu item is added to the right.

To make it simple, the application will display all the events in the calendar book whenever the MyCalender form is loaded. This requires adding an event handler to the form's `onload` event. The following code illustrates how to export the events to the `GetCalendar` application:

```
...
Using Microsoft.WindowsMobile.PocketOutlook;
...

private void MyCalendar_Load(object sender, EventArgs e)
{
    //Establish a new Outlook session
OutlookSession CalSess = new OutlookSession();

//Get the collection of appointments by calling outlookSession.Appointments.Items
    AppointmentCollection CalCol = CalSess.Appointments.Items;

    //Add each appointment to the ListView
    foreach (Appointment apt in CalCol)
{
    //Create one new ListView object for each appointment
        ListViewItem aLVItem = new ListViewItem();

        //Make the appointment date the text property of this ListView
        aLVItem.Text = apt.Start.Date.ToString();

        //Make other appointment property as the subitem of this ListView
        aLVItem.SubItems.Add(apt.Subject);

        //You can also add a field for the appointment location
        //aLVItem.SubItems.Add(apt.Location);

        //Add ListViewItem to the ListView
        CalView.Items.Add(aLVItem);
    }
}
```

This code first creates a new `OutlookSession` object and then retrieves the collection of the `Appointment` objects. In the `for` loop, a new `ListViewItem` is created for every `Appointment` object. The `start date` property of an `Appointment` object is passed to the `Text` property of the `ListViewItem`, and the `subject` property of the `Appointment` object is added to the `SubItems` of the `ListViewItem` object. By doing so, the start date will be the first column of a `ListViewItem` and the `subject` will be the second column of the `ListViewItem`. At the end of the `for` loop, simply add the `ListViewItem` object to the `Items` member of the `ListView`, which corresponds to adding a new row to the `ListView` control.

Figure 8-14 shows a sample runtime result. The three events listed in the figure are exactly the same appointments stored on the Windows Mobile 5.0 device emulator using the calendar program (choose Start⇨Calendar). Of course, when you test the code on your device or emulator, you won't be able to see anything if the Calendar application does not contain any appointments.

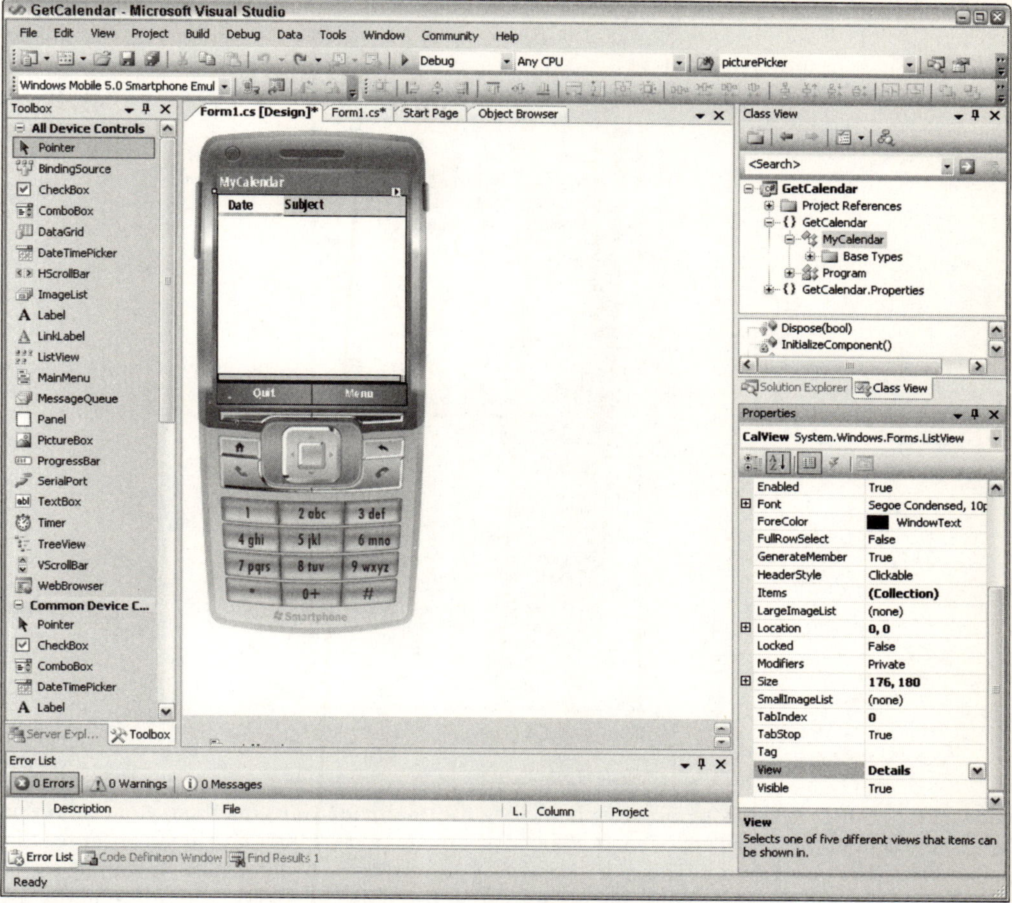

Figure 8-13

If a user has kept many events in her or his calendar, especially some recurring events such as weekly meetings and birthdays, the sample code will easily use up the memory resources and provide a long list of information that may or may not be useful. Or, in another case, if a user is trying to search all the appointments he or she has with a certain customer, it will be desirable to get a set of appointments with certain properties.

The `PimItemCollection` class in the `PocketOutlook` namespace, which is the base class of the `Appointment`, `Contact`, and `Task` collection, offers a `Restrict` method that enables users to search items in a collection. For example, the following code will pull out only those appointments with the `Categories` property set to `wrox`:

```
string query = "[Categories] == wrox";
AppointmentCollection wroxAppt = aOutlookSession.Appointment.Items.restrict(query);
```

Figure 8-14

There are two things to pay attention to when using the `Restrict` function. First, you need to put the name of the property in brackets. Second, the `Restrict` method requires a value in order for the field of the searching property to work. This is tricky when you use the <> evaluator. For example, suppose you wanted to retrieve appointments that are not located at Ball Sate University. You can certainly use a query string such as the following:

```
string query = "[Location] <> Ball State University"
```

This search will not select those appointments with an empty location field even though their locations are not set to `Ball State University`.

Using SMS

Text messaging is no doubt one of the hottest mobile applications. It can be used to keep in touch with your buddies and to vote for your favorite *American Idol* contestants. There are also some applications that can destroy a lost mobile device by sending out specially coded text messages.

The `Microsoft.WindowsMobile.PocketOutlook` namespace makes SMS programming a simple and easy process.

To send a message, you need to create an instance of the `SmsMessage` class. Then you can specify the recipient and the message, very similar to sending an e-mail message. The following snippet of code will send a text message to the phone number 1-866-436-5702 and vote for the number 2 contestant of *American Idol:*

```
OutlookSession aSession = new OutlookSession();
SmsMessage sendMsg = new SmsMessage("18664365702", "Vote to Idol #2");
aSession.SmsAccount.Send(sendMsg);
aSession.Dispose();
```

As illustrated in the preceding code, you still need to start by creating a new `OutlookSession` object. You can then construct a new SMS message by feeding the recipient's phone number and the message to the constructor of the `SmsMessage` class directly:

```
SmsMessage sendMsg = new SmsMessage("18664365702", "Vote to Idol #2");
```

Alternatively, you can construct each component of the `SmsMessage` one at a time, as follows:

```
SmsMessage sendMsg = new SmsMessage();
Recipient recv = new Recipient("18664365702");
sendMsg.Body = "Vote to Idol #2";
sendMsg.To.Add (recv);
```

The `Send()` method of the `SmsAccount` class is responsible for delivering the SMS message. In POOM, each Outlook session can have more than one e-mail account, but only one SMS message account. It makes sense because the address of a SMS account is usually the telephone number, and each Smartphone device has one unique phone number. Therefore, in SMS applications, you do not have to specify which account you are going to use, as you do in e-mail applications. Rather, use the default `SmsAccount` object of the `OutlookSession` object.

If you want users to see the SMS message compose form, call the static `DisplayComposeform()` method of the `MessagingApplication` class:

```
MessagingApplication.DisplayComposeForm(sendMsg)
```

The `Microsoft.WindowsMobile.PocketOutlook namespace` also provides a `MessageInterceptor` class that enables you to intercept a text message. To receive a text message, you must create a new instance of the `MessageInterceptor` object and register an event handler to handle the event when a new text message is received. The following code demonstrates how this can be done:

```
        //Create a new instance of SMS interceptor when form is loaded
        private void Form1_Load(object sender, EventArgs e)
        {
            msginterceptor = new MessageInterceptor();
            msginterceptor.InterceptionAction = InterceptionAction.NotifyAndDelete;
            msginterceptor.MessageReceived += new
MessageInterceptorEventHandler(msginterceptor_MessageReceived);

        }

        //Handling received message
```

```
        void msginterceptor_MessageReceived(object sender,
MessageInterceptorEventArgs e)
        {
            SmsMessage smsMsg = (SmsMessage)e.Message;
            string fullText = "Message From: "+smsMsg.From.Name;
            fullText += (" at "+ smsMsg.Received.TimeOfDay.ToString());
            fullText += (" and the message is:    "+smsMsg.Body);
            MessageBox.Show(fullText, "New Text Message !");
        }
```

In the `Form1_Load()` method, a new `MessageInterceptor` is created and the `InterceptionAction` is defined as `NotifyAndDelete`, which means the text message received will not be saved to the inbox folder of the Outlook Mobile text messaging application. Then the method `msginterceptor_MessageReceived` is registered as the event handler when SMS messages are received.

In the event handler, the event argument `e` is cast to the `SmsMessage` type. Information such as the name of the sender, the received time, and the body text can then be retrieved from the `SmsMessage` objects.

The following example demonstrates how to write an application that can send and receive text messages. Start a new Windows Mobile Smartphone device application and name the project **sms1**. Add the `Microsoft.WindowsMobile.PocketOutlook` namespace to the project, and then add two menu items to the form, Send and Quit, as shown in Figure 8-15. When a user presses the Send button, the application can display an SMS compose form with pre-composed information filled in.

Figure 8-15

Following is the full code of this simple application:

```
using System;
using System.Collections.Generic;
using System.ComponentModel;
using System.Data;
using System.Drawing;
using System.Text;
using System.Windows.Forms;
using Microsoft.WindowsMobile.PocketOutlook;
using Microsoft.WindowsMobile.PocketOutlook.MessageInterception;

namespace SMS1
{
    public partial class Form1 : Form
    {
        private MessageInterceptor msginterceptor;
        private OutlookSession aSession;

        public Form1()
        {
            InitializeComponent();
            aSession = new OutlookSession();

            //Create a new instance of SMS interceptor
            msginterceptor = new MessageInterceptor();
            msginterceptor.InterceptionAction = InterceptionAction.NotifyAndDelete;
            msginterceptor.MessageReceived += new
MessageInterceptorEventHandler(msginterceptor_MessageReceived);

        }

        //Handling received message.
        void msginterceptor_MessageReceived(object sender,
MessageInterceptorEventArgs e)
        {
            SmsMessage smsMsg = (SmsMessage)e.Message;
            string fullText = "Message From: "+smsMsg.From.Name;
            fullText += (" at "+ smsMsg.Received.TimeOfDay.ToString());
            fullText += (" and the message is:     "+smsMsg.Body);
            MessageBox.Show(fullText, "New Text Message !");
        }

        //When Send button is hit
        private void menuItem1_Click(object sender, EventArgs e)
        {
            SmsMessage sendMsg = new SmsMessage("14250010001", "Send to myself.");
            //Display the compose form
            MessagingApplication.DisplayComposeForm(sendMsg);
        }

        private void menuItem2_Click(object sender, EventArgs e)
        {
            aSession.Dispose();
```

```
                Application.Exit();
            }
        }
    }
```

A `MessageInterceptor` object and an `OutlookSession` object are added to the `Form1` class. During the initialization process of `Form1`, create a new `OutlookSession` object and a new instance of the `MessageInterceptor` object, and then register the `MessageReceived` event handler.

A simple way to verify the `MessageReceived` event handler is to send an SMS message to the device itself. On Windows Mobile device emulators, such as Smartphone device emulator and Pocket PC phone edition emulator, use the number 1-425-001-0001.

Now you can test this application. However, when you build the application, you may get an error message, as shown in Figure 8-16.

You can fix this by adding a reference to the `Microsoft.WindowsMobile` namespace to your project.

The screenshots of the runtime results are captured in Figure 8-17 and Figure 8-18, respectively.

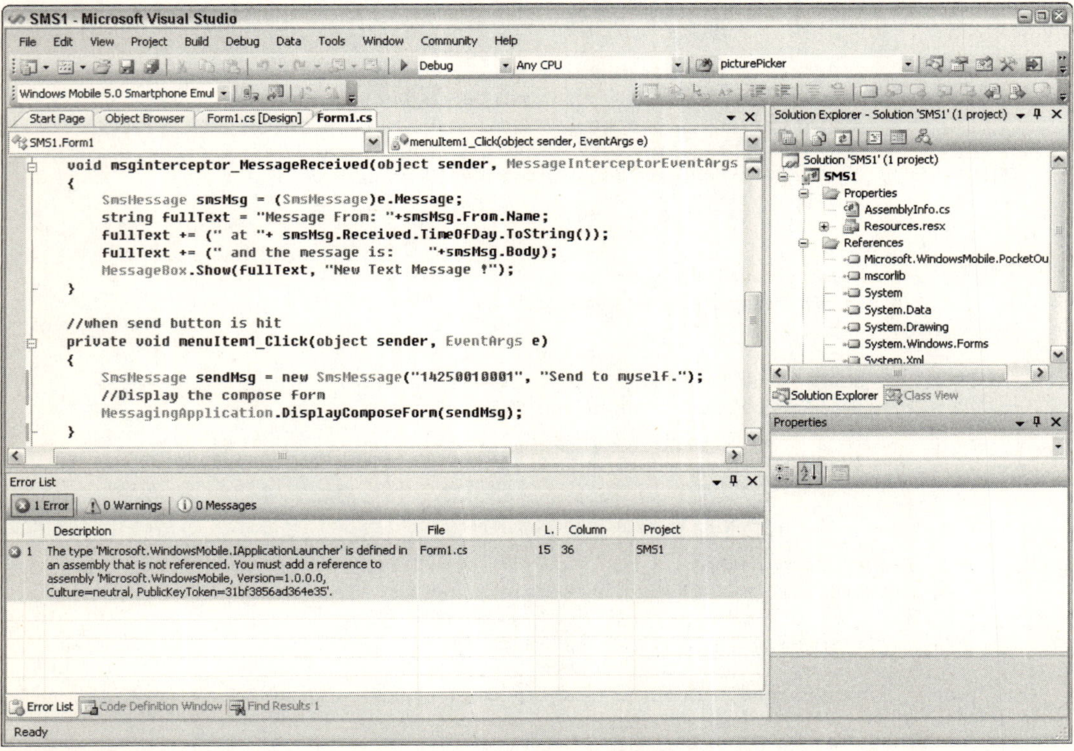

Figure 8-16

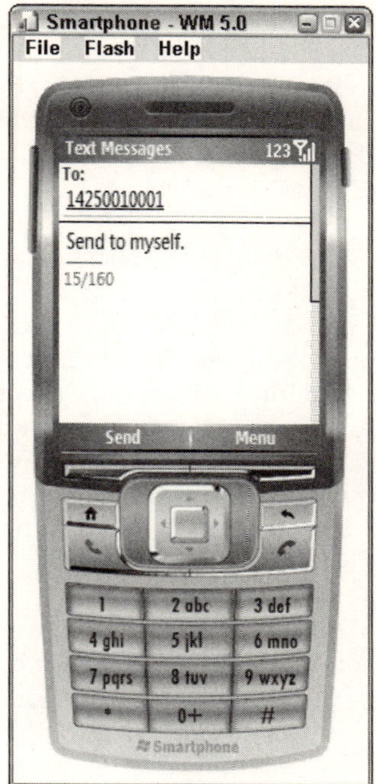

Figure 8-17

Figure 8-18

Summary

In this chapter, you have learned about how to program using the managed APIs provided in the new `Microsoft.WindowsMobile.PocketOutlook` namespace. As the name implies, it is a feature provided for Windows Mobile 5.0 devices and it is not part of the .NET Compact Framework. Before using those APIs, don't forget to add the references to your project.

Building an e-mail application, an SMS messaging application, and a PIM-data-related application starts with the instantiation of a new `OutlookSession` class. You can then access the Outlook Mobile Appointments folder, Contacts folder, and Tasks folder and manage that PIM data. To send an e-mail message or SMS message in your application, you need to first create an `EmailMessage` object or a `SmsMessage` object and fill in the related information, such as a recipient list, a subject, body text, and so on. Then you call the `EmailAccount.Send()` method or the `SmsAccount.Send()` method to deliver the message. If you want to display a user-friendly compose form before sending out the message, use the static `DisplayComposeForm()` method of the `MessagingApplication` class. The `Microsoft.WindowsMobile.PocketOutlook.MessageInterception` namespace enables you to write your own event handler when a message is received.

Compared to the unmanaged POOM APIs, the `Microsoft.WindowsMobile.PocketOutlook` namespace makes your programming job a lot easier and safer. We strongly recommend that you use those managed APIs unless you are writing applications for a previous Windows Mobile Smartphone, or the particular functions are currently not available (such as infrared) in a managed manner.

This chapter focuses on PIM data management and communications through e-mail and text messages. The next chapter expands the horizon to popular client/server applications: XML and web services.

9

XML and Web Services

XML and web services are used in a wide range of desktop applications that exchange data with each other. Smartphone applications can also take advantage of these powerful network communication mechanisms to enable users to access real-time data from a smartphone device. This chapter introduces the XML support in the .NET Compact Framework, especially the classes and types that encapsulate an XML document and data set's XML functionality. You will also learn about Visual Studio .NET's basic support for creating applications that access web services. This chapter presents everything you need to know to get started building your own XML-based network data applications and web service client applications.

The chapter covers the following topics:

- XML syntax and XML Schema syntax
- How to use the .NET Compact Framework's XML classes to read and write XML documents
- How to use Visual Studio to create applications that access a web service

Overview of XML and Web Services

eXtensible Markup Language (XML) is a W3C-recommended, general-purpose markup language capable of representing structured data with user-defined tags. Unlike HTML, which emphasizes displaying data on a web page, XML solves the problem of generating, exchanging, and interpreting text-based data between two applications. In this section, you will see that XML is frequently used as the underlying language for different purposes.

The W3C published two format standards: XML 1.1 (February 2004) and XSD 1.0 (May 2001). XSD (XML Schema Definition) describes the schema of data such as element types, relationships, constraints, and so on. XSD itself is an XML file that can be used to validate data contained in the corresponding XML data file.

A *web service* is a software system that enables machine-to-machine, text-based communication on the Internet and within an enterprise computing environment. The advantage of a web service over other communication methods such as RPC is that the two endpoints are largely independent of each other and the text messages being exchanged over HTTP conform to a standard called *SOAP*. A SOAP message is formatted as an XML file as well. A web service client sends to the web service provider some SOAP messages that contain specific requests. The web service provider receives the SOAP messages, interprets them, generates the corresponding result, wraps the result into some SOAP messages, and then sends them back to the client. Of course, the web service client must know what kind of "services" the web service can provide so that it can generate valid requests. Such a service description of a web service is described in a language named *Web Services Description Language (WSDL)*.

Several online companies offer web services as a way to expand their business by providing a platform for business partners and independent software vendors. These services are usually free within the limit of some volume of requests. For example, the Google web service that enables anybody to invoke Google web searching and other Google services is free as long as the client generates fewer than 1,000 requests per day. To facilitate web service client programming, web service providers often provide an SDK that includes some examples and class libraries.

Table 9-1 lists some well-known web services.

Table 9-1 Well-Known Web Services

Web Service	Description	Link	Cost
Amazon Web Services	A set of web services that provides access to product information, pricing, e-commerce transactions, etc.	`www.amazon.com/aws/`	Some web services are free, whereas others are pay-as-you-go or require a subscription.
eBay Developers Program	Access to eBay's marketplace to obtain item information, manage sales, conduct order and purchasing transactions, etc.	`http://developer.ebay.com`	Free
Google Web APIs	Includes Google SOAP Search API for accessing Google search services and Google Maps API for searching and displaying Google maps	`www.google.com/apis` `www.google.com/apis/maps`	Free for the first 1,000 queries per day per application
MSN Search	Web search service	`http://msdn.microsoft.com/msn/msnsearch/`	Free for the first 10,000 queries per day per machine

Web Service	Description	Link	Cost
Yahoo! Developer Network	A set of web services for building mapping applications, widget galleries, music plug-ins, search applications, and Flickr (web-based social networking) services	`http://developer.yahoo.net`	Rate limited on a daily basis per machine

XML Support in .NET Compact Framework

XML support in the .NET Compact Framework 2.0 can be described in the following categories:

- Web service client support (in combination with Visual Studio .NET)
- The `XmlDocument` class implements the W3C Document Object Model (DOM) for creating and processing XML documents. According to the World Wide Web Consortium (W3C), the Document Object Model is a platform- and language-neutral interface that enables programs and scripts to dynamically access and update the content, structure, and style of documents.
- `XmlSchema` and `XmlSchemaCollection` classes

According to MSDN, the following standards are supported by the .NET Compact Framework:

- XML 1.0 (including DTD support)
- XML namespaces (both stream-level and DOM)
- XSD schemas
- XPath expressions (XPath 1.0) for querying and navigating an XML document
- DOM Level 1 Core
- DOM Level 2 Core

The core XML classes are in the `System.Xml` and `SystemXml.Schema` namespaces. In addition, XML serialization is in the `System.Xml.Serialization` namespace, and XPath support (which is a structure query format specifying a set of XML elements) is provided in the `System.Xml.XPath` namespace.

Due to the memory constraints of mobile devices, not all classes in the `System.Xml` namespace in the .NET Framework are supported in the .NET Compact Framework. For example, the `XmlDataDocument` class, which extends the `XmlDocument` class to support structured data storage and retrieval, is not available in the .NET Compact Framework.

XML Syntax

This section uses a sample XML file, `simple.home.xml`, to introduce XML syntax. This XML file is a standard Windows Mobile home screen definition file on a Smartphone device. You can find the XML file under `\Application Data\Home` on the device emulator.

```
<?xml version="1.0"?>
<home>
    <author>Microsoft Corporation</author>
    <contacturl>http://www.microsoft.com/</contacturl>
    <title>Windows Simple</title>
    <title lang="0x0409">Windows Simple</title>
    <title lang="0x0407">Windows minimal</title>
    <title lang="0x0c0a">Windows sencillo</title>
    <title lang="0x040c">Windows simple</title>
    <title lang="0x0410">Windows semplice</title>
    <title lang="0x0816">Windows simples</title>
    <title lang="0x0416">Windows simples</title>
    <title lang="0x0809">Windows Simple</title>
    <title lang="0x0406">Windows simple</title>
    <title lang="0x0413">Windows eenvoudig</title>
    <title lang="0x041d">Windows enkel</title>
    <title lang="0x0414">Windows - enkelt</title>
    <title lang="0x040b">Windows Simple</title>
    <title lang="0x040e">Windows egyszer_</title>
    <title lang="0x0405">Windows jednoduch&#225;</title>
    <title lang="0x0418">Windows simplu</title>
    <title lang="0x041b">Windows jednoduch&#233;</title>
    <title lang="0x0415">Prosty - Windows</title>

    <version>1.0</version>
    <default target-width="240" target-height="320" font-face="Segoe Condensed"
font-size="19" padding-left="5" padding-right="5" bgcolor="transparent"
fgcolor="COLOR_HOMETEXT" padding-top="3">
    </default>
    <background bgimage="\windows\WindowsMobile.jpg" valign="bottom"
bgcolor="COLOR_TRAYGRADLEFT" />
    <scheme>
        <color name="COLOR_TRAYGRADLEFT" value="#2E97E3" />
        <color name="COLOR_HOMEHIGHLIGHT" value="#50A5E5" />
    </scheme>
    <plugin clsid="{837FC251-FE69-43ad-84E0-EBCEDEBA0884}" name="iconbar"
height="27">
        <iconbar fgcolor="COLOR_TRAYTEXT"/>
        <background gradient="title" bgcolor="COLOR_TRAYGRADLEFT" b-border-
color="COLOR_WINDOWFRAME"/>
    </plugin>
    <plugin clsid="{E09043DF-510E-4841-B652-388316977A7A}" name="carrier"
height="27">
        <label font-weight="bold">
            <text><carrier/></text>
        </label>
    </plugin>
    <plugin clsid="{44FA0F8C-082F-42b4-BE49-82559F23D5D4}" name="clock"
height="55">
        <time font-size="27"/>
        <date font-size="27" halign="right"/>
    </plugin>
    <plugin clsid="{4824B576-EFFE-45cf-BAE9-649B930CD244}" name="owner"
height="164">
        <label font-size="33" valign="bottom">
```

```
            <text><name/></text>
        </label>
    </plugin>
</home>
```

The structure of the preceding sample XML file is quite clear: A top-level home element (known as the *root element*) contains a number of sub-elements: `author`, `contacturl`, `title`, `version`, `default`, `background`, `scheme`, and various `plug-ins`. The first line of the sample XML file is an optional declaration indicating the XML version. The following can be seen as a single root element (also known as a *document element*), which may contain nested elements. An element starts with a tag in the form of `<name>` and ends with `</name>`. An element can recursively contain a number of sub-elements, thus forming a tree of elements. Between these two tags is the element's content, which can be text or some sub-elements. An element can also be empty. The following example code shows valid elements. `<!--comment -->` is used to wrap comments, which will not be parsed. An empty element can be further simplified as `<name/>`.

```
<BookTitle>Professional Microsoft Smartphone Programming</BookTitle>
<BookTitle></BookTitle> <!--empty element-->
<BookTitle/> <!--empty element-->
```

A variation of the start tag `<name>` is that it may contain attributes (name-value pairs) in the form of `<name attribute1="value1" attribute2="value2" ...>`. Attribute values must be quoted using single or double quotes. No two elements can overlap — that is, they can only be in parallel or one must contain the other. For example, the following is a valid element that contains two sub-elements:

```
<Windows Mobile Devices>
  <Smartphone>Windows Mobile on smartphone</Smartphone>
  <Pocket PC> Windows Mobile on PDA</Pocket PC>
</Windows Mobile Devices>
```

The following, however, is not a valid element:

```
<Windows Mobile Devices>
  <Smartphone>Windows Mobile on Smartphone
  <Pocket PC> Windows Mobile on PDA</Smartphone></Pocket PC>
</Windows Mobile Devices>
```

Like HTML, XML syntax does not care about indention and carriage returns; the XML parser will simply skip them. They are used only for better code formatting.

Some elements, in turn, have sub-elements. For example, a plug-in element may have sub-elements such as time and date, which can have further sub-elements. This flexible, recursive, tree-like structure makes it easy to describe structured data.

An XML Example: Customizing the Home Screen

This section walks through the sample XML file, `simple.home.xml`, to illustrate how to modify it to customize a Windows Mobile home screen. A Smartphone's home screen is defined in an XML document. A cell phone running Windows Mobile comes with several home screens: Windows Default, Windows

Basic, Windows Simple, MSN Default, Large Font, and a carrier-customized home screen. Their XML files reside in the following directory `\Storage\Application Data\Home`. To change home screens, select Start⇨Setting⇨Home Screen.

On the Smartphone emulator, the preceding sample XML file defines the home screen shown in Figure 9-1. The configuration of the screen includes the default font format and screen layout, the background image, a scheme that defines the color and gradient for the entire device, several plug-ins for the icon bar on the top, carrier information ("Fake Network" on the emulator), the clock, and the owner's name.

Figure 9-1

Each home screen plug-in uses a unique class ID. For example, the following plug-in is copied from `pocketmsn.home.xml` in the same directory as `simple.home.xml`:

```
    <plugin clsid="{865A354A-4A96-4687-B001-C155DC0DBE76}" name="calendar"
height="50">
        <background>
            <format state="selected" bgcolor="COLOR_HOMEHIGHLIGHT"/>
        </background>
        <label h="22">
            <text><subject/></text>
        </label>
        <label h="22" y="23">
            <text><time/> <location/></text>
        </label>
    </plugin>
```

This plug-in specifies the display of the `subject`, `time`, and `location` of the next appointment, the label's height (the `h` attribute) and vertical position (the `y` attribute), as well as the plug-in's `clsid`, `name`, and `height`. By adding this plug-in to the Windows simple home screen XML (simply include this piece of code as a sub-element to the `<home>` element), the next appointment will show on the home screen. If the owner information exceeds the screen, reduce its `height` attribute to a smaller value (such as 50). Figure 9-2 shows the modified Windows Simple home screen.

Figure 9-2

XML Schema

An XML schema is used to describe valid elements and attributes for an XML document. Thus, an XML schema document can be used to validate an XML document so that different parties can agree on an XML data structure while exchanging XML data documents. A schema not only defines a list of elements and attributes that can appear in the XML document, it also defines the number and order of sub-elements, data types of elements or attributes, and default and fixed values for elements and attributes. An XML schema can be embedded into the XML document or saved into a separated XML file with the .xsd filename extension.

As an XML document, an XML schema document has a root element, `<schema>`, an example of which follows:

```
<xs:schema xmlns="" xmlns:xs="http://www.w3.org/2001/XMLSchema">
<xs:element name="book">
  <xs:complexType>
    <xs:sequence>
      <xs:element name="Title" Type="xs:string">
      <xs:element name="Price" Type="xs:string">
      <xs:element name="Author" Type="xs:string">
      <xs:element name="PublicationDate" type="xs:date"/>
    </xs:sequence>
  </xs:complexType>
</xs:element>
</xs:schema>
```

The `xmlns` attribute can define a default namespace that applies to elements with no prefixes, and some number of namespaces that apply to the element with that specific namespace. A namespace defines a set of valid element names and data types. In the preceding example, `xmlns=""` specifies the default namespace (which is empty); the attribute `xmlns:xs=http://www.w3.org/2001/XMLSchema` specifies the namespace with prefix "`xs`" (including the schema element itself). The `xs` namespace defines valid names and types, such as `element`, `schema`, `string`, that correspond to XML element entries, XML schema entries, and strings in an XML file, respectively. Simple types include `element`, `attribute`, and `restriction` for defining a data range or a list of valid values for elements with specific characteristics. Complex data types can contain other elements or attributes. For example, the preceding example defines a complex type for a sequence of elements. Valid data types include `xs:string`, `xs:decimal`, `xs:integer`, `xs:boolean`, `xs:date`, and `xs:time`.

For more information about XML schema data types, see the official W3C document at `www.w3.org/TR/xmlschema-2/`. You can also find a good tutorial at `http://msdn.microsoft.com/library/default.asp?url=/library/en-us/dnxml/html/XMLSchemaComplex.asp`.

XML schema can be embedded into an XML file. In fact, different elements in an XML file can specify different XML schemas. These are generally called *inline schemas*. The following code demonstrates using an inline schema in an XML file. Schema `"foo"` is used to validate the `EmployeeName` element:

```
<?xml version="1.0" encoding="utf-8" ?>
<root>
  <SomeElement>
    <someContent/>
    <xsd:schema xmlns:xsd="http://www.w3.org/2001/XMLSchema"
        targetNamespace="foo">
        <xsd:element name="EmployeeName" type="xsd:string"/>
    </xsd:schema>
  </SomeElement>
  <f:EmployeeName xmlns:f="foo">
    ... this will be validated with schema "foo" ...
  </f:EmployeeName>
</root>
```

By default, the `System.XML` namespace is able to process inline schemas. To enable them, you need to set the `ProcessInlineSchema` flag:

```
XmlReaderSettings Xmlsettings = new XmlReaderSettings();
Xmlsettings.ValidationFlags |= XmlSchemaValidationFlags.ProcessInlineSchema;
XmlReader valReader = XmlReader.Create(@"MyXML.xml", Xmlsettings);
```

With an XML schema, an application can validate an XML data stream or an XML document, ensuring that the data is in good form and can be further used. As you will see in the following section, the .NET Compact Framework has implemented this kind of functionality in some classes.

XML-Related Classes

This section describes how to use .NET Compact Framework classes to read and write XML files, as well as how to exchange XML-based data over the web. Table 9-2 shows a list of XML-related classes. To use any of the XML-related classes in the .NET Compact Framework, you must import the `System.XML` namespace to your class, as shown here:

```
using System.XML;
```

Because of the loose syntax rules of XML regarding white space, order of attributes or elements, and so on, it is difficult to generate the same digital signature of two files that indeed contain the same piece of data. There must be a way to normalize the lexical form of an XML document. The W3C defines canonical XML, *which is a normalized lexical form for XML that removes allowed variations and imposes strict rules for the uniform representation of a specific piece of data. .NET Framework classes such as* `XmlTextReader` *and* `XmlTextWriter` *perform this canonicalization while reading and writing a raw XML document.*

Table 9-2 XML Classes in .Net Compact Framework

Classes	Description
`XmlReader`	An abstract class that provides fast, forward-only, read-only access to an XML document stream.
`XmlTextReader`	The canonical `XmlReader` implementation for fine-grained, text-based XML processing.
`XmlNodeReader`	A public class that provides node navigation across an in-memory XML DOM tree or subtree. A *node* represents a single object in an XML DOT tree.
`XmlWriter`	An abstract class that provides an interface for generating an XML document stream.
`XmlTextWriter`	The canonical `XmlWriter` implementation.
`XmlNodeWriter`	A public class that produces an in-memory DOM tree. The *DOM* is an in-memory (cache) tree representation of an XML document.
`XmlDocument`	The .NET Compact Framework's DOM implementation that models an XML DOM tree.

The following sections describe how to use these classes to process an XML document stream. The discussion centers around the following tasks:

- Reading and writing XML elements to and from an XML document stream
- Transforming XML DOM data into a relational data structure

XmlDocument and XmlTextReader

The `XmlDocument` class is used to load an XML file into memory. The XML file is parsed such that a tree representation is created in memory. Just call the `Load()` method with the XML filename (and its path) and you are ready to explore the XML file.

The constructor of the `XmlDocument` class does not need any parameters. After creating an `XmlDocument` object, you can call either `Load()` or `LoadXML()` to read an XML file. The `Load()` method is overloaded with the following four variances:

- `void Load(Stream)` —Loads the XML document from a `System.IO.Stream` object, a `System.IO.NetworkStream` object, or a `System.IO.FileStream` object.
- `void Load(String)` —Loads the XML document from a URL or a filename.

- void Load(TextReader) — Loads the XML document from a System.IO.TextReader object, a System.IO.StreamReader object, or a System.IO.StringReader object.
- void Load(XmlReader) — Loads the XML document from a System.Xml.XmlReader object. An XmlReader object is a forward-only, read-only scanner for an XML document. Note that XmlReader is an abstract class, so you have to use one of its subclasses for the read operation: XmlTextReader, XmlValidationReader, or XmlNodeReader. The XmlTextReader class is the fastest way to read an XML file, but it does not perform any schema validation. You can use the XmlValidationReader class to validate XML data against an XML schema document. It has a Schemas property that you can use for XmlSchemaCollection objects. The next section covers XmlNodeReader.

The following code snippets shows some examples of the Load() method:

```
Using System.Xml;

XmlDocument xmldoc = new XmlDocument();
// The following method reads an XML file specified in a filename
xmldoc.Load(@"\My Documents\sales.xml");
// The following shows an example of using XmlTextReader in XmlDocument.Load()
XmlTextReader xmlRdr = new XmlTextReader(@"\Storage Card\books.xml");
xmlRdr.WhiteSpaceHandling = WhiteSpaceHandling.None;  // Don't read white spaces
xdoc.Load(xmlRdr);
```

The XmlTextReader class provides fine-grained control over the read operation. Like file I/O, XmlTextReader maintains the position of the current node, which can be a comment block, a processing instruction, an element, a sub-element, and so on.

The Read() method will read the next node from the file, which could be any type of tags. The order in which nodes are read from an XML document is *depth first*. When you proceed to a sub-element, XmlTextReader will update the "depth" of the current node.

You can use its MoveToContent() method to skip any non-data elements. At the beginning of a read operation, a call to MoveToContent() will move the current node to the root node of the XML file.

You can also use the Skip() method to skip all subnodes of the current node. You can use the HasAttributes property to determine whether the current node has one or more attributes. If the HasAttributes property is true, you can use a for loop to call MoveToAttribute(), and read each attribute value using the Name and Value properties:

```
public void DisplayAttributes(XmlReader reader)
{
  if (reader.HasAttributes)
  {
    Console.WriteLine("Attributes of <" + reader.Name + ">");
    for (int i = 0; i < reader.AttributeCount; i++)
    {
      reader.MoveToAttribute(i);
      //Consume current node: reader.Name and reader.Value
    }
```

```
        reader.MoveToElement(); //Moves the reader back to the element node
    }
}
```

Now let's take a look at a complete example. Suppose you want to display all the entities in an XML document. The following is the `books.xml` file that can be downloaded along with sample code from the book's website:

```
<?xml version='1.0'?>
<!-- Some of Wrox books in the book database -->
<bookstore>
  <book Section="XML" PublicationDate="2004" ISBN="0-7645-7077-3">
    <title>Beginning XML, 3rd Edition</title>
    <author>
      <first-name>David</first-name>
      <last-name>Hunter</last-name>
    </author>
    <price>39.99</price>
  </book>
  <book Section="Java" PublicationDate="2004" ISBN="0-7645-6874-4">
    <title>Ivor Horton's Beginning Java 2, JDK 5 Edition</title>
    <author>
      <first-name>Ivor</first-name>
      <last-name>Horton</last-name>
    </author>
    <price>49.99</price>
  </book>
  <book Section="Database" PublicationDate="2005" ISBN="0-7645-7950-9">
    <title>Beginning MySQL</title>
    <author>
      <first-name>Robert</first-name>
      <last-name>Sheldon</last-name>
    </author>
    <author>
      <first-name>Geoff</first-name>
      <last-name>Moes</last-name>
    </author>
    <price>39.99</price>
  </book>
</bookstore>
```

This XML file describes three books. Each book element has three sub-elements: `title`, `author`, and `price`. Because a book may have more than one author, a book element may have more than one `author` sub-element. In addition, a book element has three attributes: `Section`, `PublicationDate`, and `ISBN`.

The following code shows an example of using `XmlTextReader` to navigate the `books.xml` file. The example writes the output to a text file. The `StreamWriter` object is created as the output specifier. The first parameter of the `StreamWriter` constructor we used is the filename, whereas the second is a Boolean value indicating whether the data will be appended to the file. Then an `XmlTextReader` object is created. In the main `while` loop, the reader keeps reading from the XML document and formats the output according to the type of each node. Although this sample XML file does not contain many non-element nodes, the following sample code uses a single `switch-case` statement to define how to process each possible type of node found in an XML document:

```
            XmlTextReader reader = null;
            StreamWriter writer = new StreamWriter(@"\Storage
Card\output.txt",false);

            try
            {
                reader = new XmlTextReader(@"\Storage Card\books.xml");

                while (reader.Read())
                {
                    switch (reader.NodeType)
                    {
                        case XmlNodeType.XmlDeclaration:
                            FormatOutput(writer, reader, "XmlDeclaration");
                            break;
                        case XmlNodeType.ProcessingInstruction:
                            FormatOutput(writer, reader, "ProcessingInstruction");
                            break;
                        case XmlNodeType.DocumentType:
                            FormatOutput(writer, reader, "DocumentType");
                            break;
                        case XmlNodeType.Comment:
                            FormatOutput(writer, reader, "Comment");
                            break;
                        case XmlNodeType.Element:
                            FormatOutput(writer, reader, "Element");
                            break;
                        case XmlNodeType.Text:
                            FormatOutput(writer, reader, "Text");
                            break;
                        case XmlNodeType.Whitespace:
                            break;
                    }
                }
            }
            catch (XmlException xmlEx)
            {
                writer.WriteLine(xmlEx.Message);
            }
            reader.Close();
            writer.Close();
```

The following code shows the `FormatOutput()` method in the `Form1` class (which is part of the `XmlDemo` sample project). We want to show all the attributes of an element, so this method has a `while` loop to print all the attribute names and attributed values. Another notable aspect of this method is that it uses the `XmlReader::depth` property to determine the number of tabs to be printed before the content of a node. This makes the output look like a tree, with proper indentation for sub-elements:

```
        private static void FormatOutput(StreamWriter writer, XmlReader reader,
String nodeType)
        {
            for (int i = 0; i < reader.Depth; i++)
            {
                writer.Write('\t');
```

```
            }
            if(reader.Name != String.Empty)
                writer.WriteLine(reader.Prefix + nodeType + "<" + reader.Name + ">:
" + reader.Value);
            else
                writer.WriteLine(reader.Prefix + nodeType + ": " + reader.Value);

            // Display the attributes values for the current node
            while (reader.MoveToNextAttribute())
            {
                for (int i = 0; i < reader.Depth; i++)
                    writer.Write("\t");
                writer.WriteLine("Attribute: " + reader.Name + " = " +
reader.Value);
            }

        }
```

The following shows the output file (`output.txt`):

```
XmlDeclaration<xml>: version='1.0'
Attribute: version = 1.0
Comment:  Some of Wrox books in the book database
Element<bookstore>:
    Element<book>:
        Attribute: Section = XML
        Attribute: publicationdate = 2004
        Attribute: ISBN = 0-7645-7077-3
        Element<title>:
            Text: Beginning XML, 3rd Edition
        Element<author>:
            Element<first-name>:
                Text: David
            Element<last-name>:
                Text: Hunter
        Element<price>:
            Text: 39.99
    Element<book>:
        Attribute: Section = Java
        Attribute: publicationdate = 2004
        Attribute: ISBN = 0-7645-6874-4
        Element<title>:
            Text: Ivor Horton's Beginning Java 2, JDK 5 Edition
        Element<author>:
            Element<first-name>:
                Text: Ivor
            Element<last-name>:
                Text: Horton
        Element<price>:
            Text: 49.99
    Element<book>:
        Attribute: Section = Database
        Attribute: publicationdate = 2005
        Attribute: ISBN = 0-7645-7950-9
        Element<title>:
```

```
            Text: Beginning MySQL
        Element<author>:
            Element<first-name>:
                Text: Robert
            Element<last-name>:
                Text: Sheldon
        Element<author>:
            Element<first-name>:
                Text: Geoff
            Element<last-name>:
                Text: Moes
        Element<price>:
                Text: 39.99
```

The `XmlDocument` class has another method, `LoadXml()`, which will load an XML string (not the XML filename).

The `Load()` and `LoadXml()` methods will throw an `XmlException` if an error occurs while loading an XML file. The errors may be caused by XML syntax error.

Sometimes you want to query an XML document with a specific string. In this case, you don't need to manually navigate the entire XML document (e.g., an `XmlDocument` object); rather, you can use the `SelectNode()` method or `SelectSingleNode()` method of an `XMLNode` object. You can simply pass a string expression that represents an XPath into the underlying XML document.

The following example uses an XPath expression to query the `XmlDocument` object. The XPath expression actually specifies a set of books that has an author with a last name of "Sheldon":

```
XmlDocument doc = new XmlDocument();
doc.Load("book.xml");
XmlNodeList nodeList;
XmlNode root = doc.DocumentElement;

nodeList=root.SelectNodes("/BookStore/Book[author/last-name='Sheldon']");
foreach (XmlNode book in nodeList)
{
  //Access each book using the "book" object
}
```

XmlNodeReader and DataSet

In many cases, you will want to extract data from an XML file and convert it into a relational data structure. That way you can take advantage of the rich support for the relational data structure of ADO.NET in the .NET Compact Framework to simplify XML data handling. This is often done using the `DataSet` class in the `System.Data` namespace.

A `DataSet` represents an in-memory cache of a relational database. You can fill a `DataSet` with data from a local relational database, an XML stream, or an XML document. Furthermore, you can merge

XML data with existing data in a DataSet. The DataSet class also provides methods to write the data into an XML stream or document, and can even generate XML schema for the data. For more information about the DataSet class, see Chapter 6.

To load XML data into a DataSet, use the ReadXml() method. Like the LoadXml() method of the XmlDocument class, the ReadXml() method of the DataSet class has been overloaded. It can accept a single parameter of a Stream object of the specified file, a String object of the filename, a TextReader object, an XmlReader object, or a combination of these three objects and an object of an enum type XmlReadMode. The XmlReadMode object determines the mode used to read the XML stream or document. Valid XmlReadMode types include Auto, DiffGram, Fragment, IgnoreSchema, InferSchema, InferTypedSchema, and ReadSchema. Usually, setting the XmlReadMode to Auto will suffice, as it will perform the most appropriate action with respect to the data being read. However, if the application encounters performance problems with this method, you might want to look into other options of the read mode for optimization. For example, if schema-based validation is not a big concern, you can use IgnoreSchema to skip inline schema in the XML document.

The following code uses the DataSet class with XML data as input:

```
DataSet ds = new DataSet();
ds.ReadXml(new StreamReader(@"\Storage Card\books.xml"), XmlReadMode.Auto);
```

This DataSet is filled with data from the books.xml file on the storage card of a Smartphone device. Note that the first argument of the ReadXml method can be the filename (@"\Storage Card\books.xml") without using the StreamReader construct. Alternatively, you can use the XmlNodeReader class with a DataSet while calling ReadXml().XmlNodeReader is a subclass of XmlReader. It can read XmlNode type, which can be, for example, an XmlDocument object or an XmlAttribute object. Note that an XmlTextReader object does not read the XmlNode type.

The following example uses DataSet with XmlDocument and XmlNodeReader:

```
XmlDocument doc = new XmlDocument();
try
{
     doc.Load(@"\Storage Card\books.xml");
}
catch (XmlException ex)
{
     MessageBox.Show(ex.Message);
     return;
}
XmlNodeReader reader = new XmlNodeReader(doc);
DataSet ds = new DataSet();
ds.ReadXml(reader);
```

One advantage of using DataSet with XML is that a DataSet is able to infer the XML from the XML data. In many cases, the XML data will be placed into multiple tables represented by a DataTable collection in the DataSet object.

A DataSet (or a DataTable) can also dump data to an XML format using the WriteXml() method. The first parameter of this method is the output specifier, which can be a stream or a filename. The second parameter is the XmlWriteMode. Following are the three modes:

- `IgnoreSchema` — Default value, no schema will be written.
- `WriteSchema` — Writes the relational structure as inline XML schema. To write the XML schema only, use the `WriteXmlSchema()` method.
- `DiffGram` — An XML format used to describe the difference between data after some update.

For example, the following line shows a `DataSet` object saved into an XML document with inline schema:

```
MyDataSet.WriteXML(@"\My Documents\results.xml", XmlWriteMode.WriteSchema);
```

Specifically, some of the `DataTable`'s `WriteXml()` overloaded methods require the third parameter, a Boolean value indicating whether the current table's descendant tables will be saved.

An XML Processing Sample Application

This section uses a sample application to illustrate the usage of these classes. We employ a `DataGrid` object to display a `DataTable` in a `DataSet`. The `DataGrid` object has a `DataSource` property for the `DataTable` with which it is associated. For example, suppose `dg` is a `DataGrid` object and `ds` is a `DataSet` that has three tables. To associate the first table in the `DataSet` object with the `DataGrid` object, you can simply use the `DataSource` property of the `DataGrid` object, as follows:

```
// Display the table
dg.DataSource = ds.DataTables[0];

// Display the table in a special view (sorted by the Name column ascending)
ds.DataTables[0].DefaultView.Sort("Name DSC");
dg.DataSource = ds.DataTable[0].DefaultView;
```

The sample application reads the XML file `books.xml` from the directory `\Storage Card\` and displays the data in the form of two tables using a `DataGrid` object. We intentionally use an XML file that cannot be placed into a single `DataTable`. Why use two tables instead of one? Because a book may have multiple authors that may not be placed into a single table along with other unique book properties such as title and price. Therefore, author information is placed into a separate table that has a foreign key relationship with the other major table.

To allow the Smartphone Emulator to access a faked storage card, you must put your files into a directory on your development machine that runs the Smartphone emulator, and configure the emulator to access the shared folder. On the Smartphone emulator, select File⇨Configure, and then enter the path to the shared folder (see Figure 9-3).

The data contained in the `books.xml` document is obvious: three books as elements, each with two attributes (`PublicationDate` and `ISBN`) and some sub-elements. Each book element has a price sub-element. The first two book elements have an author sub-element, whereas the last book element has two authors. The sample program uses `XmlNodeReader` with a `DataSet` to load the XML document. A `DataGrid` object is used to display table data. In addition, three menu items are created: View Record, View Author Table, and View Book Table. If the View Record menu item is selected, a message box containing the current data record in the selected table will appear on the screen.

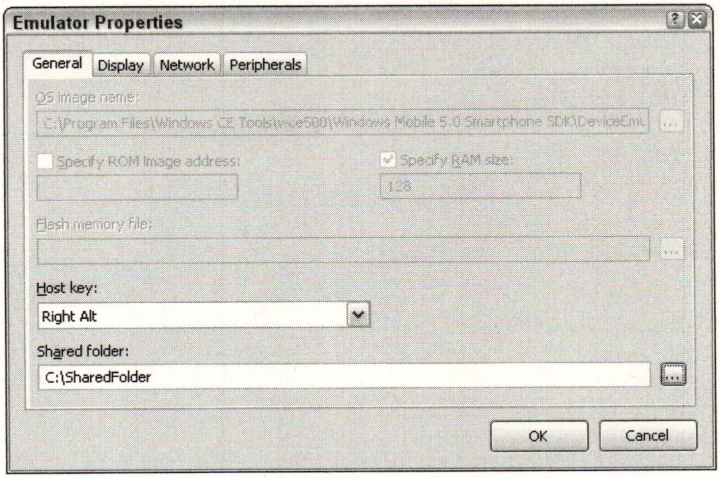

Figure 9-3

When the program starts to run and the XML file is read, the `DataSet` object contains two `DataTable` objects: one containing all book information except author names, and the other storing author names only. Initially, the first one (`ds.DataTable[0]`) will be displayed.

The main form class, which is derived from the `System.Windows.Forms.Form` class, has four data fields, as follows:

```
DataTable dt1 = null;
DataTable dt2 = null;
DataSet ds = null;
int CurrentTable = 0;
```

Two `DataTable` objects and a `DataSet` object are used to store and process XML data read from the XML file. The `CurrentTable` variable indicates which table (the major table or the author table) is about to be displayed.

The following private method will be called in the form's `Load` (in this case, `Form1_Load()`) method:

```
private void LoadData()
{

    XmlDocument doc = new XmlDocument();
    try
    {
        doc.Load(@"\Storage Card\books.xml");
    }
    catch (XmlException ex)
    {
        MessageBox.Show(ex.Message);
        return;
    }
```

```
            XmlNodeReader reader = new XmlNodeReader(doc);
            DataSet ds = new DataSet();
            ds.ReadXml(reader);
            reader.Close();
            dt1 = ds.Tables[0];
            dt2 = ds.Tables[1];

            booksData.DataSource = ds.Tables[0].DefaultView;
            CurrentTable = 0;
            label1.Text = "Book Table:";
        }

        private void Form1_Load(object sender, EventArgs e)
        {
            LoadData();
        }
```

Note that we don't have an XML schema for this document; rather, the `DataSet` class can infer the schema from this document. The result is a set of two tables: one table includes the Section column, the publicationdate column, the ISBN column, the title column, and an ID column identifying the book. The other table consists of two columns: an ID column and an Author column, with the third book including two records for the two authors.

When the form is loaded, the `DataGrid` control looks like Figure 9-4.

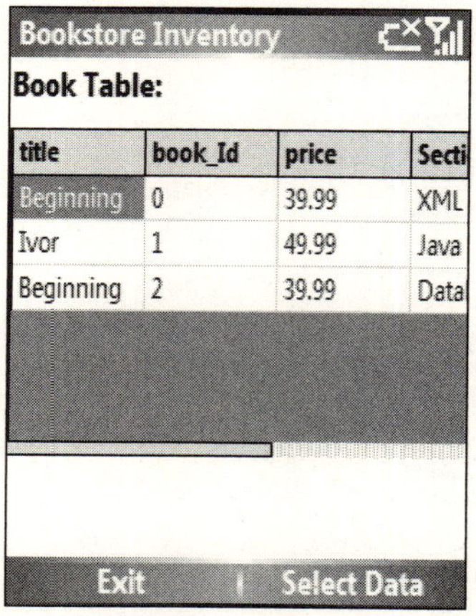

Figure 9-4

Each book has an ID starting from 0. Users can use the navigation pad to move around in the data grid. At any time, they can select the Select Data menu at the bottom of the screen, and then select View Record to display the current record. The following code shows the event handler for the menu item View Record:

```
        private void DisplayDataRow(int bookIndex)
        {
            String line = String.Empty;
            DataTable dt = CurrentTable == 0 ? dt1 : dt2;
```

```
        DataRow dr = dt.Rows[bookIndex];

        int col = 0;
        foreach (Object value in dr.ItemArray)
        {
            line += (dt.Columns[col].ColumnName + ":  " +
                value.ToString()+"\r\n");
            col++;
        }
        MessageBox.Show(line);
    }

    private void menuItem3_Click(object sender, EventArgs e)
    {
        int bookIndex = booksData.CurrentRowIndex;
        DisplayDataRow(bookIndex);
    }
```

An example of the record screen is shown in Figure 9-5.

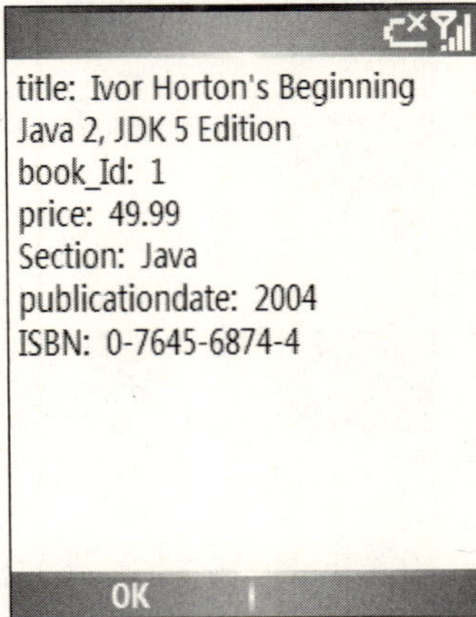

Figure 9-5

If the View Author Table menu item is selected (➪Select Data ➪View Author Table), the `DataGrid` control will switch to the other data table in the data set. The following code snippet shows the tabling switch operation when a user selects either View Author Table or View Book Table. The `SwitchTable()` method handles both cases.

```
    private void SwitchTable()
    {
        if (CurrentTable == 0)
        {
            booksData.DataSource = dt2.DefaultView;
            CurrentTable = 1;
            label1.Text = "Author Table:";
        }
        else
        {
```

```
            booksData.DataSource = dt1.DefaultView;
            CurrentTable = 0;
            label1.Text = "Book Table:";
        }
    }

    private void menuItem5_Click(object sender, EventArgs e)
    {
        SwitchTable();
    }

    private void menuItem4_Click(object sender, EventArgs e)
    {
        SwitchTable();
    }
```

An example of this author table screen is shown in Figure 9-6. Note that the third book (ID #2) has two records in the table.

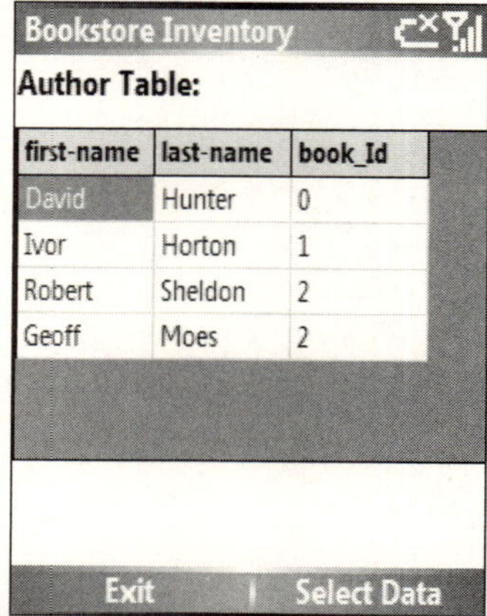

Figure 9-6

When a user selects Exit, the data in the data set is exported as two separate XML documents to the folder on the storage card. This is done by calling the `WriteXml()` and `WriteXmlSchema()` methods of the `DataSet` class:

```
    private void menuItem1_Click(object sender, EventArgs e)
    {
        ds.WriteXml(@"\Storage Card\bookstore.xml",XmlWriteMode.WriteSchema);
        ds.WriteXmlSchema(@"\Storage Card\bookstore.xsd");
        Application.Exit();
    }
```

Here we use `XmlWriteMode.WriteSchema` as the second parameter of `WriteXml()`, indicating that the schema of the data table will be saved along with the same XML document. The `WriteSchema()` method requires only an output specifier (a filename in the example).

The following example code shows the output XML document BookStore.xml. The first part of this document is a schema section, followed by the book data as shown in the original XML data file books.xml:

```
<?xml version="1.0" standalone="yes"?>
<bookstore>
  <xs:schema id="bookstore" xmlns="" xmlns:xs="http://www.w3.org/2001/XMLSchema"
xmlns:msdata="urn:schemas-microsoft-com:xml-msdata">
    <xs:element name="bookstore" msdata:IsDataSet="true"
msdata:UseCurrentLocale="true">
      <xs:complexType>
        <xs:choice minOccurs="0" maxOccurs="unbounded">
          <xs:element name="book">
            <xs:complexType>
              <xs:sequence>
                <xs:element name="title" type="xs:string" minOccurs="0"
msdata:Ordinal="0" />
                <xs:element name="price" type="xs:string" minOccurs="0"
msdata:Ordinal="2" />
                <xs:element name="author" minOccurs="0" maxOccurs="unbounded">
                  <xs:complexType>
                    <xs:sequence>
                      <xs:element name="first-name" type="xs:string" minOccurs="0" />
                      <xs:element name="last-name" type="xs:string" minOccurs="0" />
                    </xs:sequence>
                  </xs:complexType>
                </xs:element>
              </xs:sequence>
              <xs:attribute name="Section" type="xs:string" />
              <xs:attribute name="publicationdate" type="xs:string" />
              <xs:attribute name="ISBN" type="xs:string" />
            </xs:complexType>
          </xs:element>
        </xs:choice>
      </xs:complexType>
    </xs:element>
  </xs:schema>
  <book Section="XML" publicationdate="2004" ISBN="0-7645-7077-3">
    <title>Beginning XML, 3rd Edition</title>
    <price>39.99</price>
    <author>
      <first-name>David</first-name>
      <last-name>Hunter</last-name>
    </author>
  </book>
  <book Section="Java" publicationdate="2004" ISBN="0-7645-6874-4">
    <title>Ivor Horton's Beginning Java 2, JDK 5 Edition</title>
    <price>49.99</price>
    <author>
      <first-name>Ivor</first-name>
      <last-name>Horton</last-name>
    </author>
  </book>
  <book Section="Database" publicationdate="2005" ISBN="0-7645-7950-9">
    <title>Beginning MySQL</title>
    <price>39.99</price>
```

```
        <first-name>Robert</first-name>
        <last-name>Sheldon</last-name>
      </author>
      <author>
        <first-name>Geoff</first-name>
        <last-name>Moes</last-name>
      </author>
    </book>
  </bookstore>
```

This sample program demonstrates how to combine a data set with XML classes to handle structured data. The XML document used by the program is a local file on the storage card. If the data is on a remote site, you must have a way to transport the data, as well as its schema, across the network. A solution would be to use an XML web service. Recall that the `XmlDocument` and `XmlNodeReader/XmlNOdeWriter` classes both support reading and writing an XML stream. You can use these classes to write another layer of wrapper classes that may provide even a simplified interface to other developers—they don't even have to know the underlying XML data exchange and scheme inference; instead, the web service may just provide some web reference classes that work as regular assembly classes. That way, developers can focus on taking advantage of the service, rather than details about how to consume the service. These web reference classes are usually packaged in a web service SDK. The following section describes how to use the MSN Search SDK to build a Smartphone application.

Building a Smartphone XML Web Service Application

A Microsoft Smartphone application can access web services on the Internet using the .NET Compact Framework's web service support. Visual Studio .NET provides an easy way to add a web reference to your Smart Device project so that classes in the project can directly use the exposed classes and consume the web service. The web service can be either a common web service provided by sites such as MSN.com, Amazon.com, or Google.com, or a web service within an enterprise network.

Adding a Web Reference

Just as you would add a reference to an assembly in a Smart Device project, you need to add a reference to the web service exposed by an Internet site so that the interface exposed by the service is made available in the project. For example, the MSN Search web service API reference is exposed at `http://soap.search.msn.com/webservices.asmx?wsdl`.

An HTTP request sent to this URL will generate a WSDL file for the API and return it to the client. This piece of information includes web service classes, data types, and any public properties, methods, parameters, return types, and so on. Other web service APIs (such as Google API) do not use this web-based WSDL provision; instead, along with the SDK, they provide a WSDL file that essentially contains the same type of information as the web-based method. Communication is performed via SOAP, an XML-based mechanism for exchanging typed information.

To add a web reference to your Visual Studio project, select Project➪Add Web Reference. In the resulting Add Web Reference screen, enter the URL to the web service you want to invoke (see Figure 9-7). The URL actually points to a WSDL file, such as `http://soap.search.msn.com/webservices.asmx?wsdl` or `http://api.google.com/GoogleSearch.wsdl`. You can view the actual data of the WSDL file by entering the appropriate URL in your web browser.

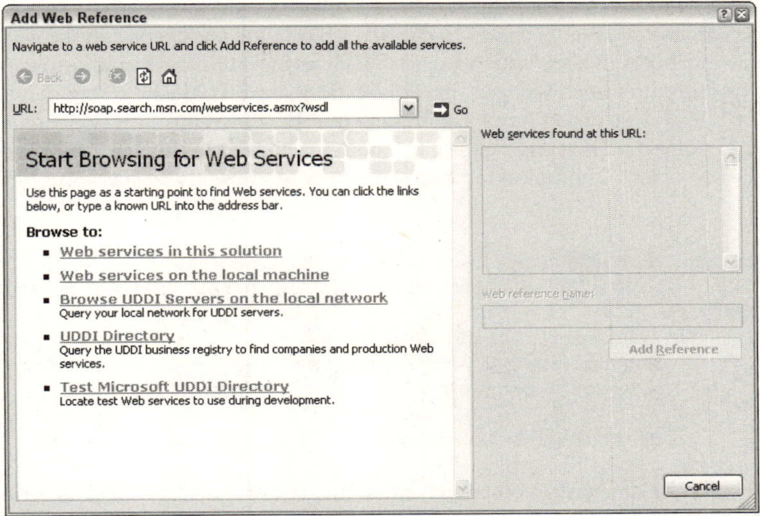

Figure 9-7

Just as you add an assembly reference to your project, adding a web reference will add a namespace to your project. In the case of adding MSN Services, the namespace `com.msn.search.soap` contains the following classes: `Location`, `MSN Search Service`, `Result`, `SearchRequest`, `SearchResponse`, `SourceRequest`, `SourceResponse`, and so on.

Another required step before developing your web service client is to obtain an application ID from the service provider. The application ID uniquely identifies an application, not the developer. In the case of the MSN Search web service, you can register your application and get the application ID at `http://search.msn.com/developer`.

Consuming Web Services

After adding the web reference, you can start to program your Smart Device project in Visual Studio just like any other smartphone application that uses a reference to some assembly. Depending on how the web service is exposed (as described in its WSDL file), simply follow the process of creating some class objects, and then invoke specific methods to consume a web service on the web. After that it's just a matter of data exchange between your application and the web service provider. Your application may simply need to retrieve some data from the web service provider, such as obtaining a map for a given address, or your application may want to perform some transactions using a web service.

The following code snippet is the core to launching the MSN Search web service. First an `MSNSearchService` object is created. Its `Search()` method will eventually be called to invoke the service. Before that, however, you have the option to indicate what sources (e.g., web, online advertisements, spell checker, etc.) will be searched. For example, you can send a string to the web service and it will tell you which word you spelled wrong. You can also send your query and obtain online advertisements about that query. The sources are defined in the `SourceRequest.SourceType` enum type.

In addition, you can use `SourceRequest` to specify what fields are needed in the search results. You can also specify some attributes of your search request, such as the culture information (en-us for English, etc.), application ID (required for each web service application), and the query string in a `SearchRequest` object. The `SearchRequest` object has a `Request` property, which is an array of some different types of `SourceRequest` objects. The `SearchRequest` object is passed to the `Search()` method of the `MSNSearchService` object. The following code shows how to use these classes to perform a search operation:

```
                MSNSearchService s = new MSNSearchService();
                //If your development computer is behind a web proxy, use the
MSNSearchService.Proxy
                //Property to indicate this
                WebProxy proxy = new WebProxy("192.168.1.100:8080");
                s.Proxy = proxy;

                SearchRequest searchRequest = new SearchRequest();

                //Create a SourceRequest to indicate what fields of search results
are needed
                SourceRequest[] sr = new SourceRequest[1];

                sr[0] = new SourceRequest();
                sr[0].Source = SourceType.Web;

                // To return all fields, use the following
                //sr.ResultFields = ResultFieldMask.All;

                // To return the Title and URL fields of search results:
                sr[0].ResultFields = ResultFieldMask.Title | ResultFieldMask.Url;


                // Set required CultureInfo string to "en-US"
                searchRequest.CultureInfo = "en-US";
                searchRequest.Query = textBox1.Text;
                searchRequest.AppID = "D4EC1031F772A8BD3BBDDA26E11B3A6ABCD597F6";
                searchRequest.Requests = sr;

                SearchResponse searchResponse = s.Search(searchRequest);
```

The `search()` method returns a `SearchResponse` object, which contains a collection of `SourceResponse` objects. Each `SourceResponse` object corresponds to a `SourceRequest` object associated with the `SearchRequest`. For example, if there are two `SourceRequest` objects of type `Web` and `Spelling`, respectively, there will be two `SourceResponse` objects for each of them. By iterating the result set of these `SourceResponse` objects, you can obtain every search result. The following code shows how to obtain the search result using the `SourceResponse` object:

```
            foreach (SourceResponse sourceResponse in searchResponse.Responses)
            {
                    Result[] sourceResults = sourceResponse.Results;
                    foreach (Result sourceResult in sourceResults)
    {
                    //Result is formatted into some fields such as
sourceResult.Url, sourceResult.Title, etc.
                    }
            }
```

Figures 9-8 and 9-9 show a simple Smartphone application that takes advantage of the MSN Search web service. Two forms are created: The first, shown in Figure 9-8, contains an input box and a Search button; the second, shown in Figure 9-9, displays the search results. The source request specifies the type of Web, meaning that only a web search is performed. For each result, its title and URL are shown on the second form. Note that both of these settings can be easily modified with the `SourceRequest` objects.

Figure 9-8

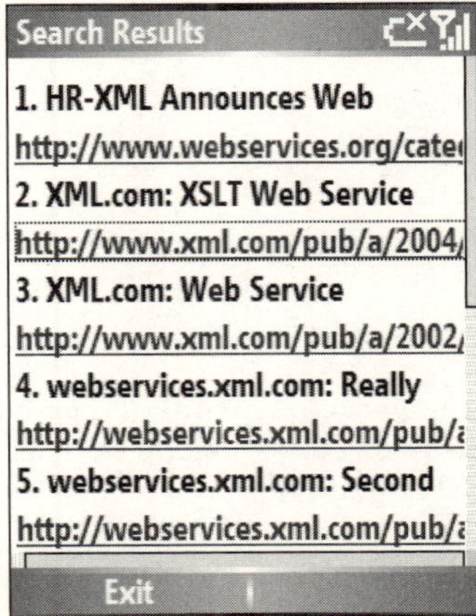

Figure 9-9

The search result form does not have any static controls. Instead, when the Search button on the first form is selected, the form is loaded with the search result, appearing as some dynamically generated controls. Each result item's title is used to populate a Label control, and each URL is used to populate a LinkLabel control. The number of controls depends on how many search result items are returned, which is determined by the value of the `Count` field of the `SourceRequest` object you passed to the `SearchRequest` object. The default value of `Count` is 10.

The following code snippet shows the `ShowResults()` method that displays search results (the `searchResponse` object, which is a member of the `form` class):

```
        void ShowResults()
        {
            int cnt = 1;
            int left = 10;
            int top = 10;
            this.SuspendLayout();
            foreach (SourceResponse sourceResponse in searchResponse.Responses)
            {
                    Result[] sourceResults = sourceResponse.Results;
                    foreach (Result sourceResult in sourceResults)
                    {

                        Label l = new Label();
                        l.Width = this.Width - 10;
                        l.Text = cnt + ". " + sourceResult.Title + " " +
sourceResult.Description;
                        l.Left = left;
                        l.Top = top;
                        LinkLabel ll = new LinkLabel();
                        ll.Text = sourceResult.Url;
                        ll.Left = left;
                        ll.Top = top + 25;
                        ll.Width = this.Width - 10;
                        this.Controls.Add(l);
                        ll.Click += new EventHandler(ll_LinkClicked);

                        this.Controls.Add(ll);

                        cnt++;
                        top += 50;
                    }
            }
            this.ResumeLayout();

        }
```

The outer `foreach` loop iterates through each `SourceResponse` object corresponding to each `SourceRequest` object, whereas the inner `foreach` loop obtains each search result within that `SourceResponse` object. The search result data is obtained by looking at the data fields of `Result` objects.

Summary

XML Web services have become a primary vehicle for exchanging data on the Internet. A number of Internet service companies provide web services that enable anyone to access their database and the software platform. This has been a clear trend in online business, as many individual developers and companies are utilizing these web services for their enterprises. It is important to understand that many of these web services are also exposed to mobile applications so that users can access these services on the go.

The .NET Compact Framework contains a handful of classes for XML document, schema, and node navigation. In addition, you can use data classes such as `DataSet` and `DataTable` to parse XML data as a relational data structure, thereby enabling an application to easily retrieve, process, and exchange XML data with other online applications.

The next chapter explains how to build customized components using managed and unmanaged APIs, and how to invoke Win32 APIs directly from within managed applications. These features greatly extend application capability and provide you with more programming options.

10

Platform Invoke

This chapter covers Platform Invoke (P/Invoke), which enables you to call unmanaged libraries from managed code. With P/Invoke, you do not have to rewrite your unmanaged code and package it into a managed fashion. You can also import functions from the Windows CE library files, which are written in unmanaged code.

Specifically, this chapter discusses the following:

- The differences between managed code and unmanaged code
- How to use Platform Invoke (P/Invoke) in the .NET Compact Framework
- How to marshal data in P/Invoke
- How to optimize P/Invoke performance

Managed and Unmanaged Code

In .NET, managed code is normally written with C#, Microsoft Visual Basic .NET, or other common language infrastructure (CLI) languages. Managed code is compiled to Intermediate Language (IL), rather than the operating system–specific assembly so that it can be executed by the common language runtime (CLR). One advantage of IL is that it can run on any operating system that supports the .NET Framework. Managed code also enjoys the core services provided by the .NET Framework, such as memory management, error exceptions, and security information. Therefore, it is much easier to write a well-behaved application with managed code.

Managed code organizes data in a managed fashion, which is referred to "managed data" in some literature. A typical feature of managed data is that the memory heap will be automatically allocated and de-allocated by the garbage collection process of the .NET runtime.

Code written and running for a non-.NET environment is considered unmanaged code. It is normally compiled to an operating system–specific assembly and is executed directly by the operating system. Unmanaged code runs outside of the CLR and cannot take advantage of managed data. As a result, when writing unmanaged code, developers have to explicitly call the de-allocator function or destructor method to de-reference the data and return allocated memory space back to the operating system. Typical programming languages that are geared toward unmanaged code include C and C++. Note however, you can write managed C++ with the help of its managed extension for C++.

Table 10-1 summarizes the differences between managed code and unmanaged code.

Table 10-1 Managed Code vs. Unmanaged Code

	Managed Code	Unmanaged Code
Portability	Executable files do not contain processor-specific code, which is to be compiled on demand by JIT.	The executable files contain processor-specific code.
Memory	Memory is managed by the .NET runtime.	Each program has to explicitly allocate and free memory.
Safety	Features such as array boundary protection prevent memory overwrites.	Each program has to implement its boundary protection features.

As you can see, managed code is often considered portable: .NET executables run on any platform that supports the CLR. It is also safe compared to unmanaged code in that implicit pointers and automatic memory management eliminate memory leaks, and array boundary protection prevents memory overwrites. If you have the luxury of choice for your software development, it is always highly recommended to go with managed code for the following reasons:

- **Managed code has better memory management.** As mentioned earlier, the .NET Compact Framework allocates and de-allocates memory automatically for managed code (whereas for unmanaged code, you have to clean up memory explicitly in your code). For example, to create a bitmap by calling the Win32 function `CreateBitmap`, you must later call on the `DeleteObject` function; otherwise, a portion of memory will not be freed by the application, resulting in memory leaks. For mobile devices that usually have only 32MB or 64MB RAM, memory leaks would cause the system to slow down or even crash very soon.
- **.Managed code has better portability.** A managed application is more portable because it has fewer dependencies on platform-specific libraries, such as Win32 libraries. For example, if you write managed code that runs on the .NET Compact Framework, your application can be ported to non-Windows CE/Mobile devices as long as the .NET Compact Framework is installed on those devices.
- **Managed code is safer to run.** The CLR will examine the IL code and determine whether it is safe. Unsafe code may be prevented from execution if the security settings of a mobile device block it.
- **Managed code has better reusability.** You can easily reuse managed code because it is immediately available in Visual Studio 2005 by adding a reference to the Solution Explorer and the appropriate namespace reference. To expose an unmanaged function in managed code, however, you have to go through complex procedures before you can even use the function.

In short, we recommend writing functions in a managed manner. However, in some cases you may have to write your code in an unmanaged manner due to concerns such as performance constraints. In addition, despite a variety of managed functions that are supported in the .NET Compact Framework, you may run into situations where a function you need is not in the .NET Compact Framework but is available in a Win32 DLL.

In both cases, you need to call into unmanaged code from your managed program. The .NET Compact Framework has an interoperability layer that allows managed code to call into Windows DLLs or interact with COM objects. This layer, termed *P/Invoke*, essentially connects the managed code with unmanaged code.

In the next section you will learn more about unmanaged code, especially unmanaged DLL, in order to gain a better understanding of P/Invoke.

Building Unmanaged DLLs

Microsoft eMbedded Visual C++ enables you to write your own DLLs for mobile devices using C++. Before introducing P/Invoke, it's worth taking a brief look at how to build your own unmanaged DLLs. To start a new Windows CE DLL project in Microsoft eMbedded Visual C++, click File⇨New⇨Projects⇨ WCE Dynamic-Link Library, as shown in Figure 10-1. Because neither IL nor the JIT compiler is available for unmanaged code, you need to specify the processor type to build one DLL for each platform. For Smartphone applications, you may also want to build a DLL for an Intel X86 CPU so that it can be executed on a Smartphone emulator from your PC.

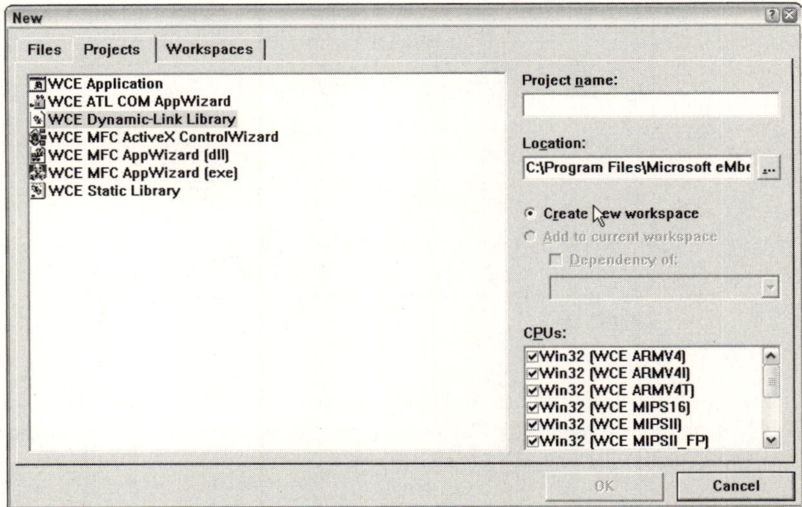

Figure 10-1

Note some complications, however. The C++ overload feature is great for developers because several functions can share a single name with different parameters, but when a source code file is compiled to an object file, each one of the overloaded functions should have a unique identifier. This is done by the compilers using a technique called *name mangling*. However, each compiler follows different conventions to mark the overloaded name. To ensure that the DLL functions can be exported correctly, you can use the Dumpbin utility to display the function names from the compiler's point of view.

By default, the 32-bit version of the `Dumpbin.exe` utility is stored at `C:\Program Files\Microsoft Visual Studio 8\VC\Bin` (assuming Visual Studio 2005 is installed; the Dumpbin utility is also available in Visual Studio 2002 and 2003). To run this program, however, you need to either specify the PATH in the environment variable or use the fully qualified filename to run the program from the command line.

To add an environment variable in Windows, right-click My Computer and choose Properties. On the Advanced tab, click the Environment Variables button (see Figure 10-2). You can then select the Path variable from the System Variables list and add the corresponding path to the Path variable, as shown in Figure 10-3.

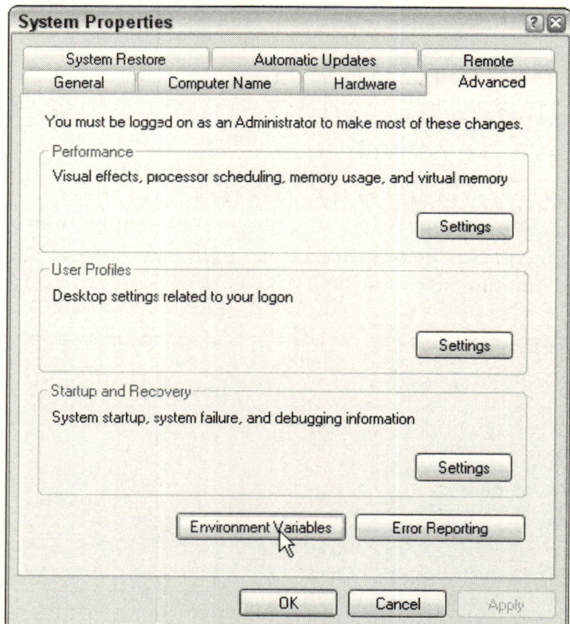

Figure 10-2

Another way to set the environment variables is to run the `vsvars32.bat` batch file shipped with Visual Studio 2005. The default location of this batch file is `C:\Program Files\Microsoft Visual Studio 8\Common7\Tools`.

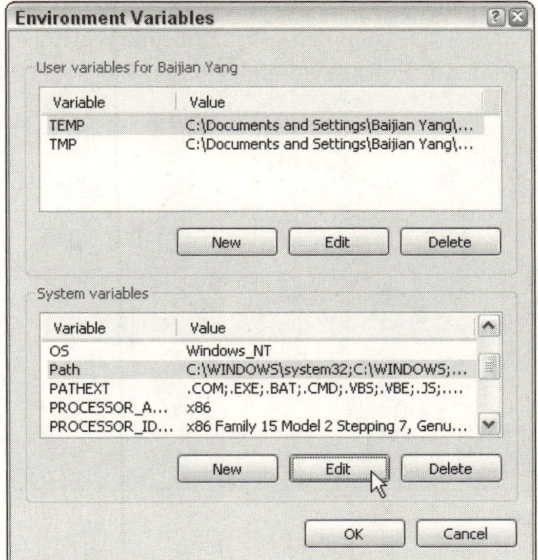

Figure 10-3

The Dumpbin utility reveals how the Visual Basic or Visual C# compiler sees a DLL. For example, if you want to find out the external names of the "coredll" library, you can use the following command:

```
C:\Program Files\Microsoft Visual Studio 8\VC\Bin>dumpbin /EXPORTS "C:\Program
Files\Windows CE Tools\wce500\Windows Mobile 5.0 SmartphoneSDK\Lib\ARMV4I\ coredll.
lib" | more
```

Figures 10-4 and 10-5 show the first and last page of the library information, respectively.

```
C:\WINDOWS\system32\cmd.exe
Microsoft (R) COFF/PE Dumper Version 8.00.50727.42
Copyright (C) Microsoft Corporation.  All rights reserved.


Dump of file C:\Program Files\Windows CE Tools\wce500\Windows Mobile 5.0 Smartph
one SDK\Lib\ARMV4I\coredll.lib

File Type: LIBRARY

     Exports

       ordinal    name

          1572    ??0exception@std@@QAA@ABV01@@Z (public: __cdecl std::exception
::exception(class std::exception const &))
          1571    ??0exception@std@@QAA@PBD@Z (public: __cdecl std::exception::e
xception(char const *))
          1570    ??0exception@std@@QAA@XZ (public: __cdecl std::exception::exce
ption(void))
          1568    ??0type_info@@AAA@ABV0@@Z (private: __cdecl type_info::type_in
fo(class type_info const &))
          1574    ??1exception@std@@UAA@XZ (public: virtual __cdecl std::excepti
on::~exception(void))
          1562    ??1type_info@@UAA@XZ (public: virtual __cdecl type_info::~type
_info(void))
          1095    ??2@YAPAXI@Z (void * __cdecl operator new(unsigned int))
-- More --
```

Figure 10-4

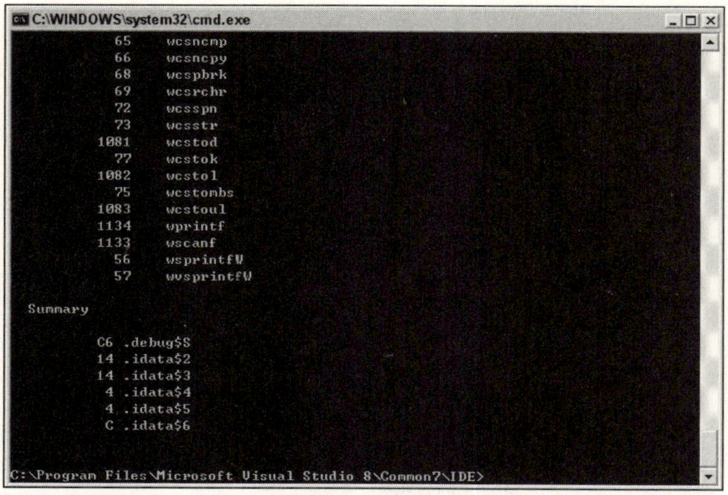

```
C:\WINDOWS\system32\cmd.exe
          65    wcsncmp
          66    wcsncpy
          68    wcspbrk
          69    wcsrchr
          72    wcsspn
          73    wcsstr
        1081    wcstod
          77    wcstok
        1082    wcstol
          75    wcstombs
        1083    wcstoul
        1134    wprintf
        1133    wscanf
          56    wsprintfW
          57    wvsprintfW

  Summary

          C6 .debug$S
          14 .idata$2
          14 .idata$3
           4 .idata$4
           4 .idata$5
           C .idata$6

C:\Program Files\Microsoft Visual Studio 8\Common7\IDE>
```

Figure 10-5

In short, the Dumpbin utility helps programmers export functions from a DLL using the correct name used by the compiler.

Using P/Invoke in the .NET Compact Framework

The P/Invoke functions and attributes are located in the `System.Runtime.InteropServices` namespace. Basically, you perform P/Invoke on the .NET Compact Framework just as you would on the full .NET Framework. However, the .NET Compact Framework supports only one-quarter of the P/Invoke functions available in the full .NET Framework. In addition, the following features are unique in the .NET Compact Framework environment:

- **Calling convention** — The full .NET Framework supports three calling conventions, such as Cdecl, StdCall, and ThisCall. Each calling convention mandates who cleans the stack and how the arguments are passed. You can change the calling convention by setting the value of the `CallingConvention` property of the `DllImport` attribute. For example, the following directive can set the calling convention to Cdecl:

```
[DllImport("coredll.dll", CallingConvention=CallingConvention.Cdecl)]
```

In the .NET Compact Framework, however, only the `Winapi` value is supported in the `CallingConvention` enumeration. Note that the `Winapi` value is not an actual calling convention; rather, it refers to the default platform convention. On Windows CE .NET, `Winapi` is referred to `Cdecl`, whereas on the full .NET Framework it defaults to the `StdCall` convention.

- **Character coding** — The character set encoding on the full .NET Framework can be set to ANSI, Unicode, or Auto through the `CharSet` property in the `DLLImport` attribute. If `CharSet` is not specified, then the default encoding is ANSI on the full .NET Framework. The .NET Compact Framework, however, supports only Unicode encoding. As a result, if a DLL function takes an ANSI string, then you need to convert the encoding before calling. You can perform this conversion by calling the `GetBytes()` method of the `ASCIIEncoding` class, as follows:

```
using System.Text;
...

    //Unicode string
    String unicodeStr = "The value of Pi (\u03a0) is 3.14159265.";

    //A new instance of ASCIIEncoding
    ASCIIEncoding ascii = new ASCIIEncoding();

    //Convert the unicode string to an ASCII encoded byte array
    Byte[] ByteArray = ascii.GetBytes(unicodeString);

...
```

- **Unidirection** — The full .NET Framework supports callbacks through delegates. This allows a DLL function to call managed code at the address of the delegate. Such a callback feature is missing in the .NET Compact Framework. Nonetheless, parameters can be passed to a DLL function by value or by reference, thereby enabling data to be returned to the .NET Compact Framework application.
- **Exceptions** — The .NET Compact Framework throws different exceptions while errors occur. If the function cannot be found, then the .NET Compact Framework throws a `MissingMethodException` exception, whereas the full .NET Framework throws an `EntryPointNotFoundException` exception. If a function is declared incorrectly, a `NotSupportedException` will be thrown on the .NET Compact Framework, rather than the `ExecutionEngineException` on the full .NET Framework.
- **Windows messages** — In the .NET Framework, the `Handle` (`hWnd`) property of the `Form` class can be exposed so that you can pass the handle to a window or a function. You can also override the `DEfWndProc()` method to customize the processing of messages sent by the operating system. The .NET Compact Framework does not support either of these members. It does, however, support the `MessageWindow` and `Message` classes in the `Microsoft.WindowsCE.Forms` namespace. You can use the `SendMessage()` and `PostMessage()` methods of the `MessageWindow` class to send messages to other windows.

To call the unmanaged code (such as Win32 APIs) from managed code, you must first declare an unmanaged code function.

Declaring and Calling an Unmanaged Code Function

A P/Invoke declaration exports a function from a DLL. The declaration is like a regular function declaration in the sense that it has a return type, takes zero or more parameters, and should be made inside a class. Conversely, a P/Invoke declaration requires three additional elements: the `DllImport` attribute, the `static` keyword, and the `extern` keyword. For example, to determine whether a key is up or down, you can export the `GetAsyncKeyState()` function from the `coredll.dll`, as follows:

```
using System.Runtime.InteropServices;
...
[DLLImport("coredll.dll")]
public static extern short GetAsyncKeyState (int nKey)
```

As you can see, the `extern` keyword is necessary in the declaration because it indicates that the function's body is not within the current assembly. The compiler should look for that external function from the `coredll.dll` specified in `DLLImport` attributes. The function is declared as `static` because it is a class method. The `coredll.dll` is stored on every Windows Mobile–based Smartphone device. It exposes many Windows CE APIs to the programmers, and functions like `kernel32.dll` and `user32.dll` on a desktop PC. If you need to call a DLL that is not yet stored on the mobile devices, you need to copy the DLL file to the device as well when you deploy your code.

Calling an unmanaged code function is the same as calling a managed code function. In C#, you can call the `GetAsyncKeyState` function as follows:

```
int i = 2;
short nkeyState = GetAsyncKeyState(i);
```

Usually, you can easily export a function from unmanaged code using the approach mentioned above. However, the signature of the exporting function might conflict with other functions or even reserved keywords in your managed code. An easy workaround is to use the `EntryPoint` property of the `DLLImport` attribute to import the function from the DLL and then rename it in the declaration. The following snippet of code exports the `GetAsyncKeyState` function from the `Coredll.dll` and makes it appear as `GetMyKeyState` in the managed code:

```
using System.Runtime.InteropServices;
...
[DLLImport("coredll.dll", EntryPoint="GetAsyncKeyState")]
public static extern short GetMyKeyState (int nKey)
...
int i = 2;
short nkeyState = GetMyKeyState(i);
```

Error Handling

The DLLs and the functions imported from the DLLs are loaded during runtime, not at compile time. The compiler is therefore not able to check the existence of the DLL, nor can it locate the entry point. To make your application more robust, it is highly suggested that you catch the exceptions with a `try...catch` block whenever you make calls to unmanaged code. For example, the following code will catch the `MissingMethodException` when calling the native `GetAsyncKeyState()` functions:

```
try
{
    GetAsyncKeyState(i);
}
catch (MissingMethodException e)
{
```

```
    MessageBox.Show("MissingMethodException for GetAsyncKeyState:" + e.toString()
);
}
```

Usually, the `NotSupportedException` indicates a mismatch between the declaration and the actual definition of the DLL function. You need to check the declaration to make certain it indeed matches the function it calls. In the case of `MissingMethodException`, it could be one of the following reasons:

- The DLL being called using P/Invoke does not exist, cannot be located, or is corrupted.
- The DLL being called is dependent on other DLLs, which are missing.
- The name of the function is incorrect.
- The parameters passed to the functions are wrong.

Another technique that can help you with the debugging and error handling of P/Invoke is to turn on the `SetLastError` property (which is set to `false` by default). This will cause the CLR to call the Windows CE `GetLastError` function and cache the returned error value. By doing so, the error value will not be overridden by other functions, and you can then safely retrieve the error information by using the `Marshal.GetLastWin32Error` method. In the following code snippet, `SetLastError` is set to `true` when importing the `GetAsyncKeyState` API to the managed code. After making an invalid call to the API, the error code can be retrieved and printed on the screen:

```
using System.Runtime.InteropServices;
using System.Windows.Forms
...


[DLLImport("coredll.dll", EntryPoint="GetAsyncKeyState", SetLastError=ture)]
Extern public static extern short GetMyKeyState (int nKey)
...

//Make an invalid call
short nkeyState = GetMyKeyState (987654321);

int lastError = Marshal.GetLastWin32Error();
string errMesg ="The error message is "+Convert.toString(lastError);

MessageBox.Show(errMesg);
```

Marshaling Data

When calling unmanaged libraries from managed code, you have to be careful about parameter passing. As mentioned previously, unmanaged libraries can only access unmanaged data, whereas managed code accesses managed data by default. The process of converting between managed data and unmanaged data is known as *marshaling*. For simple data types or objects, marshaling is automatically handled in the .NET Framework. For complex data types, you can use the `Marshal` class to copy managed data to an unmanaged memory space or copy unmanaged data to a managed memory space.

> *Marshaling an object is also known as* serializing *an object (or* deflating *an object), and unmarshaling an object is known as* deserializing *an object (or* inflating *an object).*

Marshaling Value Types

Passing parameters by values is normally easy because value types usually take only a few bytes of memory space and can be pushed and popped on the stack. Another category of data types known as *blittable* types refers to those data types that have the same sizes and data representations in both managed code and unmanaged code. Because blittable types have the identical internal "look," marshaling those blittable types requires only a simple memory block copy.

Most of the value types defined in the .NET Compact Framework are blittable types, as illustrated in Table 10-2.

Table 10-2 Common .NET Compact Framework Blittable Value Types

.NET Compact Framework Type	C# Keyword	Native C/C++ Keyword	Size (bits)
`System.Byte`	`byte`	`unsigned char`	8
`System.SByte`	`Sbyte`	`signed char`	8
`System.Int16`	`short`	`short`	16
`System.UInt16`	`ushort`	`unsigned short`	16
`System.Int32`	`int`	`int`	32
`System.Char`	`char`	`WCHAR (wchar_t)`	16

Because marshaling blittable value types is automatically handled by the .NET Compact Framework, you can simply marshal those blittable value types as if you were passing parameters to managed code.

For example, the following C/C++ function takes two integers and calculates the sum:

```
EXTERN_C
__declspec(dllexport)
int IntAdd2(int x, int y)
{
    return (x+y);
}
```

If this function is compiled and built into the library `MarshalTypeDll.dll`, it can then be declared and called from managed code. The following C# code calculates the sum of 10 and 6 and displays the result in a message box:

```
using System.Runtime.InteropServices;
...
[DllImport("MarshalTypeDll.dll")]
extern static int IntAdd2(int a, int b);
...
int sum = IntAdd2(10, 6);
MessageBox.Show( String.Format("Sum is {0}", sum) );
```

In the full .NET Framework, `long` *types (64-bit integer) and* `floating-point` *types (*`float` *and* `double`*) can be passed by value and marshaled into unmanaged code. In the .NET Compact Framework, however, this is no longer true. You should pass them by reference in order to pass those values into unmanaged code.*

Besides passing value types by value, the .NET Compact Framework also supports passing value types by reference. When parameters are passed by reference, the pointer to the data, rather than the data itself, is passed to the unmanaged code.

The C# language offers two ways to pass value types by reference. The `out` parameters are used only to pass back a value from a function. The `ref` parameters can be used to pass a value to a function and to retrieve a value from a function, but you must assign a value to a `ref` parameter before using it.

The following C/C++ function takes two `double` numbers, calculates the arithmetic mean, and stores the result to a third `double` parameter. All the `double` parameters in this example are passed by reference through pointers:

```
EXTERN_C
__declspec(dllexport)
void DoubleMean2(double *x, double *y, double *mean)
{
    *mean = (*x + *y) / 2.0;
}
```

Again, assuming the function is complied and built in the library `MarshalTypeDll.dll`, it can be accessed from managed code, as follows:

```
[DllImport("MarshalTypeDll.dll")]
extern static void DoubleMean2(
    ref double a,
    ref double b,
    out double mean);
...
double a = 10.0;
double b = 6.0;
double mean;
DoubleMean2(ref a, ref b, out mean);
MessageBox.Show(
    String.Format("Mean is {0}", mean)
);
```

Another thing you need to consider is the constants. Traditionally, constants are defined as macros in C/C++. For example, the following code defines the numeric value of each date:

```
#define SUNDAY      0
#define MONDAY      1
#define TUESDAY     2
#define WEDNESDAY   3
#define THURSDAY    4
#define FRIDAY      5
#define SATURDAY    6
```

When translating those constants in managed code, use the `const` keyword, as follows:

```
const int SUNDAY    = 0;
const int MONDAY    = 1;
const int TUESDAY   = 2;
const int WEDNESDAY = 3;
const int THURSDAY  = 4;
const int FRIDAY    = 5;
const int SATURDAY  = 6;
```

Alternately, you can simply define the constant values as an enumerated type, as follows:

```
enum   Dates : int
{
    SUNDAY    = 0,
    MONDAY    = 1,
    TUESDAY   = 2,
    WEDNESDAY = 3,
    THURSDAY  = 4,
    FRIDAY    = 5,
    SATURDAY  = 6,
}
```

Marshaling Reference Types

Generally, reference types are more complex to process because the CLR memory management system may store data differently from C and C++. In addition, the CLR may move data around in a way that is transparent for the managed code but not for C and C++.

Next, you will learn how to marshal reference types, such as arrays, strings, structures, and classes.

Passing Arrays

Arrays are stored as a contiguous space in memory in C/C++. When marshaling an array from managed code to unmanaged code, the .NET Compact Network can map all the elements to a format that is consistent with the C/C++ representation. The following example demonstrates how this can be done. The C function `MeanArray` can calculate the arithmetic mean of an array of positive integers:

```
EXTERN_C
__declspec(dllexport)
void int MeanArray(int *pItem, int len)
{
    if (len < 1)
        return -1; // Empty

    int Sum = 0;

    for ( int i= 0; i < len; i++)
        Sum += pItem[i];

    return Sum / len;
}
```

Assuming the function is again compiled and built in `MarshalTypeDll.dll`, you can call this function with P/Invoke as follows:

```
[DllImport("MarshalTypeDll.dll")]
extern static int MeanArray(int[] pItem, int len);

...

int[] stuScores = new int[] { 78, 85, 51, 92, 81, 96, 65};
int mean = MeanArray(stuScores, stuScores.Length);
MessageBox.Show(
    String.Format("The class average of final exam is {0}", mean)
);
...
```

Passing String Variables

You can use both the `System.String` and `System.Text.StringBuilder` classes to pass data to Unicode character arrays in the native code. The marshaling runtime of the .NET Compact Framework will append a null terminating character (`\0`) to the end of a Unicode character array so that the resulting array conforms to the C string format.

The `string keyword` *in C# is an alias of the* `System.String` *class. The data type of each Unicode character is* `System.Char` *in the .NET Compact Framework, which is equivalent to the* `WCHAR` *(or* `wchar_t`*,) data type in C/C++ (refer to Table 10-2).*

In the .NET Compact Framework, the `System.String` objects are immutable by design, which means they cannot be changed at runtime. If the content of a string needs to be changed, such as appending a few characters to an existing string, the .NET runtime will return a new object to hold the content of the new string. As a result, you should not pass a `String` object to unmanaged code that will modify the string.

The `StringBuilder` class in the `System.Text` namespace represents a mutable array of Unicode characters. You can therefore marshal a `StringBuilder` object to the unmanaged code if the string will be modified. When creating a new instance of the `StringBuilder` object, the best practice is to specify the maximum number of characters it can hold. You can do this by specifying the `Capacity` property in the `StringBuilder` constructor. Note that you need to ensure that the `Capacity` property is sufficient to hold all possible results.

For example, the following unmanaged function `StrConcat` concatenates `pStr1` and `pStr2`, and passes the resulting string to `pStr3`:

```
   EXTERN_C __declspec(dllexport)
void StrConcat(WCHAR * pStr1, WCHAR * pStr2, WCHAR * pStr3)
{
    size_t len1 = wcslen(pStr1);
    size_t len2 = wcslen(PStr2);
    size_t k;

    for (k = 0; k < len1; k++)
    {
```

```
            pStr3[k] = pStr1[k];
        }

        for (k = 0; k < len2; k++)
        {
            pStr3[len1+k] = pStr2[k];
        }

        pStr3[len1+len2] = '\0';
    }
```

If the preceding unmanaged code is compiled and built in `MarshalTypeDll.dll`, the following managed code shows how to P/Invoke the `StrConcat` function:

```
[DllImport("MarshalTypeDll.dll")]
extern static void StrConcat(string inStr1, string inStr2, StringBuilder outStr);
...

string inStr1 = "Happy ";
string inStr2 = "Smartphone Programming";

StringBuilder outStr = new StringBuilder(inStr1.Length + inStr2.Length);

StrConcat(inStr1, inStr2, outStr);
MessageBox.Show(
        String.Format("The concatenated string is: '{0}'", outStr)
        );
...
```

Note that in this example, you can also use the `StringBuilder` objects to pass the first two parameters, but you cannot pass the third parameter as a `String` object because the unmanaged code will modify the data it contains.

Note also that even though you can use either `StringBuilder` objects or `String` objects to pass the first two parameters, using two `String` objects is recommended because the unmanaged code is not supposed to modify the first two parameters; passing two immutable `String` objects is safer.

Passing Structures and Classes with Blittable Fields

If all the fields in a structure or a class are blittable, the .NET Compact Framework can automatically marshal the structure or the class by sequentially laying out the fields in memory in the same order as they appear in the unmanaged code.

Consider a structure in C/C++ that defines a point by *x* and *y* coordinates, and a function to calculate the distance between two points:

```
struct Point
{
    double x;
    double y;
};

EXTERN_C
```

```
__declspec(dllexport)
void GetDistance ( Point * p1, Point * p2, double * pDist)
{
    double dx = p1->x - p2->x;
    double dy = p1->y - p2->y;
    *pDist = sqrt(dx*dx + dy*dy);
}
```

To call the unmanaged `GetDistance()` function from the managed code, first declare the managed version of the `Point` structure in your C# code, as follows:

```
public struct Point
{
    public double x;
    public double y;
    public Point (double x, double y)
    {
        this.x = x;
        this.y = y;
    }
};
```

Assuming the unmanaged function is in the `MarshalTypeDll.dll` library, the following example shows how to P/Invoke the function from your managed code:

```
...
[DllImport("MarshalTypeDll.dll")]
extern static void GetDistance (
        ref Point p1,
        ref Point p2,
        out double dist);
...

double result;

Point point1 = new Point(0.0, 0.0);
Point point2 = new Point(8.0, 6.0);

GetDistance (ref point1, ref point2, out result);
MessageBox.Show(
    String.Format("Distance is {0}", resultDistance)
    );
...
```

The `StructLayout` attribute in the .NET Compact Framework 2.0 enables you to specify the physical layout of the data fields of a structure or a class. The layout options are defined in the `LayoutKind` enumeration, which supports the following three options:

- **Auto** — The runtime automatically choose the layout in unmanaged memory. This is the default setting when `LayoutKind` is specified.

- **Explicit** — You can explicitly set the position for each member of an object by using the `FieldOffset` attribute. For example, the unmanaged memory space for the `X` and `Y` members starts from byte 0 and byte 4, respectively:

```
[StructLayout(LayoutKind.Explicit)]
public struct IntPoint
{
    [FieldOffset(0)] public int X;
    [FieldOffset(4)] public int Y;
}
```

- **Sequential** — The members of an object will be laid out sequentially in unmanaged memory according to the order in which they appear in the structure or class declaration.

In the `GetDistance` example, you can also specify the structure layout as sequential as follows:

```
[StructLayout(LayoutKind.Sequential)]
public struct Point
{
    public double x;
    public double y;
    public Point (double x, double y)
    {
        this.x = x;
        this.y = y;
    }
};
```

Setting the layout to sequential is optional in this example, because data members `X` and `Y` are both blittable. The .NET Compact Framework runtime can correctly handle the marshaling without the help of the `StructLayout` attribute.

Passing Structures and Classes with Non-Blittable Fields

Prior to version 2.0, the .NET Compact Framework has very limited support to marshal complex structures or classes that contain non-blittable data members. As a result, marshaling some complex structures or classes is simply impossible. The .NET Compact Framework 2.0 now supports the `MarshalAs` attribute, which enables you to indicate how to marshal complex data between managed code and unmanaged code.

To identify the format of the unmanaged data, you can use the `MarshalAs` attribute followed by the `UnmanagedType` enumeration. For example, the following C# code can pass a string variable as a two-byte null-terminated Unicode character array to unmanaged code:

```
Void PInvokeAnCFunction ( [MarshalAs(UnmanagedType.LPWStr)] string s);
```

Table 10-3 lists several common members in the `UnmanagedType` enumeration.

Refer to MSDN online at `http://msdn2.microsoft.com/en-us/library/system.runtime.interopservices.unmanagedtype.aspx` *to get a full list of the members.*

Table 10-3 Common Members in the UnmanagedType Enumeration

Name	Size (Bytes)	Description
AnsiBStr	1	An ANSI character string
Bool	4	A Boolean type (Win32 BOOL type)
ByValArray	N/A	An array of value type data
ByValTstr	N/A	An array of characters
I1	1	Signed integer
I2	2	Signed integer
I4	4	Signed integer
LPStr	1	Null-terminated ANSI character string
LPWStr	2	Null-terminated Unicode character string

The `MarshalAs` attribute makes it possible to marshal a complex structure or a class. For example, the following C structure `ComplexStruct` contains an integer array, a character array, and a pointer to a string:

```
struct ComplexStruct
{
    int   intAry[10];
    char  charAry[80];
    WCHAR  *pStr;
};
```

This next bit of code illustrates how to define this structure in C#:

```
struct ComplexStruct
{
    [MarshalAs(UnmanagedType.ByValArray, sizeConst=10)]  int[]  intAry;
    [MarshalAs(UnmanagedType.ByValTStr,  sizeConst=80)]  string str1;
    [marshalAs(UnmanagedType.LPWStr)]      string str2;
};
```

Note that when using `ByValArray` or `ByValTStr`, you must set `sizeConst` to indicate how many elements are in the array.

An Example of a P/Invoke Application

Chapter 4 described how to change the text input mode by using the `InputMethod` and `InputModeEditor` classes in the `Microsoft.WindowsCE.Form` namespace. This chapter presents an alternative solution: calling the Win32 APIs with P/Invoke. By walking through this example, you will learn how to leverage the native APIs in your managed Smartphone device applications.

To change the input mode in native code, use the `SendMessage()` API, which can send various messages to a window or windows. The signature of this function is as follows:

```
LRESULT SendMessage(
    HWND       hWnd,
    UINT       Msg,
    WPARAM     wParam,
    LPARAM     lParam
);
```

The `SendMessage()` function takes four parameters. The `hWnd` parameter is the handle to the window that will receive the message `Msg`. The `wParam` and the `lParam` parameters provide additional message-specific information.

To set the input mode, the `Msg` parameter must be `EM_SETINPUTMODE`, which is defined as `0x00DE` in `winuserm.h` in the `Include\Armv4i` directory of the Windows Mobile 5.0 SDK. You also need to indicate which input mode is in the `lParam` parameter. Following are the supported input modes and their numeric values:

- `EIM_SPELL` — `0`, specifies the Spell input mode (also called *multi-tap*)
- `EIM_AMBIG` — `1`, specifies the Ambiguous input mode (also called *T9*)
- `EIM_NUMBERS` — `2`, specifies the Numbers input mode
- `EIM_TEXT` — `3`, specifies the user preferred input mode, which is the user's last Spell or Ambiguous selection

To set the input mode for a TextBox control, the handle of the Textbox must be passed to the `hWnd` in the `SendMessage()` function. The approach in this example to get the handle is to call the Win32 native API `GetFocus()`, which returns the window handle of the currently focused control.

After going through the related native Win32 APIs, you are ready to build a Smartphone device application that can change the input mode calling the Win32 APIs using P/Invoke.

Start a new Smartphone device application by choosing File⇨New⇨Project in Visual Studio 2005, and name the project **InputMode**. From Solution Explorer, rename the default `Form1.cs` to **InputMode.cs** by right-clicking `Form1.cs` and choosing Rename. Then make the Form Designer the currently active window by pressing Shift+F7. From the Properties window, set both the `name` property and the `text` property of the form to `InputForm`.

Next, add controls to the form and set their properties as described in Table 10-4.

Table 10-4 Controls in the InputMode Sample Application

Name	Type	Text	Location	Size	Add To
lbName	Label	Input Last Name	0,0	152,22	InputForm
txtName	TextBox		0,25	152,22	InputForm
lbPhone	Label	Input Phone Number	0,60	152,22	InputForm

Name	Type	Text	Location	Size	Add To
txtPhone	TextBox		0,85	152,22	InputForm
lbNotes	Label	Input Comments	0,120	152,22	InputForm
txtComment	TextBox		0,145	152,22	InputForm
mnuDone	MenuItem	Done			mainMenu1
mnuInputMode	MenuItem	Input Mode			mainMenu1
mnuSpell	MenuItem	Spell			mnuInputMode
mnuT9	MenuItem	T9			mnuInputMode
mnuNumber	MenuItem	Number			mnuInputMode
mnuText	MenuItem	Text			mnuInputMode

Figure 10-6 shows the UI interface.

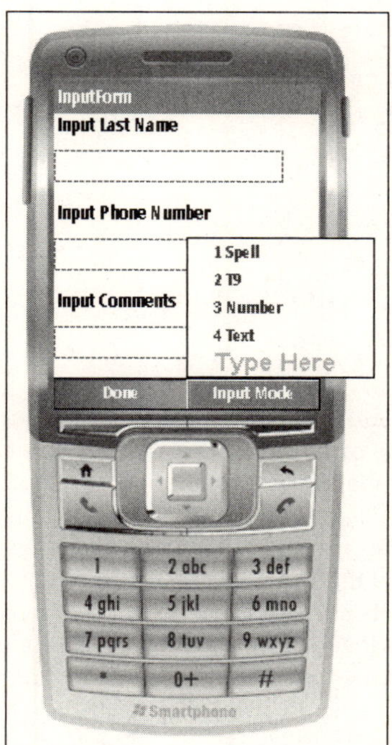

Figure 10-6

To call the native code, first include the `System.Runtime.InteropServices` namespace, as follows:

```
using System.Runtime.InteropServices;
```

Both the `GetFocus()` function and the `SendMessage()` function are located in `coredll.dll` and can be imported to C# code as follows:

```
    [DllImport("coredll.dll", EntryPoint = "GetFocus")]
    public static extern IntPtr GetFocus();

    [DllImport("coredll.dll", EntryPoint = "SendMessage")]
    public static extern int SendMessage(IntPtr hWnd, uint Message, uint wParam,
uint lParam);
```

Next, define the Text input mode constant so that it can be used by the `SendMessage()` function to set the input mode for a Textbox. The following code defines related constants:

```
    //Define text mode constant
    public const uint EM_SETINPUTMODE = 0xDE;

    public const uint EIM_SPELL = 0;     //Multi-tap
    public const uint EIM_AMBIG = 1;     //T9
    public const uint EIM_NUMBERS = 2;   //Number
    public const uint EIM_TEXT = 3;      //Text
```

Because there are three TextBox controls on the form, it makes sense to add a method for changing the input mode:

```
//Set the input mode of ctrl to MODE
private void SetInputMode(Control ctrl, uint MODE)
{
...
    ctrl.Focus();
    IntPtr hWnd = GetFocus();
    SendMessage(hWnd, EM_SETINPUTMODE, 0, MODE);
...
}
```

As shown above, the `SetInputMode()` method first sets the focus of the form to `Control ctrl`. Then the window handle of `ctrl` can be retrieved from the `GetFocus` function imported earlier. Next, call the Win32 API `SendMessage` and set the input mode of `ctrl` to `MODE`.

Having defined the `SetInputMode()` method, there are two ways you can use it. First, during the initialization process, you can call this method and set the TextBox controls to an appropriate input mode. For instance, TextBox `txtPhone` expects users to input numbers. You can set the input mode of `txtPhone` to `EIM_NUMBERS` when the form is loaded. The following code pre-sets the input mode of `txtName` to `EIM_SPELL`, `txtPhone` to `EIM_NUMBERS`, and `txtComment` to `EIM_TEXT`:

```
//Preset input modes of each TextBox
SetInputMode(txtName, EIM_SPELL);
SetInputMode(txtPhone, EIM_NUMBERS);
SetInputMode(txtComment, EIM_TEXT);
```

Second, you can add event handlers for those input mode menu items. For example, when users select T9 mode, set the currently focused control to T9 Input mode. To find out which control is currently focused, check the `Focused` property of each control in the form, as follows:

```
// Find the focused control in current form
private Control getFocusedCtrl(Form form)
{
    foreach (Control c in form.Controls)
    {
        if (c.Focused) return c;
    }
    //Return null if no control is focused
    return null;
}
```

Following is the event handler when the menu item `mnuT9` is clicked:

```
//Set input mode to T9
private void mnuT9_Click(object sender, EventArgs e)
{
    Control fc = getFocusedCtrl(this);

    //Set input mode to current focused control
    if (fc != null) SetInputMode(fc,EIM_AMBIG);

    //Set input mode to TextBox Tb_name if no control is currently focused
    else SetInputMode(txtName,EIM_AMBIG);
}
```

In the preceding click event handler, when the currently focused control cannot be found, it uses the `txtComment` as a fallback control and sets the input mode for `txtComment`.

Add similar click event handlers for `mnuName`, `mnuPhone`, and `mnuText`. Also add a click event hander to `mnuDone` to close the application when the left soft key is pressed.

Following is the full listing of the code in `InputForm.cs`:

```
using System;
using System.Collections.Generic;
using System.ComponentModel;
using System.Data;
using System.Drawing;
using System.Text;
using System.Windows.Forms;

using System.Runtime.InteropServices;

namespace InputMode
{

    public partial class InputForm : Form
    {
        //Import native GetFocus API
        [DllImport("coredll.dll", EntryPoint = "GetFocus")]
        public static extern IntPtr GetFocus();

        //Import native SendMessage API
```

```
        [DllImport("coredll.dll", EntryPoint = "SendMessage")]
        public static extern int SendMessage(IntPtr hWnd, uint Message, uint
wParam, uint lParam);

        //Define text mode constant
        public const uint EM_SETINPUTMODE = 0xDE;
        public const uint EIM_SPELL = 0;     //Multi-tap
        public const uint EIM_AMBIG = 1;     //T9
        public const uint EIM_NUMBERS = 2;   //Number
        public const uint EIM_TEXT = 3;      //Text

        //Set the input mode of ctrl to MODE
        private void SetInputMode(Control ctrl, uint MODE)
        {
            //Set the focus, obtain the window handler, and then set the mode
            try
            {
                ctrl.Focus();
                IntPtr hWnd = GetFocus();
                SendMessage(hWnd, EM_SETINPUTMODE, 0, MODE);
            }
            catch (MissingMethodException e)
            {
                MessageBox.Show("Error: " + e.ToString());
            }
        }

        public InputForm()
        {
            InitializeComponent();

            //Preset input modes of each TextBox
            SetInputMode(txtComment, EIM_TEXT);
            SetInputMode(txtPhone, EIM_NUMBERS);
            SetInputMode(txtName, EIM_SPELL);

        }

        // Find the focused control in current form
        private Control getFocusedCtrl(Form form)
        {
            foreach (Control c in form.Controls)
            {
                if (c.Focused) return c;
            }
            //Return null if no control is focused
            return null;
        }

        //Set input mode to spell (Multi-tap)
        private void mnuSpell_Click(object sender, EventArgs e)
        {
            Control fc = getFocusedCtrl(this);

            //Set input mode to current focused control
```

```
            if (fc != null) SetInputMode(fc, EIM_SPELL);

            //Set input mode to TextBox Tb_name if no control is currently focused
            else SetInputMode(txtName, EIM_SPELL);
        }

        //Set input mode to T9
        private void mnuT9_Click(object sender, EventArgs e)
        {
            Control fc = getFocusedCtrl(this);

            //Set input mode to current focused control
            if (fc != null) SetInputMode(fc,EIM_AMBIG);

            //Set input mode to TextBox Tb_name if no control is currently focused
            else SetInputMode(txtName,EIM_AMBIG);
        }

        //Set input mode to Numeric
        private void mnuNumber_Click(object sender, EventArgs e)
        {
            Control fc = getFocusedCtrl(this);

            //Set input mode to current focused control
            if (fc != null) SetInputMode(fc,EIM_NUMBERS);
            //Set input mode to TextBox Tb_name if no control is currently focused

            else SetInputMode(txtName,EIM_NUMBERS);

        }

        //Set input mode to Text
        private void mnuText_Click(object sender, EventArgs e)
        {
            Control fc = getFocusedCtrl(this);
            //Set input mode to current focused control
            if (fc != null) SetInputMode(fc,EIM_TEXT);
            else SetInputMode(txtName,EIM_TEXT);

        }

        //
        private void mnuDone_Click(object sender, EventArgs e)
        {
            this.Close();
        }
    }
}
```

This application demonstrates how you can use P/Invoke native APIs in your managed code. You can apply the same approach when writing your own P/Invoke applications.

Optimizing P/Invoke Performance

In many situations, exporting native Windows CE APIs through P/Invoke certainly makes a developer's life a lot easier. You should realize, however, that you are sacrificing performance for the convenience. As some reports have pointed out, calling and executing unmanaged DLLs from the .NET Compact Framework can be more than five times slower than executing a managed implementation. This is really not a big deal on a powerful PC running the full .NET Framework, but the penalty may be significant for mobile devices such as Smartphone. Naturally, you may wonder where the overhead comes from and how to minimize the negative impact.

The performance penalties mainly come from two sources. The first overhead is marshaling managed data to unmanaged data. Obviously, the more parameters you need to pass to unmanaged code, the more computational time and memory will be consumed. In addition, non-blittable marshaling will certainly affect the performance much more than blittable marshaling where data representations are the same. The second extra cost is due to .NET runtime management. As discussed earlier in this chapter, managed data is under the control of the .NET runtime. Processes such as garbage collection are always monitoring this data. When calling native APIs from managed code, however, the .NET Compact Framework first has to unregister that data from garbage collection so that unmanaged data will not be controlled by the garbage collection process. Once the unmanaged code finishes execution, the .NET runtime needs to reclaim the data as managed data so that it can be governed by the runtime environment. This extra process cost appears to be nontrivial as well.

Now that you know why P/Invoke adversely affects performance, let's look at what you can do to limit the penalties as much as possible.

The first technique results from the fact that non-blittable marshaling is much more complex than blittable marshaling. If you have a choice, always choose blittable marshaling when performing interop calls. The next thing you should consider when optimizing P/Invoke performance is to group those calls to bigger trunks. This is because each interop call has a fixed cost. By cutting the number of interop calls, it is highly likely to reduce the time spend on interop management.

Specifically, when calling from managed code to unmanaged code, keep in mind the following guidelines suggested by the Microsoft .NET Compact Framework 2.0 team:

- P/Invoke calls are normally handled more quickly when the parameters are basic blittable types or simple types. This includes the following:
    - All blittable value types except the `long` type, which has a size of 64 bits and is more efficient to pass by reference, rather than by value
    - Simple data types such as `String` and `Array`, which are not blittable but which the .NET Compact Framework can marshal pretty quickly
    - `struct` or `class` types that contain only blittable or simple data types
- Using `in` and `out` attributes for arguments can speed up the marshaling process.
- `Marshal.Prelink` and `Marshal.PrelinkAll` can be used in the .NET Compact Framework 2.0 to provide earlier initialization of the native functions when the application first starts. Therefore, users probably will wait longer for applications to launch but they can enjoy a faster response once the application is loaded.

Summary

The .NET Compact Framework supports a subset of P/Invoke services available in the full .NET Framework. This feature enables you to call in to the old-fashioned yet still convenient Win32 system libraries. You can also use the P/Invoke service to interact with other unmanaged libraries you have manually created.

The typical steps to perform P/Invoke are as follows:

1. Import the DLL that contains the native APIs by using the `DllImport` attributes. This requires the inclusion of the `System.Runtime.InteropServices` namespace.
2. Declare the calling function in your managed code.
3. Call the native APIs using the format defined in your managed code.

Blittable data types have the same size and memory representations in both managed code and unmanaged code. You can simply pass them by value without any problems. Complex data types such as arrays and strings can be passed by reference. For structure and class types, the `LayoutKind` attribute and the `MarshalAs` attribute can be used to indicate how each data field is mapped to the unmanaged memory.

Calling into native APIs using P/Invoke may incur performance penalty and is therefore recommended only as a last resort. In the next chapter you will learn how to handle errors and how to debug in Smartphone applications.

11

Exception Handling and Debugging

This chapter is about dealing with exceptional conditions in your code and discovering problems in your code. The .NET Compact Framework offers a number of exception classes that are used by other classes when an error occurs. Applications can choose to handle some of these exceptions programmatically, leaving others to the .NET Compact Framework CLR. Applications can also define their own exception types for specific abnormal cases. The proper use of exception handling can enhance the robustness of your application.

Debugging a Smartphone application enables you to trace into the code at runtime and discover an application's unexpected behaviors. Visual Studio 2005 can debug managed applications running on the emulator or on a Smartphone device. Depending on the debugging setting of the application, you can perform source-code-level debugging or assembly-level debugging. The debugger provides a set of windows for viewing and controlling objects and threads.

Topics covered in this chapter include the following:

- Exception and exception handling constructs in C#, and exception classes in the .Net Compact Framework
- Debugging support in Visual Studio 2005
- Best practices for exception handling and application debugging

Exceptions and Exception Handling

Exceptions are errors that occur while an application executes. Hardware-based exceptions such as page faults can be handled by computer hardware. Software-based exceptions are handled either by the application or by the default exception handling mechanism of the runtime environment or the operating system.

Exceptions

The purpose of exception handling can be summarized as follows:

- Exception handling enables applications to proceed in an expected fashion; application developers should take into account potential exceptions pertaining to the program flow, and handle them proactively. Leaving these exceptions to the runtime or the operating system is not a suggested approach as it often breaks the program logic and even crashes the application.
- Exception handling enables applications to define their own exceptions as a way to facilitate special situations in program flow. Applications can throw exceptions at desired places and catch them wherever needed.
- Unlike return code, exceptions are never ignored — either the application or the .NET Compact Framework runtime will catch the exception.

From the standpoint of the application, there are handled exceptions and unhandled exceptions. As the name implies, exceptions that are handled by the application are known as *handled exceptions*, whereas those left to the runtime or the operating system are known as *unhandled exceptions*. Therefore, it is critical to identify which exceptions should be handled by the application. Usually, application developers can obtain this information from the runtime framework. In the .NET Compact Framework, many classes have built-in support for exceptions. For example, the `System.Xml.XmlReader::Read()` method will throw a `System.Xml.XmlException` when an error occurs while parsing an XML data stream. Applications using this method should capture this exception and notify the caller of this method via error code.

The try...catch Statement

When an exception occurs, an exception handler defined by the application or the system will be called. The basic construct of exception handling in C# is as follows:

```
try
{
    // Statement that may throw exceptions
}
catch(ExceptionType e)
{
    // This catch block will handle only the indicated exception type
}
```

You can associate zero or more `catch` blocks with a `try` block. A `catch` block can handle a specified type of exception. When you need to handle multiple types of exceptions, the `catch` blocks should start from specific exception types and proceed to more general types. An exception type filter must be placed before its base exception type filter. Thus, once an exception occurs, it will go through the filters (the `catch` blocks). In addition, a `catch` block can have no argument, meaning that it will catch all exceptions that come through the `catch` block filters. However, only one exception can be placed into a single `catch` block.

Exceptions are handled according to the call stack of functions, meaning that an exception will unwind the call stack until a handler for that type of exception is found. If an exception falls through the list of `catch` blocks, the .NET Compact Framework runtime will catch it and handle it using some built-in default handlers. In this case, we call these exceptions *unhandled* because the application code relies on the runtime to handle them.

Here is an example of I/O exception handling when a file is created on the device:

```
StreamWriter sw = null;
try
{
      sw = new StreamWriter(@"\Storage Card\output.log", false);
}
catch (DirectoryNotFoundException e)
{
      // The file path is incorrect
}
catch (IOException e)
{
      // IO exception such as failure to create the file
}
```

The MSDN documentation states that the constructor of `StreamWriter` used in this example (`StreamWriter::StreamWriter(string path, bool append)`) may throw one of the following seven exceptions:

- `UnauthorizedAccessException`
- `ArgumentException`
- `ArgumentNullException`
- `DirectoryNotFoundException`
- `IOException`
- `PathTooLongException`
- `SecurityException`

Depending on the setting of the file to be written, the caller may need to catch one or more exceptions. In this example, you are concerned with the case in which the file path specified does not exist, leaving all other I/O-related exceptions to the `IOException` filter. In fact, `DirectoryNotFoundException` is a subclass of `IOException`, which is also a base class of several other exception classes, including `FileNotFoundException` and `PathTooLongException`.

The finally Statement

In some cases, even when an exception occurs and has been handled in a `catch` block, there may still be some clean-up code that needs to be executed after the `catch` block is executed. Although you can, of course, put this part of clean-up code in every `catch` block, it would be more efficient to have another code block that will be always executed, no matter what exceptions occurred and are being handled. In C#, this block is the `finally` block. Here is an (incorrect!) example of using `finally` for the preceding example:

```
try
{
    sw = new StreamWriter(@"\Storage Card\output.txt", false);
    sr = new StreamReader(@"\Storage Card\input.txt");
    string line = sr.ReadLine();
    while (line != null)
    {
        sw.WriteLine(line);
        line = sr.ReadLine();
    }
}
catch (DirectoryNotFoundException e)
{
    // The file path is incorrect
    Console.WriteLine(e.Message);
}
catch (IOException e)
{
    // IO exception such as failure to create the file
    Console.WriteLine(e.Message);
}
finally
{
    sw.Close();
    sr.Close();
}
```

The `try` blocks contain calls to the constructors of `StreamReader` and `StreamWriter`, as well as the call to `StreamReader::ReadLine()`. The `catch` blocks catch `DirectoryNotFoundException` and `IOException`. Finally, the `finally` block simply closes the previously created `StreamReader` and `StreamWriter` objects. The problem with this `finally` block is that if any one of the two objects threw an exception, regardless of whether it has been handled or not, the `Close()` call will throw a `NullReferenceException` because the object is not initialized properly. For example, if any of the file paths specified in the two constructor calls do not exist, you will hit the `NullReferenceException` in the `finally` block. The solution to this problem is to determine whether the object is null before closing it:

```
finally
{
        if(sw != null)
            sw.Close();
        if(sr != null)
            sr.Close();
}
```

The earlier example code uses a `Console.WriteLine()` call in the `catch` blocks. When developing Smartphone applications, you can use this method to output some debugging information in the Output window of Visual Studio 2005 in debug mode.

The throw Statement

You can explicitly throw an exception in your code as an indication of an error to other parts of the program. You can also rethrow an exception in your `catch` block and let the outer `catch` block catch it. Here is an example of using the `throw` statement:

```
static void Main(string[] args)
{
    try
    {
        FileCopy();
    }
    catch (IOException e)
    {
        // This line will catch the re-thrown DirectoryNotFoundException
        Console.WriteLine(e.Message);
    }
}

static void FileCopy()
{
        try
        {
            sw = new StreamWriter(@"\Storage Card\output.txt", false);

            // We intentionally put a non-existent file path
            sr = new StreamReader(@"\Storage Card1\input.txt");
            string line = sr.ReadLine();
            while (line != null)
            {
                sw.WriteLine(line);
                line = sr.ReadLine();
            }
        }
        catch (DirectoryNotFoundException e)
        {
            // The file path is incorrect
            Console.WriteLine(e.Message);
            throw new DirectoryNotFoundException("Re-throw: " + e.Message);
        }
        catch (IOException e)
        {
            // IO exception such as failure to create the file
            Console.WriteLine(e.Message);
        }
        finally
        {
            if (sw != null)
                sw.Close();
            if (sr != null)
                sr.Close();
        }

}
```

The rethrown `DirectoryNotFoundException` will be caught by the `catch` block in the `Main()` method. The purpose of doing this "rethrowing" is that the method `FileCopy()` may need to indicate to the caller when a specific exception occurs. Aside from using a return value, you can use `throw` to easily pass the exception information to the caller, rather than use a defined method return value. This way, you can separate common program flow (which will return normally) from abnormal code paths in which exceptions may occur and be handled and rethrown if needed.

Exception Stack Trace

As stated earlier, exceptions are handled by unwinding the function stack call. If an exception is not handled in the current function or method, it will be passed to the caller of this method, and so on, until a handler for this exception catches it. Once the exception is handled, the program will continue to execute from that point. A handler can rethrow the exception it is dealing with, or it can throw a different one that wraps the underlying exception. Such a technique is called *exception chaining* or *exception wrapping*. It is the responsibility of the caller of the current method to handle these exceptions; it is impossible for a managed application to handle all possible exceptions. A good practice is to handle those exceptions that are most likely to occur, leaving other exceptions to the .NET runtime.

All exception classes derived from `System.IO.Exception` have a `StackTrace` property containing the stack trace. More details of these classes are discussed in the next section.

The following example shows three different scenarios when an exception is handled along with the function call stack. The code is part of a Smartphone console application. Although the `Console::WriteLine()` call does not output anything, you can still see a message regarding unhandled exceptions on the emulator or target device:

```
    class ClassA
    {
        public static void MethodA(int i)
        {
            if (i == 0)
                throw new DivideByZeroException("DivideByZeroException in
MethodA()");
            else if (i == 1)
                throw new FileNotFoundException("FileNotFoundException in
MethodA()");
            else
                throw new NullReferenceException("NullReferenceException in
MethodA()");
        }
    }
    class ClassB
    {
        public static void MethodB(int i)
        {
            try
            {
                ClassA.MethodA(i);
            }
            catch (FileNotFoundException e)
```

```
            {
                Console.WriteLine(e.StackTrace);
                Console.WriteLine(e.Message);
                throw new IOException();
            }

        }
    }
    class ClassC
    {
        public static void MethodC()
        {
            try
            {
                // Depending on the parameter passed to the following call, an
exception will be thrown
                // If the parameter is 0, MethodA throws DivideByZeroException,
which is un-handled in this program
                // If the pamameter is 1, MethodA throws FileNotFoundException,
which is handled in MethodB(). MethodB() re-throws IOException, which is handled in
MethodC()
                // Otherwise, MethodA throws NullReferenceException, which is
handled in MethodC()
                ClassB.MethodB(0);
            }
            catch (IOException e)
            {
                Console.WriteLine(e.StackTrace);
                Console.WriteLine(e.Message);
            }
            catch (NullReferenceException e)
            {
                Console.WriteLine(e.StackTrace);
                Console.WriteLine(e.Message);
            }
        }
    }
```

In this example, `ClassC::MethodC()` calls `ClassB:MethodB()`, which in turn calls `ClassA::MethodA()`. Depending on the parameter passed to `ClassA::MethodA()`, three exceptions may be thrown and handled or unhandled:

- `DividedByZeroException` — The code does not handle this exception; thus, the .NET CLR will handle this exception after popping up a message box called Exception Assistant (shown in Figure 11-1) in debug mode. If the application is run without debugging, you will see a message on the emulator, as shown in Figure 11-2. Note that the name of the application executable is `ExceptionHandling.exe`.
- `FileNotFoundException` — `ClassB::MethodB()` handles this exception and then throws an `IOException`, which is handled in `ClassC::MethodC()`.
- `NullReferenceException` — This exception is handled in `ClassC:MethodC()`.

By default, if an exception handler is available, the code will not break when the exception is thrown. However, if you want to break whenever a type of exception occurs, choose Debug⇨Options, and then specify an exception from the Category list or add your own user-defined exception to the list.

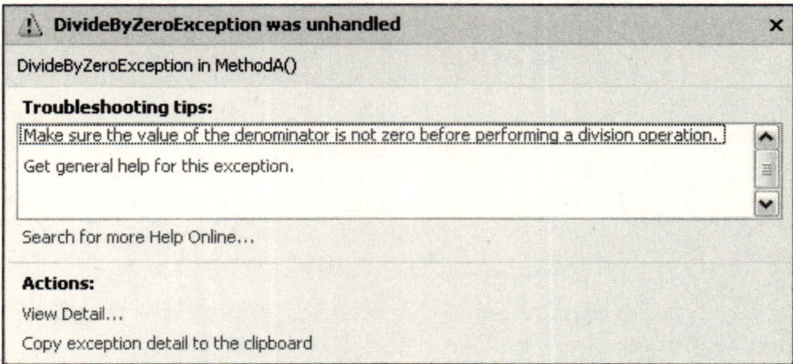

Figure 11-1

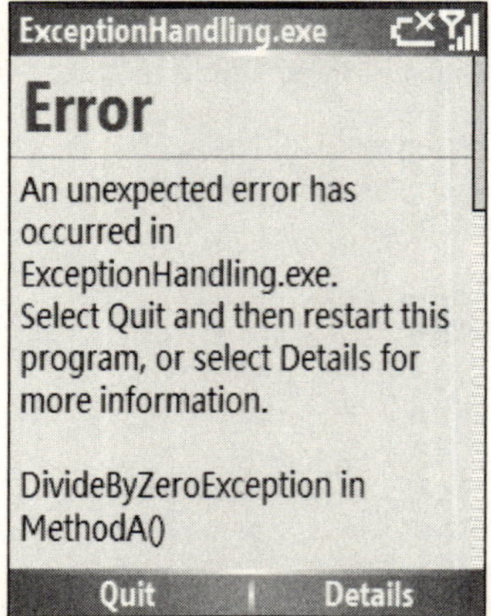

Figure 11-2

The Exception Class

Now let's look at the base exception class, `System.Exception`. All other exception classes are derived from this class, including user-defined exceptions. In fact, in C#, you can throw any object derived from the `System.Object` class as an exception. However, it is a good practice to always throw an exception object that inherited from `System.Exception`, as it has defined almost all the necessary properties and methods. Subclasses do not need to provide any additional properties and methods.

Table 11-1 lists the most commonly used properties of the `System.Exception` class in the .NET Compact Framework. The second column shows whether the property has a `Get` accessor, a `Set` accessor, or both.

Table 11-1 System.Exception Properties

Property	Accessor	Description
`Message`	`Get`	A string that describes the exception. Subclasses can override this property.
`StackTrace`	`Get`	A string representation of the call stack when the current exception was thrown. Subclasses can override this property.
`InnerException`	`Get`	An `Exception` instance that was wrapped in the current exception. This is used in exception chaining when an exception is thrown along with a reference to an earlier exception (passed in the constructor).
`HResult`	`Get/Set`	A 32-bit value that uniquely identifies an exception.

Table 11-2 lists two frequently used methods in `System.Exception` in the .NET Compact Framework.

Table 11-2 System.Exception Methods

Method	Return Type	Description
`GetBaseException()`	Exception	The root cause of the current exception in an exception chain. For all the exceptions on the exception chain, there is only one base exception: the first exception whose base exception is null.
`ToString()`	String	Returns a string representation of the current exception that may include the name of the class that throws the exception, the `Message` property, the return value of `ToString()` of the inner exception, and the result of calling `Environment.StackTrace`, which represents the current stack trace. Subclasses can override this method.

Applications are not likely to use `System.Exception` directly; rather, either some specific exception types or user-defined exceptions are used to cover possible program errors. The following section introduces the exception hierarchy.

The SystemException Class

The `System.SystemException` class is thrown by the .NET CLR when unhandled exceptions occur in the application. These exceptions are mainly runtime check errors. Applications do not need to catch a `SystemException`; instead, only derived classes of `SystemException` need to be caught. Table 11-3 lists the classes derived from `SystemException`.

Table 11-3 Classes Derived from SystemException

Class	Description
`System.AccessViolationException`	Thrown when a protected memory address is being accessed.
`System.ArgumentException`	Thrown when at least one parameter of a method is invalid. This exception class further extends to `System.ArgumentNullException` and `System.ArgumentOutOfRangeException`.
`System.ArithmeticException`	Thrown when an application contains arithmetic errors, such as divided by zero (`DividedByZero Exception`), number overflow (`Overflow Exception`), or floating number infinity (`NotFiniteNumberException`).
`System.IndexOutofRangeException`	When an index used to access an array is out of the range.
`System.IOException`	The base class for I/O exceptions. Classes derived from `IOException` include `DirectoryNot FoundException`, `EndOfStreamException`, `FileNotFoundException`, `FileLoadException`, and `PathTooLongException`.
`System.NotImplementedException`	Thrown when a request method is not implemented.
`System.NotSupportedException`	Thrown when a request method does not exist.
`System.NullReferenceException`	Thrown when a null object is being accessed.
`System.OutOfMemoryException`	Thrown when the runtime is running out of memory.
`System.Security.SecurityException`	Thrown when a security error occurs.
`System.Xml.XmlException`	Thrown when an XML processing error occurs.

The ApplicationException Class

`ApplicationException` is the base class of all user-defined exceptions. Like `SystemException`, `ApplicationException` is derived from the `Exception` class but does not add any new functionality. It is used to distinguish system exceptions thrown by the .NET CLR from those defined by the application.

In most cases, a user-defined exception derived from `Exception` or `ApplicationException` does not need to add any new functionality. It is recommended that your user-defined exception class should simply override the constructors that call the base constructors. However, it is also possible to add some additional information to the exception. It is also suggested that your user-defined exception should be named with the word "Exception." The following code is an example of a user-defined exception:

```
class InvalidZipCodeException : ApplicationException
{
    public InvalidZipCodeException()
```

```
        {

        }
        public InvalidZipCodeException(string message) : base(message)
        {

        }
        public InvalidZipCodeException(string message, Exception inner)
            : base(message, inner)
        {
        }
        public override string ToString()
        {
            return "InvalidZipException: " + this.Message + this.StackTrace;
        }
    }
```

The `InvalidZipCodeException` defines the following three constructors:

- A default constructor
- A constructor taking a string to be added to the `Message` property of the class
- Another constructor taking a string for the `Message` property and an inner exception for the `InnerException` property

The default constructor is added because we want to create an `InvalidZipCodeException` without using any parameter. The other two constructors simply call the corresponding base class (`ApplicationException`) methods. In addition, we override the `ToString()` method just to demonstrate that we can add some meaningful information when the method is called.

To use `InvalidZipCodeException`, throw an instance of it at the appropriate place. For instance, the following example uses a simple character-by-character check to verify whether a string indeed contains a zip code:

Note that this example can certainly use a return code rather than an exception to notify the caller of any invalid zip code. The intent is to demonstrate how a user-defined exception can be used.

```
        public static bool ProcessZipCodeInput(string zip)
        {
            char[] zipCode = zip.ToCharArray();
            if (zip.Length != 5)
                throw new InvalidZipCodeException();
            foreach (char x in zipCode)
            {
                if (!(x >= '0' && x <= '9'))
                {
                    // One of the three constructors of the user-defined exception
class may be used
                    throw new InvalidZipCodeException("User-defined exception:
Invalid zip Code. ", new ArgumentException());
                }
            }
            return true;
        }
```

When throwing a user-defined exception, you have the option to use one of the available constructors. In the preceding example code, you can use either the default constructor, or the one that accepts a string parameter (`User-defined exception: Invalid Zip Code`), or the one that accepts the string parameter and an exception instance (`new ArgumentException()`). For example, the following two throw calls can also be used in the preceding example:

```
                    //throw new InvalidZipCodeException();
                    //throw new InvalidZipCodeException("User-defined exception:
   Invalid Zip Code. ");
```

Any code that calls `ProcessZipCodeInput()` should be put into a `try-catch` block. For example, if a string `"239d3"` is passed to the `ProcessZipCodeInput()` method, an exception will be thrown and should be handled:

```
            try
            {
                ProcessZipCodeInput("239d3");
            }
            catch (InvalidZipCodeException e)
            {
                Console.WriteLine(e.ToString());
            }
```

Best Practices of Exception Handling

As shown in the sample code discussed so far, exception handling can act as a method for execution event notification across method/function calls. Note the following issues regarding when, where, and how to use exception handling to improve the robustness of an application:

- **Exceptions vs. condition checking.** Some errors can be discovered by checking whether the variable truly holds the expected value or by checking the return value of a method. Without using exceptions, the code can simply return some error code in the case of a specific type of error. The key is to determine whether this kind of error occurs quite rarely in the code path. On the one hand, if the error is indeed exceptional, using exceptions is recommended because you won't waste CPU cycles on that condition-checking code in normal cases. On the other hand, if the error is almost certain to occur every time in the current method, you may prefer condition checking to exception handling because the latter usually results in significant overhead of stack unwinding and exception creation and deletion.
- **Exceptions vs. return error code.** A method that may incur some error can either return a specific error code or raise some exception. The general guideline is to return a common error code (for example, `null` for an object, `-1` for an integer, `false` for a Boolean) when a common error occurs, or define your return codes for your application logic, but throw an exception when an unusual, critical error occurs.
- **Exception classes vs. user-defined exception classes.** Use the .NET Compact Framework's exception classes for all general exceptions. Write your user-defined exception classes only for a special error in your application. For example, if the application requires numeric input following a pattern, then a user-defined exception can be created for invalid input that does not conform to the pattern.

Debugging in Visual Studio 2005

Chapter 3 talked briefly about debugging Smartphone applications. This section first describes the debugging features in Visual Studio 2005, and then we turn to some advanced topics regarding debugging support in the .NET Compact Framework.

Debugging Windows

Basic debugging functions supported by Visual Studio 2005 include the following: place breakpoints (F9 to toggle the breakpoint), step into (F11), step over (F10), or step out of the code (Shift+F11), and attach to a process on a device or the emulator. In addition, you can use a handful of debug windows to view and change variables, and to evaluate expressions and function calls.

> **Visual Studio 2005 debugger's "Attach to process" support is disabled by default for processes running on the emulator or Smartphone devices. To enable this feature, add the following DWORD key to the registry of the device or the emulator:**
>
> ```
> HKLM\Software\Microsoft\.NetCompactFramework\Managed
> Debugger\AttachEnabled = 1
> ```
>
> **You can use the WinCE Remote Registry Editor (in Visual Studio Remote Tools of the Visual Studio 2005) to do this.**

The following debugging windows are listed under Debug⇨Windows:

- **Immediate window** — You can enter an expression or a statement to inspect or change its value. Two modes are supported in the immediate window. The first is *command mode,* in which you can enter a Visual Studio command prefixed by a > character. For example, to open the commands window, enter **>cmd**. The other mode is *immediate mode,* in which you can enter a variable or a statement. To evaluate an expression, prefix the expression with a `?` (such as **? 3+5**).
- **Watch windows** — These windows enable you to enter and edit a variable, object, or expression, which will be evaluated automatically while debugging. The display of an object enables you to quickly view the properties and fields.
- **Locals window** — This window automatically displays local variables and objects. You can change the values as well.
- **Autos window** — This window automatically displays local variables, objects, and expressions of the current line and preceding lines. You can change the values as well.
- **Call stack window** — This window displays the method call stack, including all methods, parameters, and return values.
- **Modules window**–This window lists loaded modules.
- **Process window** — This window lists currently running processes in the debugger.
- **Threads window** — This window lists all the threads of the application.
- **Breakpoint window** — This window lists all breakpoints. You can set conditional breakpoints and configure a macro or print a message for a breakpoint.

In addition, when you hover the mouse over the variable or object while the code breaks at a statement or an exception, you will see the DataTip window, which enables you to inspect and change complex data types. You can even expand properties or fields, which themselves are other types.

Note that managed code based on the .NET Compact Framework does not support disassembly debugging.

Debugging Setting

The C# compiler (`csc.exe`) supports a `/debug` option, which you can set to one of the following two values:

- `full` — This setting allows source code debugging when the program is run in a debugger and when a debugger is attached to the running program. This is the default setting for a project's "Debug" configuration.
- `pdbonly` — This setting allows source code debugging when the program is run in a debugger but only enables assembly-level debugging when a debugger is attached. This is the default setting for a project's "Release" configuration.

To change this setting in Visual Studio 2005, go to Project⇨Project Properties⇨Build⇨Advanced⇨Debug Info. On the command line, you can also use `/debug+` or `/debug` to get the same result as `/debug:full`. Conversely, `/debug-` will disable debugging, just as `/debug` is not included in the compiler options. The following line is an example of using `/debug:full` for the file `MyExample.cs`:

```
csc /debug:full MyExample.cs
```

Using `/debug:full` will have some impact on the performance of the program because the JIT code size will increase and the .NET Compact Framework CLR will take longer time to JIT-compile the MSIL code. The programmatic way to control these settings is to use `System.Diagnostics.DebuggableAttribute` in your code.

A related compiler option is `/optimize` (or `/o`), which specifies whether to optimize the code for better performance. Usually, code under debugging should not use this option. The same setting in Visual Studio 2005 is available under Project⇨Project Properties⇨Debug.

Deploying and Debugging in Visual Studio

In Chapter 3, you learned how to deploy your Smartphone application onto a device or an emulator. Basically, select Project⇨Project Properties⇨Device and select the device from the Target device box. Or, if you have the Device toolbar selected, you can directly select the device there. After this step, debugging an application running on the device or the emulator can be as easy as debugging a desktop application: You will be able to perform source code debugging, put breakpoint in the code, and so on. Of course, the user's input must be done on the device or the emulator, but anything else is done on your development PC. See Chapter 3 for a detailed discussion of the device debugging features in Visual Studio.

Defining Symbols

Sometimes conditional compilation is needed to quickly separate code for debugging purposes from release code. The common way to do this is to use a conditional check on a symbol, as shown in the following example:

```
#ifdef DEBUG
// Some debug code goes here
#endif
```

This piece of code checks whether the `DEBUG` symbol is defined either in your code (`#define DEBUG`) or on the compiler command line (`csc /define:DEBUG MyExample.cs`). If so, statements following `ifdef` will be compiled. Otherwise, the compiler will skip the `ifdef-endif` enclosed code. You can set the `DEBUG` symbol as a compiler option with `/define:DEBUG`. The code segment can be placed anywhere in the code.

Another symbol that is often used for debugging and tracing is `TRACE`. The compiler option is `/define:TRACE`. In fact, you can define whatever symbols you like using `/define:` (or `/d:`), followed by your symbol on the compiler command line. Note that in C#, you can't assign a value to a defined symbol. Multiple symbols can be defined and separated by commas. In Visual Studio 2005, these settings are found under Project⇨Project Properties⇨Build.

Limitations of Debugging

Despite the powerful functionality of Smartphone application debugging, it has the following limitations (as of the .NET Compact Framework 2.0 and Visual Studio 2005):

- **Just-My-Code debugging is not supported.** Just-My-Code debugging is a feature that, once enabled, allows the programmer to see only the user code while debugging; library class methods and system code will not be displayed. The debugger will search for symbols files to determine whether a piece of code is "My Code." If symbols for the code exist, it is considered "My Code." This feature is not available for .NET Compact Framework debugging.
- **Edit-and-Continue is not supported.** In the .NET Framework application development, you can edit the code and continue to run in a debugging session, thus saving the time for recompilation. This feature is not supported in .NET Compact Framework debugging.
- **The `next` statement is not supported.** You cannot set the instruction point when debugging .NET Compact Framework applications.

AppVerifier

Microsoft provides another free tool, called AppVerifier for Windows Mobile 5.0, that you can use to test Smartphone applications against common native coding mistakes such as memory leaks, handle leaks, GDI resource leaks, and some heap corruptions. For managed code debugging, it is also useful for testing problems in intensive I/O applications that may stem from the .NET Compact Framework CLR.

For more information about AppVerifier for Windows Mobile, visit `www.microsoft.com/downloads/details.aspx?FamilyId=D275348A-D937-4D88-AE25-28702C78748D&displaylang=en` or search for "Application Verifier Tool for Windows Mobile" at download.Microsoft.com.

Multithreaded Debugging

Applications can have multiple threads to perform different tasks in parallel and in sync. The .NET Compact Framework has done a good job encapsulating multithreading details into many classes such that you don't need to create threads yourself. For example, the `BeginInvoke()` method of a `Control` internally uses the `ThreadPool` thread to perform the specified task asynchronously. The `BeginRead()` and `BeingWrite()` methods of the `Stream` class are two other examples of multithreading in the .NET Compact Framework.

There are still cases where you need to create and manage multiple threads in a Smartphone application. For instance, for an application that retrieves web pages from the Internet and caches them locally, network access, local file access, and UI updates can be performed with three threads simultaneously; thus, one will not block others. A major debugging topic involves multiple threads running concurrently and interoperating with each other. Common issues in multithreaded applications include *race condition* (the execution of multiple threads depending on the timing of events), *deadlock* (two threads waiting for each other to release a resource), and *access violations* (a thread accessing a resource that has been released). AV (access violation) can be fairly easy to detect, as the .NET Compact Framework runtime will throw exceptions in these cases. For the other two problems of concurrency, you need the debugger to help.

Managed Threads

A managed thread in the .NET Compact Framework CLR is not directly mapped to an operating system thread. The CLR may schedule some managed threads using a single operating system thread. A managed thread may be migrated from one operating system thread to another, but to application developers this is completely transparent. If you have debugged applications that use multiple operating system threads in Visual Studio .NET, you will find that debugging managed multithreaded applications is very similar. Before discussing the details, let's go over the threading support in the .NET Compact Framework.

To create a managed thread, use `System.Threading.Thread`. You need to define a thread procedure and pass it to the constructor of the `Thread` object. Table 11-4 lists some notable properties and methods of the `Thread` class.

Table 11-4 Thread Class Members

Member	Description
`Thread::Start()`	Starts the thread.
`Thread::Abort()`	Terminates the thread. This method will throw a `ThreadAbort Exception`.
`Thread::Join()`	Blocks the calling thread until the thread being joined terminates.
`Thread.CurrentThread`	Returns the current running thread.
`Thread.Sleep()`	Put the current thread into sleep.
`Thread::ManagedThreadId`	Returns the unique thread ID.
`Thread::Name`	Gets or sets a thread name. Once set, `Name` cannot be changed.

Member	Description
`Thread::Priority`	Gets or sets thread priority, which is one of the values defined in the `ThreadPriority` enumeration.
`Thread::IsBackground`	Gets or sets a value indicating whether the thread is a background thread. This setting determines whether the process can terminate. A process cannot terminate until all its foreground threads terminate. Once all the foreground threads have terminated, the CLR terminates the process and stops all background threads of the process.

The following code snippet shows how to create and start a thread:

```
        Thread newThread = new Thread(ThreadProc);
        // Set the Name property of the new thread
        newThread.Name = "A new thread other than the main thread";
        newThread.Start();
    ...
    // Thread procedure
    private static void ThreadProc()
    {
        // Current thread's Name
        string name = Thread.CurrentThread.Name;
        // Sleep for 1 second
        Thread.Sleep(1000);

    }
```

As shown in this example, in the .NET Compact Framework 2.0, you can pass the method name for the new thread to the constructor of a `Thread` object. You can certainly also use the "old" scheme — that is, pass a new `ThreadStart` delegate that is created with the thread method to the constructor:

```
        Thread newThread = new Thread(new ThreadStart(ThreadProc));
```

If the thread method is fairly simple, you can put it inline in the constructor call, as follows:

```
        Thread newThread = new Thread(delegate()
             {
             // ThreadProc statements
             }
        );
```

The .NET Compact Framework CLR provides another facility for multithreaded applications: the thread pool. The *thread pool* consists of a set of worker threads (background threads) managed by the CLR for each application. You can post asynchronous I/O tasks and short callbacks to the thread pool using the `ThreadPool.QueueUserWorkerItem()` method or the `ThreadPool.RegisterWaitForSingleObject()` method. That way, you don't need to create and manage a new thread yourself. The disadvantage of using a thread pool worker thread is that the tasks can't take too long to finish. Otherwise, the thread pool may become fully occupied and can't accommodate new worker thread requests. In addition, you can't change a thread pool thread's priority, and they are all background threads.

Race Condition

Multiple threads may need to change the same object. If the code that accesses the object is not properly protected, you will see garbled object data. This is often called a *race condition*. The code block that modifies the shared object is called a *critical section*. Applications must ensure that at any given time there is only one thread in the critical section.

Critical sections can be protected by a lock. Only a thread that acquires the lock can enter the critical section. Other threads waiting for the lock will be blocked. The locking mechanism is implemented as a `lock` construct in C# that is built on top of the `Monitor` class in the .NET Compact Framework. If there are multiple resources to protect, a mutex can be used for each resource. A *mutex* is a named synchronization object that provides exclusive access to a resource. Once a `System.Threading.Mutex` object is created, a thread can call `WaitOne()` to obtain the mutex, and any other threads calling WaitOne() will be blocked. A thread should call `ReleaseMutex()` when it finishes the access.

You can use the `lock` construct of C# as follows:

```
lock(lockObject)
{
    // Enter critical section
.....
}
```

The `lockObject` can be a simple object type. It is suggested that this object should be a private member of the class. Locking on a public type may lead to deadlock because other code may also lock this type.

The following `RaceCondition` class demonstrates a first-come-first-take procedure of a number of work items. The class manages a private variable `numItems` that can be changed by the public method `TakeWorkItem()`. In the `TakeWorkItem()` method, we check to see if the number of items are greater than zero. If so, the method will let the underlying thread sleep for some random time to simulate the time for that work item. After the sleep, it will check the number of work items again. If multiple threads are executing in this method, there will be a chance that when the thread comes out of sleep, the number of work items have already be decremented by another thread. An exception is thrown when a thread wakes up and identifies zero or a negative number of work items.

```
class RaceCondition
{
    private Object myLock = new object();
    private int numItems = 0;
    Random r = new Random();

    public RaceCondition(int items)
    {
        numItems = items;
    }
    public void TakeWorkItem()
    {
        if (numItems > 0)
        {
            // There are still work items; so take one.
            Thread.Sleep(r.Next(100,1000));
```

```
                if(numItems <= 0)
                    throw new ApplicationException("Race condition: negative number
of items!");
                numItems--;
            }
            else if (numItems == 0)
            {
            }
            else
            {
                throw new ApplicationException("Race condition: negative number of
items!");
            }
        }
    }
```

You can test the race condition situation in the preceding class by simply creating a `RaceCondition` object with *x* work items and then creating more than *x* threads that execute the `RaceCondition::TakeWorkItem()` class:

```
            RaceCondition rc = new RaceCondition(sw,5);  // 5 work items
            Thread[] workers = new Thread[10]; // 10 workers
            for (int i = 0; i < 10; i++)
            {
                Thread t = new Thread(rc.TakeWorkItem);
                workers[i] = t;
                workers[i].Start();
            }
```

The unhandled `ApplicationException` will be raised when a thread sees zero or a negative number of work items. When the application breaks, you can see all the currently running threads in the Threads window. You can view each thread's current statement by switching to that thread.

> *Note that you will not see all ten threads because some of the earlier ones are already finished when the exception is raised.*

A lock is applied to protect the `TakeWorkItem()` method. This guarantees that only one thread can enter the critical section (i.e., to change the `numItems` variable). Therefore, no thread will see zero or a negative number of work items once they have entered the critical section:

```
            lock (myLock)
            {
                if (numItems > 0)
                {
                    // There are still work items; so take one.
                    Thread.Sleep(r.Next(100, 1000));
                    if (numItems <= 0)
                        throw new ApplicationException("Race condition: negative
number of items!");
                    numItems--;
                }
                else if (numItems == 0)
```

```
                {
                }
                else
                {
                    throw new ApplicationException("Race condition: negative number
of items!");
                }
            }
```

Deadlock

A mutex guarantees mutually exclusive access to a resource. You have to be cautious, however, when using a mutex for thread synchronization. Deadlock can occur if two or more threads are holding some resource and waiting for others to unlock other resources, and all resources can't be shared among threads. Because no thread can preempt other threads to forcibly obtain a requested resource, these threads end up in a cyclical wait state.

The following code shows an example of deadlock — the famous philosopher's dinner problem. In this simplified scenario, three philosophers, John, Jack, and Joe, are sitting around a table in the middle of which is a bowl of spaghetti. As illustrated in Figure 11-3, the table has been set with a number of forks equal to the number of philosophers. Eating the spaghetti requires *two* forks, however, so a philosopher must pick up both the fork to his left and the fork to his right. If each philosopher takes the fork to his left and then waits for the fork on his right to become available, nothing can happen; therefore, deadlock occurs.

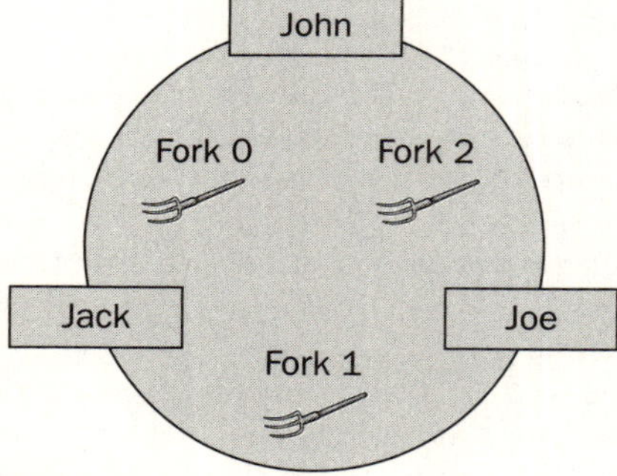

Figure 11-3

```
    class Deadlock
    {
        private StreamWriter sw = new StreamWriter(@"\Storage
Card\Philosopher.txt");
        Mutex[] forks = new Mutex[3];
        public void Dinner()
        {
            // Create three forks (mutex)
            for(int i = 0 ;  i < 3; i ++)
            {
                forks[i] = new Mutex();
```

```
            }
            // Tell the philosopher which forks to grab
            // Philosopher John = new Philosopher(sw, forks[2], forks[0]);
            // Solution to the deadlock problem: let the philosopher try the
smaller fork ID first
            Philosopher John = new Philosopher(sw, forks[0], forks[2]);
            Philosopher Jack = new Philosopher(sw, forks[0], forks[1]);
            Philosopher Joe = new Philosopher(sw, forks[1], forks[2]);

            Thread t1 = new Thread(John.Eat);
            t1.Name = "John";
            Thread t2 = new Thread(Jack.Eat);
            t2.Name = "Jack";
            Thread t3 = new Thread(Joe.Eat);
            t3.Name = "Joe";

            t1.Start();
            t2.Start();
            t3.Start();

            t1.Join();
            t2.Join();
            t3.Join();
            sw.WriteLine("Main thread exits.\n");
            sw.Close();
        }

    }
    class Philosopher
    {
        private StreamWriter sw = null;
        private Mutex lfork; // First fork to grab
        private Mutex rfork; // Second fork to grab
        public Philosopher(StreamWriter logfile, Mutex fork_left, Mutex fork_right)
        {
            sw = logfile;

            lfork = fork_left;
            rfork = fork_right;
        }

        public void Eat()
        {
            lfork.WaitOne();
            Log(Thread.CurrentThread.Name + "\t acquired " + lfork.GetHashCode());
            Thread.Sleep(1000);
            rfork.WaitOne();
            Log(Thread.CurrentThread.Name + "\t acquired " + rfork.GetHashCode());
            // The philosopher starts to eat
            Log(Thread.CurrentThread.Name + "\t is eating");
            rfork.ReleaseMutex();
            lfork.ReleaseMutex();
        }
        private void Log(string s)
```

```
        {
            sw.WriteLine(s);
        }
    }
```

Each philosopher is represented by a `Philosopher` object. The `Dinner()` method of the `Deadlock` class creates three mutexes for three forks. The `Eat()` method in the `Philosopher` class waits for the `lfork` mutex and then the `rfork` mutex that represents the two forks for a philosopher. Note that the philosopher always tries to grab the fork on the left. Thus, the assignment of `lfork` and `rfork` for each thread is done according to the seating layout. For example, John's `lfork` is Fork #2, and his `rfork` is Fork #0. A named thread is created for each `Philosopher` object to run the `Eat()` method. The main thread, which executes the `Dinner()` method, will wait for all the philosopher threads to finish. This is achieved by using the `Thead::Join()` method, which makes the calling thread wait for the completeness of the thread object.

We add the `Thread.Sleep()` call in the `Eat()` method to produce the deadlock scenario in which all three threads have sufficient time to complete the `lfork.WaitOne()` call and are waiting at the `rfork.WaitOne()` call. Without this instrumental trick, we may never see deadlock happen because the thread can quickly obtain two mutexes and finish very rapidly.

When deadlock occurs, you have to use the debugger to break the application (Debug➪Break All) if the program runs within Visual Studio 2005. If the program was launched on the device or on the emulator, you can attach the debugger to the process (after the registry key `AttachEnabled` is enabled; see the previous section for details) and then break the process. Figure 11-4 shows the Attach to Process dialog box that appears when you select Debug➪Attach to Process. The application to attach is named Chap11Threading, the process of the running assembly on the emulator. If the program is deadlocked, you need to break the application by selecting Debug➪Break All.

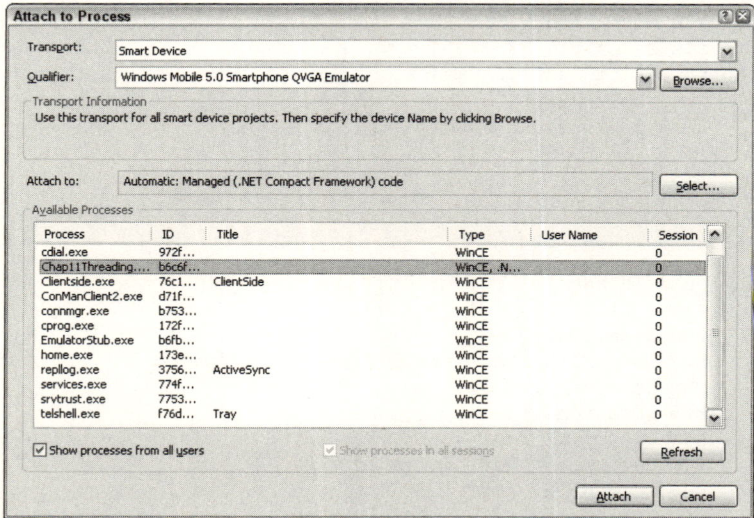

Figure 11-4

Then you can use the Threads window (Debug➪Windows➪Threads) to view the threads information, as shown in Figure 11-5. Four threads are currently running at the time we break the application: the main thread and the other three "philosopher" threads (created by the main thread). Each thread has an `ID` property and a `Name` property. The "philosopher" threads have been named with the philosopher's name. Right-click a thread and choose "Switch to thread" to see in the code window at which statement the thread is executing. Referring to the preceding code, in this example you can see that each thread waits at the same statement, `rfork.WaitOne()`, when you switch to each thread.

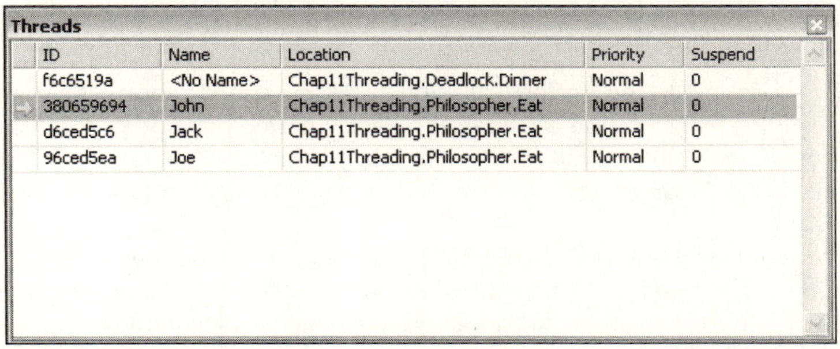

Threads

ID	Name	Location	Priority	Suspend
f6c6519a	<No Name>	Chap11Threading.Deadlock.Dinner	Normal	0
380659694	John	Chap11Threading.Philosopher.Eat	Normal	0
d6ced5c6	Jack	Chap11Threading.Philosopher.Eat	Normal	0
96ced5ea	Joe	Chap11Threading.Philosopher.Eat	Normal	0

Figure 11-5

One solution to the philosopher's dinner problem is to force the philosophers to always try the fork with a smaller number first. As shown in Figure 11-3, we know that John sits between Fork #2 and Fork #0, Jack sits between Fork #0 and Fork #1, and Joe sits between Fork #1 and Fork #2. Therefore, John always tries Fork #0 first; Jack tries Fork #0 first, and Joe tries Fork #1 first. By doing this, either John or Jack will be able to acquire Fork #0 but not both—one of them must wait on the first mutex he tried and wait for the other to finish, and other philosophers do not wait for him because he has no acquired mutex. Thus, the cyclic waiting condition of deadlock does not hold anymore. The change to the code is quite simple: just exchange the `lfork` and `rfork` mutex assignments for John, as the others' assignments are already following the "smaller mutex ID first" rule:

```
            // Philosopher John = new Philosopher(sw, forks[2], forks[0]);
            // Solution to the deadlock problem: let the philosopher try the
smaller fork ID first
            Philosopher John = new Philosopher(sw, forks[0], forks[2]);
            Philosopher Jack = new Philosopher(sw, forks[0], forks[1]);
            Philosopher Joe = new Philosopher(sw, forks[1], forks[2]);
```

The deadlock will not occur after this change. A sample output (in the file `philosopher.txt` on the storage card if a Smartphone is used or in the shared folder on the desktop machine if the Smartphone emulator is used) is shown in the following code. The hash codes of the three mutexes are `878385` (between John and Jack), `878386` (between Jack and Joe), and `878387` (between John and Joe). In this run, the eating sequence is Joe, John, and Jack. Depending on the timing, the eating sequence may vary over multiple runs.

```
Joe      acquired 878386
John     acquired 878385
Joe      acquired 878387
Joe      is eating
```

```
John     acquired 878387
John     is eating
Jack     acquired 878385
Jack     acquired 878386
Jack     is eating
Main thread exits.
```

Summary

A good computer program should perform as expected even under abnormal circumstances. In many cases, a developer's focus is on the "perfect" case where the major logic is being implemented. Understandably, many exceptional cases and errors may be completely ignored. This can be dangerous because programs can perform erroneously or crash when errors are not handled properly.

The .NET runtime provides a number of exceptions that will be raised when errors occur. As a Smartphone application developer, you need to identify the most likely exceptions in the code path and handle them programmatically. The basic programming language construct of exception handling — the `try-catch` block — can be easily embedded into exception-prone code. Unhandled exceptions will be taken care of by the .NET runtime. You should be aware of the overhead of stack unwinding when an exception is hunting for a handler.

As you know, when a program does not perform as it should, you can use the powerful Visual Studio debugger to dig into the execution of the code and pinpoint the problem. The debugger has been fully integrated with the Smartphone emulator and the device so that you can debug managed Windows Mobile code in the desktop Visual Studio 2005 environment. This chapter also covered multithreaded application debugging — including some cool features that the debugger offers to view thread execution status and control threads. Along with the discussion of multithreaded debugging, you were also introduced to key concepts such as race condition and deadlocks, as well as C#'s threading support.

Part III
Advanced Topics

12

Device and Application Security

This chapter and Chapter13 introduce the security features and security model in Windows Mobile 5.0. For software developers, it is not good enough simply to develop an application without considering security-related issues. Indeed, writing secure code and enhancing program security is not a bonus but a business requirement.

The security discussions presented in these two chapters apply to different type of applications you have learned so far: file I/O, database, networking, e-mail, etc. Although the topic of security is introduced later than those topics, it does not mean you should develop your application first and deal with security later. Research shows that the later you add security to your software development cycle, the more it will cost you.

This chapter discusses the following topics:

- Recent security threats and trends for mobile devices
- Security features supported in Windows Mobile 5.0
- Managing certificates and configuring security policy settings
- Enhancing device and application security in Windows Mobile 5.0 programmatically

Mobile Threats

The first mobile threat, Cabir.A, appeared in 2004 and soon spread to many countries. It was downloaded by many customers via Bluetooth. The virus was still in a primitive form when compared with its desktop counterparts. It wasn't long, however, before the threats grew. According to the reports released by McAfee, the number of viruses targeting Symbian OS as of 2006 increased to 120. And since the beginning of 2006, that number has increased by another 30 percent. To add some drama to the stories of mobile threats, a celebrity's Sidekick II cell phone was hacked in early 2005. Some private pictures taken with the phone's camera were stolen and posted on the web. If you think it won't happen to you because you aren't a celebrity, think again. Those hackers actually got inside the servers that save customers' private data, such as calendars, contacts, and pictures. They were capable of stealing any sensitive data from more than millions of customers! According to Mercer Management Consulting Research, the worm outbreaks on mobile devices in 2005 could infect 30 percent of the population. Not surprisingly, half of the mobile users surveyed in Japan would change service providers just to get better security.

The consequences of mobile threats are not negligible. In addition to end users suffering from lost privacy, they may be unable to communicate properly, especially in emergency situations. For a corporation, the threats are even more severe. When sensitive data is stolen, not only does a corporation lose its intellectual property, it also hurts the company's business reputation, devastates consumer confidences, and may incur severe financial crises.

It is time to face the brutal truth: Mobile devices are more prone to security threats than their desktop counterparts. Following are several contributing factors that make mobile devices more vulnerable:

- **Weak user authentication.** Most mobile devices do not require an interactive logon process. If some are equipped with Power-on-Password protection, the authentication is normally handled locally.
- **No security filesystem.** Most mobile operating systems currently do not include many security features in their file systems. You cannot audit which user accessed what file at what time. To make things even worse, some mobile OSs do not fully support advanced encryption, such as 128-bit DES and AES.
- **No role-based access control**. The design philosophy underlying a mobile device assumes a single-user scenario. Role-based access control is missing in many mobile devices. As a result, a user session cannot be established.
- **Lacking secured communications.** Mobile devices rely heavily on wireless communication technologies, such as CDMA, GSM, WiFi, InfraRed, and Bluetooth. Most of these wireless communication channels are not secure and are subject to eavesdropping.
- **Easily stolen or lost.** Mobile devices are portable, small, and lightweight, and it makes no sense to lock such a device in a room where physical access is strictly prohibited. If a device somehow falls into the wrong hands, all the sensitive data saved on that device is stolen as well.

Microsoft .NET Compact Framework 2.0 has beefed up its support for security, but it still has several key limitations. First, the .NET Compact Framework assumes an open platform and grants full trust to all code. Second, the .NET Compact Framework does not support Code Access Security (CAS), Microsoft's solution for restricting the operations an application can execute if the application is not signed with trusted certificates. You will learn more about certificates and trust in the next section. Finally, the .NET Compact Framework does not provide role-based security; therefore, you cannot use the security permission objects that are available in the full .NET Framework.

Dealing with mobile threats is not an easy task. Generally, you should apply not only software-based solutions, such as a security policy, encryption, and so on, but also some hardware-based security solutions, such as a biometric device that can authenticate users during the power-on phase. More important, you should inform end users about the threats and educate them about how to better defend themselves.

Glossary of Terms

Microsoft has defined a number of terms to describe the device and application security features in Windows Mobile 5.0 development and deployment. Having a good understanding of terms will greatly help you to develop and ship Smartphone applications with enhanced security.

Digital Signatures, Certificates, and Application Signing

When you package and deploy your application, it is critical to assure your users that the code distributed to them indeed came from you and has not been tampered with after it was published. The industry-standard solution to this problem is to include developers' information into the code. The information you want to add to the code must be able to identify you. This code, termed a *digital signature*, can be created using a public-key algorithm. The process of adding a digital signature to your code is called *application signing*.

In public-key cryptography, an entity (a person, computer, mobile device, etc.) has two keys: a public key, which is publicly available to everyone, and a private key, which is known only to the owner. Well-known PKI (public key infrastructure) encryption algorithms ensure that if some data in a message is encrypted using one of the two keys, only the entity holding the other key can decrypt it. Thus, the public-key pair can be used to check message authentication and integrity . The operation of signing uses one's private key to encrypt a hash code of the data, commonly known as *message digest*, produced by a one-way hash function. The signature and the original data are sent to the receiver, who will basically perform the same operations: use the hash algorithm (agreed on beforehand) to produce a hash code of the received data, use the sender's public key to decrypt the digital signature, and compare the result with the hash code just generated. If they match, then the data is indeed from the sender and has not been tampered with.

Note, however, a problem with the aforementioned scenario: How can the receiver obtain the genuine public key of the sender? In addition, how does the receiver map a public key to the correct identity? Most important, how can this entire procedure be automated so that it can be done completely transparently to a user (so that a user will not need to access some website to download a public key)? *Digital certificates* are designed to solve this problem. A generally trusted certification authority verifies the identity of an entity and issues a digital certificate as proof so that others can trust the certified entity. A digital certificate contains the public key of the entity, its identity, the expiration time, and the hashing algorithms used.. A digital certificate can be transferred along with the signed data or via other means, and is also signed using the CA's private key. After verifying the certificate using the CA's public key, the receiver of the data can retrieve the sender's public key from the certificate, which is guaranteed to belong to the sender.

Certificates are usually verified not with one single CA, but with a hierarchy of CAs that are chained to a root CA. Certificate verification is performed along the chain toward the root CAs. A software provider or an individual can obtain and purchase an SPC (Software Publishing Certificate) from one or more CAs in order to make its products trusted by users.

In short, a certificate is a certified digital signature. If you publish an application without signing it, that application is considered to be an *unsigned application*.

Privileged and Unprivileged Applications and Certificate Stores

Certificates saved on Smartphone devices are organized into certificate stores, the two most important of which are the *privileged certificate store* and the *unprivileged certificate store*. If an application is signed with a certificate that is saved in the privileged store, it is categorized as a *privileged application*. Conversely, if you sign an application with a certificate from an unprivileged certificate store, the application is referred to as an *unprivileged application*.

Note that a privileged certificate is not fundamentally different from an unprivileged certificate. The only difference is that the privileged certificate is saved in the privileged certificate store, whereas the unprivileged certificate is kept in the unprivileged certificate store. In addition, if an application is signed with a certificate that is not in either the privileged certificate store or the unprivileged store, then Windows Mobile 5.0 treats it as an *unsigned application*.

In addition to the privileged and unprivileged certificate stores, there are four other certificate stores. Table 12-1 summarizes all six certificate stores on Windows Mobile 5.0 devices.

Table 12-1 Certificate Stores

Certificate Store	Description
Privileged Execution Trust Authorities	Privileged certificates are saved in this store.
Unprivileged Execution Trust Authorities	Unprivileged certificates are saved in this store.
SPC	Contains Software Publishing Certificates (SPCs) for signing cabinet files.
Root	Contains root certificates and appears in the Certificates applet of a Windows Mobile 5.0 Smartphone as "Root".
CA	Contains certificates obtained from other certificate authorities.
MY	Stores certificates for an end user's personal use, and appears in the Certificates applet of a Windows Mobile 5.0 Smartphone as "Personal".

Trusted and Normal Applications

At runtime, *trusted applications* in Windows Mobile 5.0 can write all the registry keys and call all the system APIs. Conversely, *normal applications*, also termed *untrusted applications*, are barred from accessing certain system APIs and are not allowed to write certain registry keys. Those restricted registry keys and their subkeys are listed as follows:

```
HKEY_CURRENT_USER\Security
HKEY_LOCAL_MACHINE\Comm
HKEY_LOCAL_MACHINE\Drivers
```

```
HKEY_LOCAL_MACHINE\HARDWARE
HKEY_LOCAL_MACHINE\Init
HKEY_LOCAL_MACHINE\Loader
HKEY_LOCAL_MACHINE\Security
HKEY_LOCAL_MACHINE\Services
HKEY_LOCAL_MACHINE\SYSTEM
HKEY_LOCAL_MACHINE\WDMDrivers
```

Generally speaking, APIs that need to access filesystem security, database security, and user authentication are all barred from accessing it when run from normal applications. For a full list of protected system APIs, refer to the Windows Mobile 5.0 SDK document or MSDN.

"Privileged application" and "unprivileged application" are the terms used to describe what type of certificates are used to sign the application, whereas "trusted applications" and "normal applications" are characterized by what they can do during runtime.

Security Policies and Roles

Windows Mobile 5.0 has defined a number of security policy settings that enable you to specify how security is enforced on a Smartphone device. For example, if you want to prohibit an unsigned application from running on a Smartphone device, you can assign a value of 0 to the Unsigned Application Policy setting.

Of course, you don't want everyone to be able to modify those security policy settings. In Windows Mobile 5.0, security roles are defined to determine what security policy settings and what Smartphone resources one can access. *Security role* is a logical term to categorize how physical users are related to the device. Table 12-2 lists some common security roles in Windows Mobile 5.0 for Smartphone.

Table 12-2 Common Security Roles

Role	Decimal Value	Description
SECROLE_NONE	0	The message is not assigned by any security role.
SECROLE_OEM	2	OEM role. By default, this security role cannot change security settings.
SECROLE_OPERATOR	4	Mobile operator role.
SECROLE_MANAGER	8	Manager role. It is the highest level of all the security roles and can access all the security settings.
SECROLE_USER_AUTH	16	User authenticated role. It is assigned to the PIN-signed WAP push message and Remote API (RAPI).
SECROLE_USER_UNAUTH	64	User unauthenticated role. It is assigned to the unsigned WAP push message.
SECROLE_OPERATOR_TPS	128	Trusted provisioning server role. It is assigned to WAP messages that come from an authenticated push initiator.

Security policies in Windows Mobile 5.0 include the policy ID, the policy value, and the required security role. For instance, the policy ID of the Unsigned Application Policy is 4102. This policy is associated with SECROLE_MANAGER, which means only the manager role can modify this setting through an OTA message. The default value of this policy is 1, which indicates that unsigned applications are allowed to run on the device. Any value other than 1 is treated the same as 0 and will prohibit unsigned applications from running on the device.

For more information, refer to the "Security Policies" section.

Windows Mobile 5.0 Security Models

Two security models are available for Windows Mobile 5.0 Smartphone devices: a one-tier model and a two-tier model. Both models are also available in earlier platforms such as Smartphone 2002 and 2003. A Smartphone device is pre-built with either one of the security models and you cannot "flash" a one-tier device to a two-tier device.

The one-tier security model determines whether a Windows Mobile application is allowed to run by examining the certificates of the application and the device policy settings. Figure 12-1 illustrates the flowchart of this process.

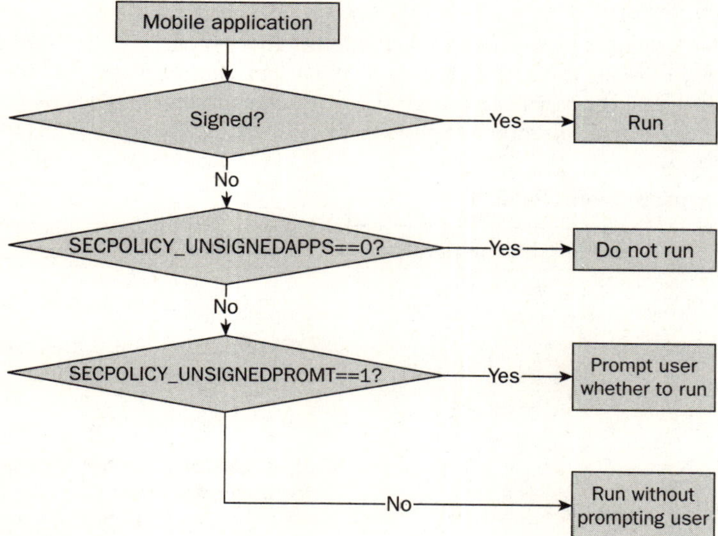

Figure 12-1

In the one-tier model, an application can run as long as it is signed with a certificate. For unsigned applications, the SECPOLICY_UNSIGNEDAPPS security policy setting is consulted to determine whether the application is allowed to run. If the policy permits unsigned applications to run, another policy setting, SECPOLICY_UNSIGNEDPROMPT, will determine whether to prompt users. For one-tier devices, applications always run in privileged mode, which means applications have full access to the devices, including restricted APIs and protected registry settings.

The two-tier security model introduces the normal execution mode into the system. This process is illustrated in Figure 12-2. An application is first checked to determine whether it is signed with a privileged certificate. It is considered a trusted application and can run in privileged mode only if the application is signed with a privileged certificate. Likewise, an application can run in normal mode if the application is signed with an unprivileged certificate. If an application is not signed, the two-tier security model goes through the same process as the one-tier model: It checks the security policy settings and determines both whether it is allowed to run and whether users are prompted before execution.

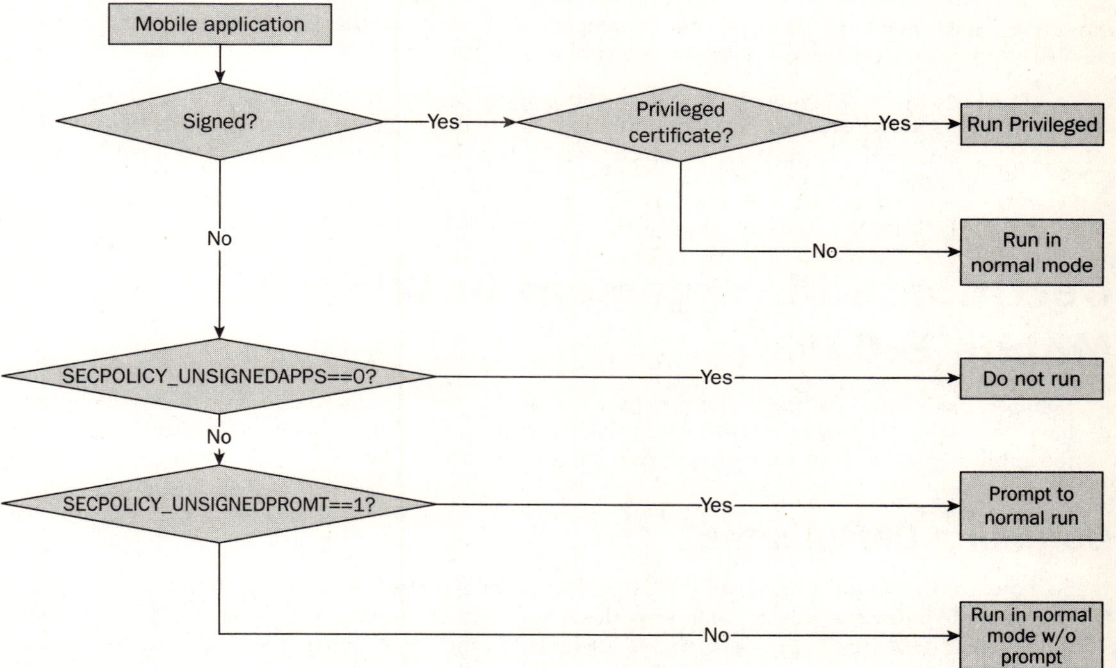

Figure 12-2

The security model and security policy settings together determine how applications are executed on a Smartphone device. Common configurations are as follows:

- **Security-Off.** By turning off corresponding security policies, an application is not required to have a certificate to run on a Smartphone device. You can use this setting to determine whether the security settings conflict with your application. In practice, the Security-Off configuration is not recommended because an attacker can easily hack into your device and take control of it.
- **One-Tier-Prompt.** In One-Tier Prompt mode, users are prompted whether to execute an unsigned application. Users have the power to say no if they suspect some applications are fishy. This certainly reduces the chances of being hacked by unidentified or unknown applications. In addition, many Windows Mobile software vendors are reluctant to sign their applications, either for marketing purposes or simply because they do not want to spend the extra time and money. With the One-Tier Prompt setting, users can still install and run these applications with ease.

- **Two-Tier-Prompt.** As with One-Tier Prompt mode, users have control over unsigned applications. Unlike One-Tier Prompt mode, however, those unsigned applications can be executed only in normal mode and therefore have no access to privileged APIs and protected registry settings. Two-Tier Prompt mode is an ideal security configuration for Smartphone devices for personal use because of the compromise between software compatibility and security.
- **Third-Party-Signed.** This mode enforces strong security policies. Applications must be signed with a valid certificate in order to run. This configuration prevents Smartphone devices from being attacked by anonymous applications and enables computer forensic investigations. It is safer than the Two-Tier Prompt mode, although users may have trouble with certificates and might be unable to operate certain applications properly.
- **Locked.** As the name suggested, the Locked configuration will prevent any third-party applications from installing. Such configurations normally target Smartphone devices for special industrial or business purposes in which software updates and system maintenance can be obtained only through device vendors.

Certificate Management in Windows Mobile 5.0

Both the one-tier and two-tier models require that applications be checked to determine whether they have been signed with valid certificates. This section describes where to obtain certificates, how to sign the applications, and how to manage those certificates.

Obtaining Certificates

So, how do you obtain certificates? For day-to-day development, you do not need to make any purchases; the Windows Mobile 5.0 SDK provides several sample certificate keys. Typically, they are stored in the folder `C:\Progam Files\Windows CE Tools\wce500\Windows Mobile 5.0 Smartphone SDK\Tools`. The `SDKSamplePrivDeveloper.pfx` certificate is particularly useful. By signing an application with this certificate, you can run the application in privileged mode, which means you can access all the system APIs and restricted registry settings.

Note that you should not ship those test-only certificates to end users. Be sure to remove those certificates from the certificate store.

How, then, to obtain certificates that can deploy your applications to Smartphone devices? You can certainly pay and obtain the certificates from various Certificates Authorities, such as GeoTrust or VeriSign. However, we have noticed many developers complaining that such certificates are either not recognized or not valid on certain devices.

To avoid any possible troubles this may incur, it is probably better to obtain certificates through Microsoft's Mobile2Market program (`http://msdn.microsoft.com/mobility/windowsmobile/partners/mobile2market/default.aspx`). Mobile2Market partners provide certificate authority specifically for Windows Mobile devices. In addition to obtaining certificates to sign your application, if you are willing to pay more, your application logo can be certified. This is *not* required to deploy your application but may be advantageous for marketing purposes. (Note that we are not necessarily advocating the Mobile2Market program; we just want you to be aware that getting the proper certificates can be a tricky process. You should certainly research whether the certificates can be deployed to the targeted Smartphone devices beforehand.)

Signing Applications with Certificates

There are two ways you can sign Windows Mobile Smartphone applications. The first way is to sign your application during the development phase through the Visual Studio 2005 IDE. First, select Project from the main menu, open the Properties of your current project, and click the Devices tab (see Figure 12-3). Then check the "Sign the project output with this certificate" option, which will enable the Select Certificate button. Click the Select Certificate button and then choose the desired certificates from the resulting dialog box. If your certificates are not present, you can click the Manage Certificates button to search for a certificate, as shown in Figure 12-4. The Manage Certificates window enables you to view detailed information about existing certificates.

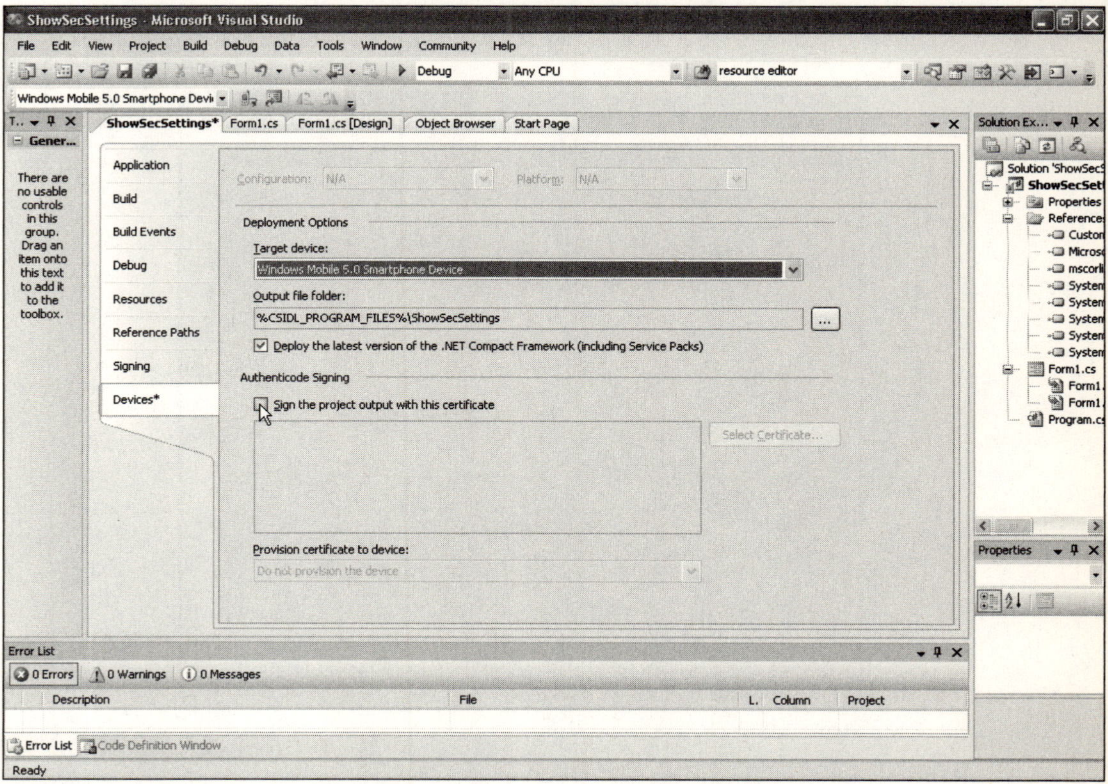

Figure 12-3

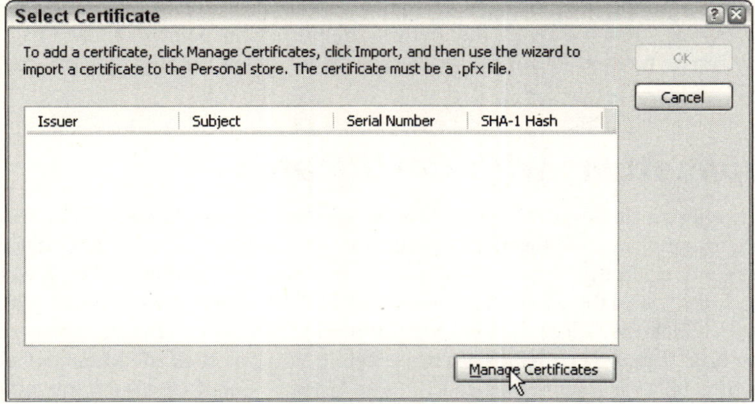

Figure 12-4

You can also import a certificate into the certificate stores on your PC, as shown in Figure 12-5.

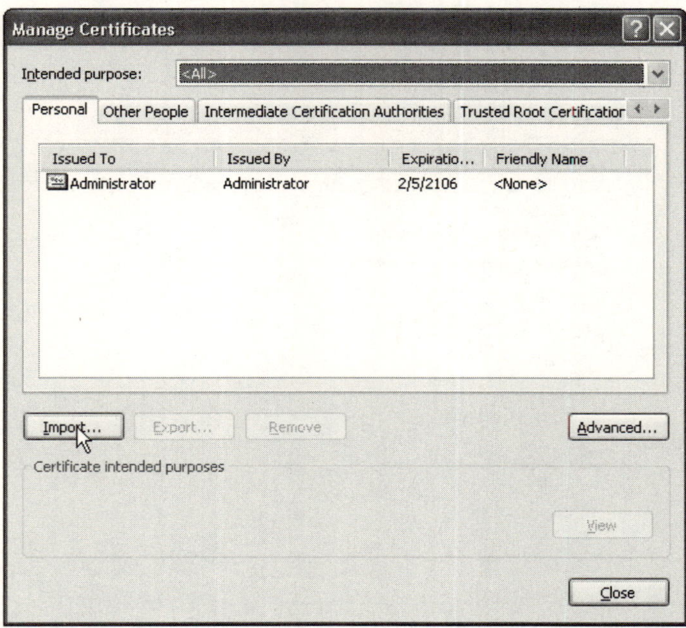

Figure 12-5

Click the Import button and select the certificates you wanted to sign. During this process, you may need to type in the password of the certificates. Generally, certificates purchased from a vendor are password protected. For certificates exported from the certificate stores or created using tools such as MakeCert.exe, you have the option to make them more secure with password protection, or easy to use without password protection. The Certificate Import Wizard, shown in Figure 12-6, asks you where to store the certificates. In most cases, you can simply let the wizard find a proper place for you automatically.

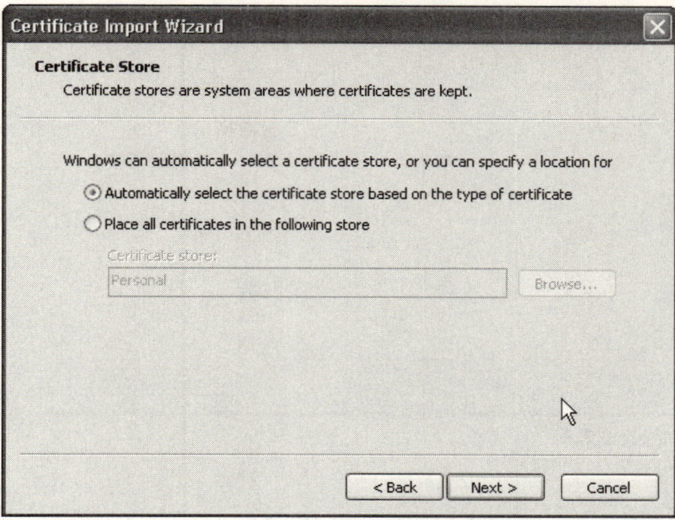

Figure 12-6

Once the certificate is imported, the Manage Certificates window will reappear with a new certificate shown in the window (see Figure 12-7). Now close the Manage Certificates window. The Select Certificate window appears again. This time, a certificate is available in the selection list to enable you to sign your application (see Figure 12-8).

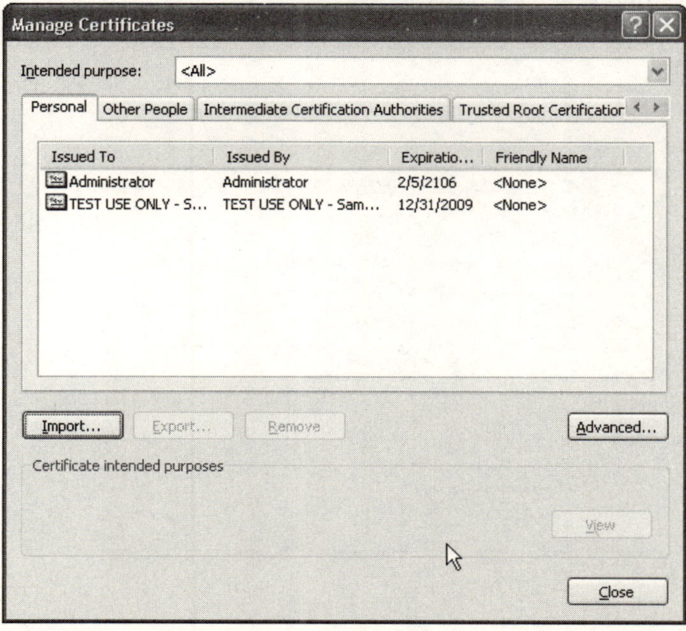

Figure 12-7

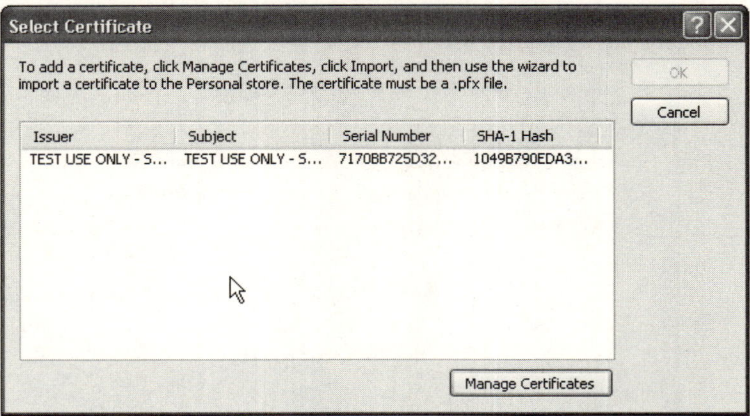

Figure 12-8

Once you have selected the certificate to sign your application, detailed information about the certificate is shown in the project's properties window, as shown in Figure 12-9. Note that the drop-down list on the bottom of the Devices tab enables you to specify how the certificate is provisioned to the device. The three options are as follows:

- Do not provision the device.
- Add the certificate to the privileged store.
- Add the certificate to the unprivileged store.

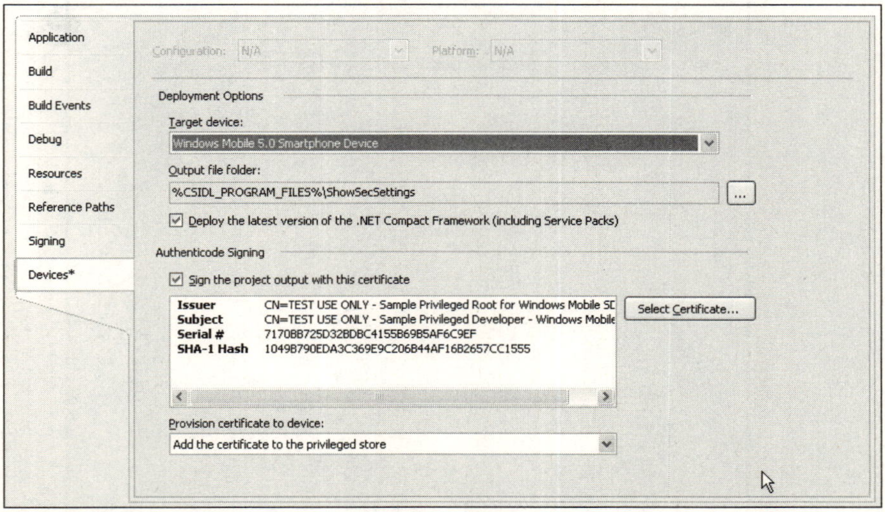

Figure 12-9

After selecting the certificate, you can leave the rest to Microsoft Visual Studio 2005; your application will be signed automatically during the build process.

Alternatively, you can sign your cabinet files with the .NET Compact Framework command utility `Signtool`. In Visual Studio 2005, by default, this tool can be found in both the `common7\tools\bin` directory and the `SDK\v2.0\Bin` directory of your Visual Studio 2005 folder. Note that you need to sign all the binary executable files. The typical syntax for using this command is as follows:

```
Signtool sign /f SDKSamplePrivDeveloper.pfx *.exe *.dll
```

After signing the binaries, you can build your cabinet file with these signed executable files and then sign the cabinet file again with the `Signtool` command, as follows:

```
Signtool sign /f SDKSamplePrivDeveloper.pfx sampleCabfile.cab
```

This command signs the cabinet file sampleCabfile.cab with the certificate `SDKSamplePrivDeveloper.pfx`.

Regardless of which tool you use to sign your application, you should always pick `.pfx` certificates rather than `.cer` certificates to sign your application. This is because in PKI architecture, files are signed with the private key and decrypted with the public key. The `.pfx` certificates contain both private and public key rings, whereas `.cer` certificates have only public keys. The differences between `cer` certificates and `pfx` certificates can be further illustrated by examining how they are created.

Visual Studio 2005 comes with two tools to help you create your own test certificate. `MakeCert.exe` is the application to use to create your public key and private key and save them separately to two files. The following command will create a private key and a public key for the common name `MyOrg` (the private key is saved as `MyPrivKey.pvk`, whereas the public key is stored as `MyPubKey.cer`):

```
MakeCert -sv MyPrivKey.pvk -n "CN=MyOrg" MyPubKey.cer
```

When the `.pvk` file and the `.cer` certificate are both available, you can create a `.pfx`file. The tool that enables this conversion is `Pvk2pfx.exe` in Visual studio 2005. The following command creates the `.pfx` certificate `MyPfx.pfx` from both the private key file `MyPrivKey.pvk` and the public key file `MyPubKey.cer`. The `.pfx` certificate `Mypfx.pfx` is also going to be protected by a password `MyPass`.

```
Pvk2pfx -pvk MyPrivKey.pvk -spc MyPubKey.cer -pfx MyPfx.pfx -PO MyPass
```

Managing Certificates

Most Windows Mobile 5.0 Smartphone devices also come with an applet that enables you to view the certificates. If you click Start➪Settings➪Security➪Choose Certificates, you can see two certificate stores, Personal and Root, as indicated in Figure 12-10. The Personal certificate store is the same as the MY certificate stores listed in Table 12-1.

The detailed information about a certificate, such as who issued it and when it expires, is also displayed on the screen, as shown in Figure 12-11.

Figure 12-10

Figure 12-11

This applet allows you to view certificates from two certificate stores only; it is not capable of installing or removing any certificates. You will need third-party tools to manage certificates of all certificate stores. In most of situations, these third-party applications need to be signed with privileged certificates.

Certificates can be loaded to or removed from Windows Mobile Smartphone in an XML-formatted document known as an *XML provisioning file*. For example, assuming the security role you obtained is sufficient to add a certificate to the privileged certificate store, the corresponding XML snippet is as follows:

```
<wap-provisioningdoc>
   <characteristic type="CertificateStore">
      <characteristic type="Privileged Execution Trust Authorities">
         <characteristic type="{hash of certificate}">
            <parm name="EncodedCertificate" value="{encoded hash of certificate}"/>
         </characteristic>
      </characteristic>
   </characteristic>
</wap-provisioningdoc>
```

In the preceding example, the XML file begins with `<wap-provisioningdoc>`, which indicates that the file is an XML device provisioning file. The first `characteristic` determines the configuration is about the certificate stores. The next `characteristic` indicates the configuration of the privileged certificate store, which is a subnode of the certificate store. Then a new certificate can be added to the privileged store by feeding the hash of the certificate and the encoded hash of the certificate.

To remove a certificate from a certificate store, you simply add `nocharacteristic` to the XML provisioning file. For example, the following XML code will remove a certificate from the SPC store based on the hash of that certificate:

```
<wap-provisioningdoc>
   <characteristic type="CertificateStore">
      <characteristic type="SPC">
         <nocharacteristic
         type="{hash of certificate}"/>
      </characteristic>
   </characteristic>
</wap-provisioningdoc>
```

Once the XML provisioning file is ready, you need to consider how to deliver it to the device. A typical method is to use RapiConfig and ActiveSync, as follows:

1. Cradle the targeting Smartphone device to a desktop.
2. Establish an ActiveSync connection to the desktop.
3. Launch the command-line window from the desktop and change the directory to the `Tools` folder of Windows Mobile 5.0 SDK.
4. Use the RapiConfig tool to deliver the file, as shown in the following example:

```
rapiconfig <provisioning.xml>
```

Note that if you are delivering the provisioning XML file through a CAB Provisioning Format (CPF) file, you must name the XML file `_setup.xml`.

Security Policies

Each security policy has a policy ID and a default value, and requires a security role in order to be modified. For example, the unsigned application policy is identified by its policy ID 4102. If the value is 1, then unsigned applications are allowed to run on the device; otherwise, if the value is not 1, which will be treated as 0, then unsigned applications are not allowed to run on the device. This policy is associated with security role SECROLE_MANAGER, which means only those who have manager role authority can change the value of this policy.

Table 12-3 summarizes the common security policies, default values, and corresponding security roles.

Table 12-3 Security Policies on Windows Mobile—Based Smartphones

Policy Setting	Policy ID	Description	Role Required to Modify Policy
Auto Run Policy	2	Indicates whether applications stored on a Multimedia Card (MMC) are allowed to run automatically when the card is inserted into the device 0 means allow 1 means restricted The default value is not defined.	SECROLE_ MANAGER
Grant Manager Policy	4119	Grants the system administrative privileges held by SECROLE_MANAGER to other security roles Possible values are: SECROLE_USER_AUTH SECROLE_NONE OPERATOS_TPS The default value is OPERATOS_TPS.	SECROLE_ MANAGER
Grant User Authenticated Policy	4120	Grants privileges held by SECROLE_ USER_AUTH to other security roles Possible values are: SECROLE_USER_AUTH SECROLE_USER_UNAUTH The default value is SECROLE_ USER_AUTH.	SECROLE_ USER_AUTH

Policy Setting	Policy ID	Description	Role Required to Modify Policy
Message Authentication Retry Number Policy	4105	Specifies the maximum number of times a user is allowed to try when authenticating with a WAP PIN-signed message Possible values are from 1 to 256. The default value is 3.	SECROLE_ MANAGER
Privileged Applications Policy	4123	Specifies whether a device is a one-tier device or a two-tier device Possible values are 0 and 1 0 indicates a two-tier device 1 indicates a one-tier device Any value other than 1 is treated as 0. The default value is 0.	SECROLE_ MANAGER
RAPI Policy	4097	Describes how Remote API (RAPI) through ActiveSync is handled. Possible values are 0, 1, or 2. 0 indicates that the ActiveSync service is shut down and therefore rejects RAPI. 1 indicates full access. RAPI calls can go through ActiveSync without restrictions. 2 indicates that access to ActiveSync is available to the User Authenticated security role. The default value is 2.	SECROLE_ MANAGER
Service Indication (SI) Policy	4109	Indicates whether a Smartphone accepts SI messages, which are sent to the device to notify users of service updates, new services, and provisioning services	SECROLE_ MANAGER
Service Loading (SL) Message Policy	4108	Determines what security roles are allowed to download new services or provision XML to the device The default value is SECROLE_PPG_ TRUSTED.	SECROLE_ MANAGER

Table continued on following page

Policy Setting	Policy ID	Description	Role Required to Modify Policy
Trusted WAP Proxy Policy	4121	Indicates the level of permissions required to create, modify, or delete a trusted proxy The default value is SECROLE_OPERATOR | SECROLE_OPERATOR_TPS | SECROLE_MANAGER.	SECROLE_ MANAGER
Unsigned Applications Policy	4102	Indicates whether unsigned applications are allowed to run on a Windows Mobile–based device 0 indicates that unsigned applications are not allowed to run. 1 indicates that unsigned applications are allowed to run. Any value other than 1 is treated as 0. The default value is 2	SECROLE_ MANAGER
Unsigned CABS Policy	4101	Indicates whether unsigned .cab files can be installed on the device 0 means no unsigned .cab files can be installed. The default value is SECROLE_USER_AUTH, which indicates that unsigned .cab files will be installed if the security role is SECROLE_USER_AUTH.	SECROLE_ MANAGER
Unsigned Prompt Policy	4122	Indicates whether a user is prompted to accept unsigned applications 0 means the user will be prompted. 1 means the user will not be prompted. Any value other than 1 is treated as 0. The default value is 0.	SECROLE_ MANAGER

Before configuring security policy settings, it is recommended that you first query the current settings. Then you can make the appropriate changes based on your needs. Both querying and configuring require you to compose an XML provisioning file. The following is an example of a query XML file:

```
<wap-provisioningdoc>
    <characteristic type="SecurityPolicy">
        <!--  query security ID 4122 -->
        <parm-query name="4122"/>
    </characteristic>
</wap-provisioningdoc>
```

This XML snippet will query the current value of policy ID 4122, the unsigned prompt policy.

Similarly, the following XML snippet sets the unsigned application policy value to `1`:

```
<wap-provisioningdoc>
    <characteristic type="SecurityPolicy">
        <!--  Allow unsigned apps to run  -->
        <parm name="4102" value="1" />
    </characteristic>
</wap-provisioningdoc>
```

The next bit of XML code will remove policy ID 4109 from the policy settings:

```
<wap-provisioningdoc>
    <characteristic type="SecurityPolicy">
        <noparm name="4109"/>
    </characteristic>
</wap-provisioningdoc>
```

An Example Code

This section puts together what you've learned so far and presents you with an example application. You will learn how to sign an application and use APIs in the `Microsoft.WindowsMobile.Configuration` namespace.

The `Microsoft.WindowsMobile.Configuration` namespace is one of the new namespaces that is introduced in Windows Mobile 5.0. It currently has only one `ConfigurationManage` class, which enables you to load the configuration XML file to Windows Mobile devices. The key method of this class is `ProcessConfiguration`, which sends the configuration XML file to the device for processing. The syntax of this method is as follows:

```
Public static XmlDocument ProcessConfiguration (XmlDocument configDoc, bool
metadata);
```

`configDoc` is the provisioning XML configuration document. If `metadata` is true, the method returns not only the original device configuration XML file, but also the processing errors in XML format. If `metadata` is false, only the original device configuration XML file is returned.

Note that by default the `Microsoft.WindowsMobile.Configuration` namespace is not visible to Windows Mobile 5.0 device applications. You need to add a reference to the namespace in your project. To add a reference in Visual Studio 2005, click Project⇨Add Reference, and select `Microsoft.WindowsMobile.Configuration` from the .NET tab.

The function of the sample code is to inform users which security configuration is currently used in a Smartphone device — that is, Security-Off, One-Tier-Prompt, Two-Tier-Prompt, Third-Party-Signed, or Locked.

Because the application touches some restricted registry settings, it needs to be signed with a privileged certificate. For testing purpose, you can use the privileged certificate SDKSamplePrivDeveloper.pfx, which comes with the SDK. The path of this certificate is usually `Program Files\Windows CE Tools\wce500\Windows Mobile 5.0 Smartphone SDK\Tools`.

When importing this certificate (following the steps described earlier in the section "Signing Applications with Certificates"), you are prompted to enter a password. For the sample certificates, just use the empty password and click Next, as shown in Figure 12-12.

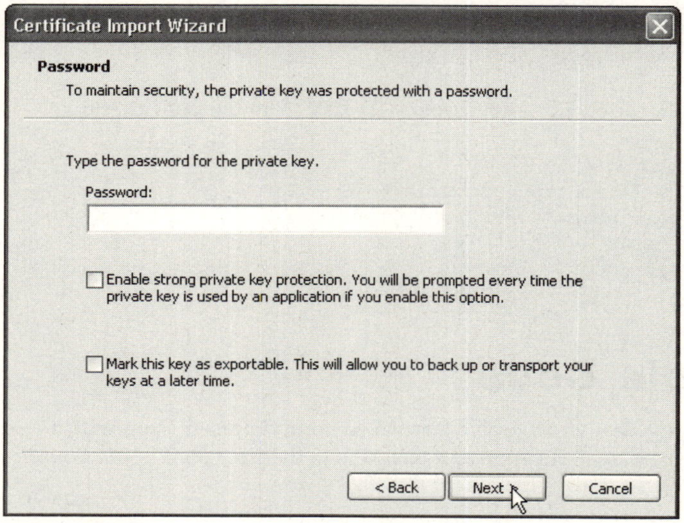

Figure 12-12

To begin, an XML provisioning file is needed to query the values of Unsigned Application Policy, Unsigned Prompt Policy, and Privileged Applications Policy. Their corresponding policy IDs are 4102, 4122, and 4123, respectively. The following XML code illustrates how to perform the query:

```
<wap-provisioningdoc>
  <characteristic type="SecurityPolicy">
    <parm-query name="4102"/>
    <parm-query name="4122"/>
    <parm-query name="4123"/>
  </characteristic>
</wap-provisioningdoc>";
```

To perform the same query in your program, a typical solution is to feed the content of the XML provisioning file to a string and construct an `XmlDocument` object from the string. The following code snippet shows how this can be done with C#:

```
            XmlDocument xmlConf = new XmlDocument();

            string strQuery = "<wap-provisioningdoc><characteristic
type=\"SecurityPolicy\">" +
                "<parm-query name=\"4102\"/><parm-query name=\"4122\"/><parm-query
name=\"4123\"/>"+
```

```
                "</characteristic></wap-provisioningdoc>";

            xmlConf.LoadXml(strQuery);
```

After the provisioning `XmlDocument xmlconf` is created, you can feed it to the `ProcessConfiguration()` method of the `ConfigurationManager` class to process the query, as follows:

```
            //Retrieve security settings
            XmlDocument xmlRslt =
    ConfigurationManager.ProcessConfiguration(xmlConf, false);
```

Once the query results are returned, you can parse the resulting XML document and retrieve the value of each policy ID. Based on the values of policy IDs, the security configurations can be mapped out, as indicated in Table 12-4.

Table 12-4 Security Configurations and Related Security Policies

	Unsigned Application Policy (4102)	Unsigned Prompt Policy (4122)	Privileged Application Policy (4123)
Two-Tier Model with Prompt On	1	0	0
Two-Tier Model with Prompt Off	1	1	0
Third-Party-Signed or Locked	0	1	0
One-Tier Model with Prompt On	1	0	1
One-Tier Model with Prompt Off	1	1	1

Following is the code for this example application:

```
using System;
using System.Collections.Generic;
using System.ComponentModel;
using System.Data;
using System.Drawing;
using System.Text;
using System.Windows.Forms;

using System.Xml;
//Need to add reference: Microsoft.WindowsMobile.Configuration
using Microsoft.WindowsMobile.Configuration;

namespace ShowSecSettings
{
    public partial class Form1 : Form
    {
        public Form1()
        {
```

```
            InitializeComponent();
        }

        // Read the security configuration
        private void Form1_Load(object sender, EventArgs e)
        {

            XmlDocument xmlConf = new XmlDocument();

            // Generate an XML query string
            // Policy ID 4102: Unsigned Applications Policy
            // Policy ID 4122: Unsigned Prompt Policy
            // Policy ID 4123: Privileged Applications Policy

            string strQuery = "<wap-provisioningdoc><characteristic
type=\"SecurityPolicy\">" +
                "<parm-query name=\"4102\"/><parm-query name=\"4122\"/><parm-query
name=\"4123\"/>"+
                "</characteristic></wap-provisioningdoc>";
            xmlConf.LoadXml(strQuery);

            //Retrieve security settings
            XmlDocument xmlRslt =
ConfigurationManager.ProcessConfiguration(xmlConf, false);

            XmlNodeList settingList = xmlRslt.FirstChild.FirstChild.ChildNodes;
            int unsigned =
Convert.ToInt32(settingList[0].Attributes["value"].Value);
            int prompt   =
Convert.ToInt32(settingList[1].Attributes["value"].Value);
            int tier     =
Convert.ToInt32(settingList[2].Attributes["value"].Value);

            string settingMsg = "not one of those predefined configurations";

            switch (tier) {
                case 0:
                    if (unsigned == 1 && prompt == 0)
                    {
                        settingMsg = "Two Tier with Prompt on";
                    }
                    else if ( unsigned == 1 && prompt == 1)
                    {
                        settingMsg = "Two Tier with prompt off";
                    }
                    else if ( unsigned == 0 && prompt == 1)
                    {
                        settingMsg = "Third Party Signed or Locked";
                    }
                    break;
                 case 1:
                    if (unsigned == 1 && prompt == 0)
                    {
                        settingMsg = "One Tier with Prompt on";
```

```
                }
                else if (unsigned == 1 && prompt == 1)
                {
                    settingMsg = "One Tier with Prompt off";
                }
                break;
            default:
                break;
        }

        MessageBox.Show("The current security setting is " + settingMsg + ".");
    }

    private void menuItem1_Click(object sender, EventArgs e)
    {
        Application.Exit();
    }
  }
}
```

Figure 12-13 shows the running result of this example application.

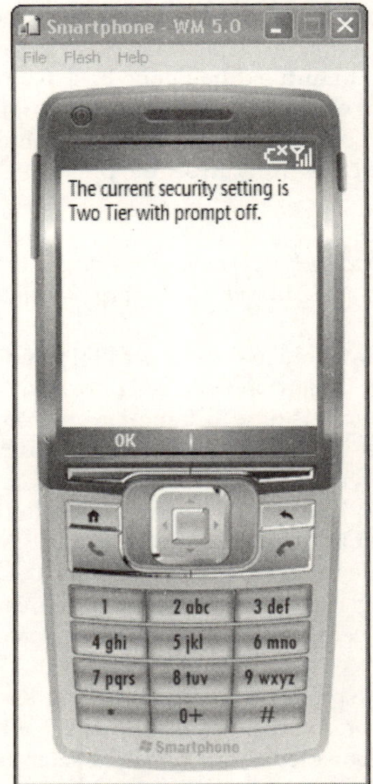

Figure 12-13

Perimeter Security

Perimeter security controls who and what application can access a mobile device, either physically or through remote access via ActiveSync or an over-the-air (OTA) message. This section briefly discusses several key features that can be used to enhance perimeter security.

Physical Access Control

Physical access control is a puzzle in the mobile community. There is really no guaranteed physical policy that can prevent an intruder from accessing the device, unless you are not using it. Therefore, a better approach is to find ways to prevent intruders from accessing the information saved on your device or spreading malicious programs on your device.

Passwords are one popular solution. They are easy to implement and can successfully block amateur intruders. On mobile devices, there are several different passwords or PINs.

Power-on-Password (POP) protection is a security feature that can prevent unauthorized users from accessing mobile devices. However, this feature is not available on a Windows Mobile–based Smartphone because it does not really help much with security. How could you receive phone calls if the device remained at the powered-off stage? Conversely, if it is left powered on all the time, then an unauthorized user can easily access all the applications and data without the need to input a password.

A more realistic security feature is to turn on the Phone Lock feature, which will lock the device when it is not in use after a certain amount of time. Users are then prompted to input the password to unlock the device. To turn on the Device Lock feature on a Windows Mobile–based Smartphone, click Start⇨Settings⇨Security⇨Enable Phone Lock. You can set the phone lock timer with your password, as shown in Figure 12-14.

Note that Phone Lock is not Keypad Lock. Keypad Lock is enforced to prevent making calls in case buttons are accidentally pressed. It should not be considered a security feature because it is very easy to bypass.

SIM PIN protection requires a user to key in the correct PIN to unlock the SIM card. A SIM (subscriber identity module) card serves as an ID card in GSM cell phones. A user's personal information, phone number, and contacts are all kept on this tiny, rectangular card. An obvious advantage of using a SIM card is that a user's phone number is not restricted to a certain device. Because a SIM card contains private information, a PIN is required to access the information stored on the card. To further enhance the access security of a SIM card, many mobile service providers also enable the SIM PIN lockout policy. For instance, if the wrong PIN is keyed in three times in a row, the SIM card is temporarily locked. A user will need to call the service provider to get the code to unlock the SIM card. Continuing to try a PIN on a temporarily locked SIM card may cause the card to be permanently disabled. This lockout feature can prevent private information from being stolen through brute-force attacks.

Finally, biometrics is another way to protect who can access the device. For example, some mobile devices provide a fingerprint scanner. A user is authenticated only if his or her fingerprints match the record saved on the device. Fingerprints, if implemented correctly, can be safer than passwords or PINs because they are unique from person to person and users do not have to remember passwords. However, some reports indicate that current fingerprinting scanners are far from being secure. One Japanese researcher has claimed he can hack a thumb scanner in less an hour with materials costing only $10. Similar implementation failures also occur in other biometric systems, such as face and voice recognition systems and iris scanners.

Figure 12-14

Antivirus Considerations

As previously mentioned, the number of mobile viruses is increasing at a dramatic rate. If possible, third-party antivirus applications, such as BullGuard Mobile Anti-virus (`www.bullguard.com/mobile`), Airscanner (`www.airscanner.com`), or McAfee VirusScan Mobile (`www.mcafee.com`) should be installed to protect your device.

In the Windows Mobile 5.0 security architecture, the two-tier security model greatly helps protect mobile devices from being infected by viruses or malicious code, because those applications are unlikely to be signed with privileged certificates. However, viruses are still able to get into your mobile devices through other channels, such as an external storage card and Remote API through ActiveSync.

Viruses and malicious code saved on an external card, such as MMC or a flash card, if executed, can copy themselves to the system memory of the device. To reduce mobile threats from external storage, you can leverage the Auto Run security policy and set the policy value so that the executables on an external card are not allowed to launch by themselves.

As indicated in Table 12-3, the Auto Run Policy has an ID of 2, and the default value is not defined. This could be a security hole for a mobile device. You can set the value of this policy to 1, which disallows an application on an external devices from running automatically. The following is a sample XML provisioning document:

```
<wap-provisioningdoc>
    <characteristic type="SecurityPolicy">
        <!--  Allow auto run  -->
        <parm name="2" value="1" />
    </characteristic>
</wap-provisioningdoc>
```

Remote APIs (RAPI) allow applications to launch from the desktop but are executed on mobile devices through Microsoft ActiveSync connections. The RAPI security policy defined in Windows Mobile 5.0 can help you restrict the access from RAPI. As listed in Table 12-3, the policy ID of the RAPI security policy is 4097. If the value of the RAPI security policy is set to 0, RAPI is working in closed mode and all RAPIs are disabled and access to ActiveSync is disabled. If the value is set to 1, the RAPI is working in open mode and full access is granted to RAPI calls and access to ActiveSync. This is not a recommended setting because intruders can upload unsigned DLL files to the mobile device and invoke them to run in privileged mode. The default setting is 2, which causes RAPI to work in restricted mode, whereby only signed applications can be uploaded to the device.

Summary

In this chapter, you have learned about two types of security models in Windows Mobile–based Smartphone devices: the one-tier model and the two-tier model. In the one-tier model, an unsigned application has the potential to access all system APIs and registry keys, posing a severe security threat. In the two-tier model, however, only applications signed with privileged certificates can access protected system APIs and restricted registry keys. This helps protect the device better because if an unsigned application is allowed to run, it can only run as normal mode and would cause less damage to the device and operating system.

Security policies are a set of policy settings that can customize the security behavior of a mobile device. Developers can code an XML provisioning file to query or change the security settings. The XML file can be uploaded to the device through ActiveSync. Alternatively, you can use the methods of the `ConfigurationManager` class in the `Microsoft.WindowsMobile.Configuration` namespace to query and update system settings. Changing security policies often requires the manager security role. For developers, it means your application must be signed with a privileged certificate to change the security policy settings.

A number of technologies can help protect perimeter security, such as various passwords and PINs, Phone Lock, and biometric recognition systems. Developers should be aware of the security attacks coming from storage cards and RAPI through ActiveSync. To better protect perimeter security, you should disable the Auto Run policy of a storage card and set the RAPI policy to restricted mode.

The security features discussed in this chapter mainly target device security and application security. Applying the technique you have learned in this chapter, mobile devices are less likely to be hit by viruses or malicious code, and your applications are more likely to be accepted by end users.

In next chapter, you will learn how to enhance data and communications security to prevent sensitive data from being lost or stolen.

13

Data and Communication Security

Chapter 12 described various methods to secure Smartphone devices and applications using application signing and policy settings. Those security measures, however, are simply not enough to prevent data theft via eavesdropping. You should certainly apply other security techniques to protect sensitive data.

This chapter introduces several key methods to protect local data stored in Smartphone devices and to protect the data exchange between Smartphone devices and servers. Encryption is widely used to protect sensitive data. If data saved on your local device is encrypted, intruders will have a hard time decoding the message even if they physically possess your device. In addition to data encryption, you should secure your communication channel so that hackers cannot simply listen in and figure out what information you are trying to communicate with your corporate servers. Even if hackers could intercept a message, you want to make sure that the message is so scrambled that it is completely beyond their comprehension.

In this chapter, you will learn the following:

- How to encrypt and decrypt messages with the classes offered in the `System.Security.Cryptography` namespace
- How to write an application that talks to an encrypted SQL Server Mobile database with password protection
- How to apply secured communication technologies such as SSL and VPNs to enhance the data communication channel
- How to secure XML Web services with user authentication

Data Protection

This section includes two major parts. "Data Encryption" introduces how to encrypt data and save it to a local file, and "Database Encryption and Password Protection" describes how to secure data stored in a SQL Server Mobile database.

Data Encryption

Unlike the desktop filesystems, such as NTFS, for which users can easily set up file permissions and access control lists (ACLs) to help protect files, the Windows Mobile filesystem does not provide the same function because it is geared more toward single-user usage. To prevent data theft on Windows Mobile devices, data encryption provides a vital and practical safeguard.

The .NET Framework supports a number of encryption classes in the `System.Security.Cryptography` namespace. Fortunately, those classes are also available in the Compact Framework 2.0. For applications targeted at .NET Compact Framework version 1.1 or earlier, you will need to call the Windows CE CryptoAPI via P/Invoke. For example, the `CryptCreateHash()` function in Windows CE can be used to initiate the hashing of a data stream; the prototype of the function is as follows:

```
BOOL CRYPTFUNC CryptCreateHash(
  HCRYPTROV  hProv,
  ALG_ID     Algid,
  HCRYPTKEY  hkey,
  DWORD      dwFlags,
  HCRYPTHASH *phHash
);
```

The following code shows how to declare this function in C# using P/Invoke:

```
[DllImport("crypt32.dll")]
  public static extern bool CryptCreatHash(
    ulong  hProv,
    uin    Algid,
    ulong  hKey,
    uint   dwFlags,
    ref ulong phHash,
  };
```

Of course, another way to implement file and data encryption is via hardware encryption. For example, Cisco offers storage cards that have built-in hardware encryption. Encrypting and decrypting data through different chips at the hardware level is actually a very popular solution because it doesn't consume much of the computing power. In addition, developers do not need to redo their work to make the application capable of encryption. However, this does not mean that hardware encryption is superior to software solutions. You may want to pick software encryption due to its lower cost.

The `System.Security.Cryptography` namespace provides three major cryptography algorithms:

- **Symmetric**—Symmetric encryption algorithms use the same key to encrypt and decrypt data. Table 13-1 summarizes four symmetric algorithms that are supported in the .NET Compact Framework 2.0.

- **Asymmetric** — Asymmetric algorithms normally require a set of keys: a public key and a private key, which only the user knows. Messages encrypted via the public key can be decrypted with the corresponding private key, and vice versa. Asymmetric algorithms, such as Rivest, Shamir, and Adleman (RSA) and Digital Signature Algorithm (DSA), are the essential building blocks for secure communications over a network. Note that asymmetric encryption usually is more computationally demanding than symmetric encryption.
- **Hashing** — Hashing is the process of mapping information to a fixed-length binary string. A typical application of hashing is the certificate discussed in Chapter 12. Unlike symmetric and asymmetric encryption, hashing functions such as MD5 and SHA1 do not really help to hide information; rather, they are used to maintain data integrity.

Table 13-1 Symmetric Algorithms in .NET Compact Framework 2.0

Algorithm	Default Key Size	Default Implementation Class
DES	64	`DESCryptoServiceProvider`
RC2	128	`RC2CyptoServiceProvider`
RinjinDael	256	`RijinDAelManaged`
TripleDES	192	`TripleDESCryptoServicProvider`

The following example uses the symmetric encryption methods offered in the `System.Security.Cryptography` namespace. The sample application first encrypts a string and saves it as a text file locally on a Smartphone device. The encrypted file can be opened for reading with or without decryption. Of course, the string is comprehensible only when the proper decryption method is present.

To begin, start a new Smartphone device project from Visual Studio 2005 and name the project **dataEncrypt**. In the Form Designer, add a menu item named **mnuQuit** to the left soft key and set the `Text` property of this menu item to `Quit`. Add another menu item named **mnuOptions** to the right soft key with the `Text` value set to `Options`. Then add two submenu items to `mnuOptions`: `mnuRead`, with the `Text` value '`Read Cipher`', and `mnuDecrypt`, with the `Text` value `Decrypt`. Figure 13-1 shows the UI of this application.

Two major functions of the sample application are encrypting a clear-text file and decrypting an encrypted file. To encrypt a file, you need to create a new instance of a cryptography algorithm. Then you can instantiate an object of the `CryptoStream` class, which links the data stream with the encryption algorithm. The following snippet shows how a string is encrypted using the default Rijndael encryption algorithm, which is also known as the Advanced Encryption Standard (AES):

```
using System.Security.Cryptography;
...
        private int bLen;
        private Encoding defEncode = Encoding.Default;
        private string filename = "SimpleEnc.txt";

        //Creates the default Rijndael encryption algorithm
        private SymmetricAlgorithm sAlg = SymmetricAlgorithm.Create();

        void EncryptText(string text)
```

```
        {
            //Convert strings to byte array
            Byte[] buff = new Byte[1024];

            buff = defEncode.GetBytes(text);
            bLen = buff.Length;

            //Write the byte array to file SimpleEnc.txt
            FileStream fo = new FileStream(filename, FileMode.OpenOrCreate);
            CryptoStream encStream = new CryptoStream(fo, sAlg.CreateEncryptor(),
CryptoStreamMode.Write);
            encStream.Write(buff,0,bLen);

            //Close stream
            encStream.Close();
            fo.Close();

        }
```

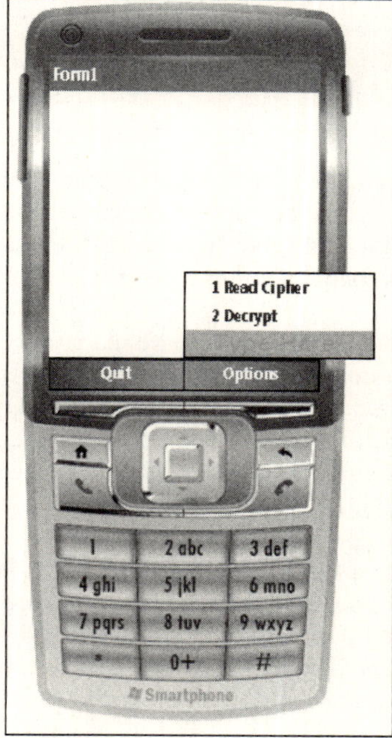

Figure 13-1

In the preceding example, a symmetric algorithm object `sAlg` is created by calling the `SymmetricAlgorithm.Create()` method, which uses the default Rijndael algorithm. When linking the data I/O stream to the `CryptoStream`, you need to create an `Encryptor` object by calling the `CreateEncryptor()` method. For the Rijndael algorithm, you can create a unique `Encryptor` by setting the key value and initialization vector (IV). If the key value and the IV are missing, a random number will be used to generate a new encryptor. In the sample code, a `CryptoStream` object, `encStream`, is created to write to a file output stream `fo` with a Rijndael encryptor:

```
CryptoStream encStream = new CryptoStream(fo, sAlg.CreateEncryptor(),
CryptoStreamMode.Write);
```

The `CryptoStream` class can read and write data in the form of byte arrays. A quick way to convert a text string to a binary array is by calling the `GetBytes()` methods of the `Encoding` class. In the sample code, a reference to the default system code is first created:

```
private Encoding defEncode = Encoding.Default;
```

A string `text` (assuming it occupies less than 1,024 bytes) can then be converted to a byte array as follows:

```
Byte[] buff = new Byte[1024];
buff = defEncode.GetBytes(text);
```

Similarity, you can use the `GetString()` method to convert a byte array to a string. For example, the following code will convert the first 10 bytes of the byte array `buff` to a string `str`:

```
string str = Encoding.Default.GetString(buff,0,10);
```

When decrypting the data, a decryptor is needed to link the `CryptoStream` with the data stream. Usually, you should feed the same key value and IV value you used in encryption to create this decryptor. Because the key and IV are not specified during the encryption process, you don't have to pass those two when creating a new decryptor. The encryption process with `CryptoStream` is illustrated in the following code:

```
    //Open a new file input stream
    FileStream fi = new FileStream(filename, FileMode.Open, FileAccess.Read);

    //Link the file input stream to the CryptoStream
    CryptoStream decStream = new CryptoStream(fi, sAlg.CreateDecryptor(),
CryptoStreamMode.Read);

    //Read data to a byte array
    Byte[] buff = new Byte[1024];
    decStream.Read(buff, 0, bLen);

    //Convert the byte array to a string value
    string decText = defEncode.GetString(buff, 0, bLen);

    //Close both streams
    decStream.Close();
    fi.Close();
```

In the sample application, the encryption process takes place when the form is loaded. When the menu item `mnuRead` is pressed, the encrypted data file will be read without decryption, which consequently displays garbled text on the screen. When the menu item `mnuDecrypt` is pressed, the encrypted file will be decrypted and displayed. The following code shows the entire sample application:

```
using System;
using System.Data;
using System.Text;
using System.Windows.Forms;
using System.IO;
using System.Security.Cryptography;

namespace DataEncrypt
{
    public partial class Form1 : Form
    {

        private int bLen;
        private string filename = "SimpleEnc.txt";
        private Encoding defEncode = Encoding.Default;

        //Creates the default Rijndael encryption algorithm
        private SymmetricAlgorithm sAlg = SymmetricAlgorithm.Create();

        public Form1()
        {
            InitializeComponent();

            //Encrypt "Guess Who I am" and save it as SimpleEnc.txt.
            string clearText = "Guess Who I am";
            EncryptText(clearText);
        }

        private void mnuQuit_Click(object sender, EventArgs e)
        {
            Application.Exit();
        }

        void EncryptText(string text)
        {

            //Convert strings to byte array
            Byte[] buff = new Byte[1024];

            buff = defEncode.GetBytes(text);
            bLen = buff.Length;

            //Write the byte array to file SimpleEnc.txt
            FileStream fo = new FileStream(filename, FileMode.OpenOrCreate);
            CryptoStream encStream = new CryptoStream(fo, sAlg.CreateEncryptor(),
CryptoStreamMode.Write);
            encStream.Write(buff,0,bLen);

            //Close stream
            encStream.Close();
```

```
            fo.Close();

        }

        //Read text without decryption
        private void mnuRead_Click(object sender, EventArgs e)
        {
            FileStream fi = new FileStream(filename,FileMode.Open,FileAccess.Read);

            Byte[] buff = new Byte[1024];
            fi.Read(buff, 0, bLen);

            string text = defEncode.GetString(buff, 0, bLen);

            fi.Close();

            MessageBox.Show("The message is:  " + text);

        }

        //Decrypt the message and show it on screen
        private void mnuDecrypt_Click(object sender, EventArgs e)
        {
            FileStream fi = new FileStream(filename, FileMode.Open,
FileAccess.Read);
            CryptoStream decStream = new CryptoStream(fi, sAlg.CreateDecryptor(),
CryptoStreamMode.Read);

            Byte[] buff = new Byte[1024];
            decStream.Read(buff, 0, bLen);

            string decText = defEncode.GetString(buff, 0, bLen);

            MessageBox.Show("The message is:  " + decText);

            decStream.Close();
            fi.Close();


        }
    }
}
```

Figures 13-2 and 13-3 display the running results of the sample application, respectively.

Figure 13-2

Using other symmetric algorithms in your application is very similar to the preceding example. For instance, if you prefer to use less complex DES encryption for faster computation, you can specify the algorithm name when creating a new `SymmetricAlgorithm` object, as follows:

```
SymmetricAlgorithm DESAlg = SymmetricAlgorithm.Create("DES");
```

You can then create the corresponding encryptor or decryptor object by calling the `CreateEncryptor()` or `CreateDecryptor()` method, respectively.

Figure 13-3

Database Encryption and Password Protection

Many Smartphone-based applications interact frequently with data saved in databases, such as SQL Server Mobile. Securing data saved in those databases is obviously very important, especially when the databases are shared at the enterprise level and contain confidential data.

Fortunately, Microsoft SQL Server 2005 Mobile Edition offers two techniques that can help you protect data saved in the database: password protection and data encryption.

Password protection in SQL Server Mobile has the following features:

- Passwords cannot be recovered. If a password is lost, the data saved in the database is inaccessible. As a software developer, you may want to design your application so that data can be regenerated when a password is forgotten. For example, allowing data synchronization between SQL Server Mobile and SQL Server may help to keep the risks to a minimum.
- The strength of the protection is determined by the length of the password, which can be up to 40 characters long. Strong passwords, which include a combination of letters, numbers, and special characters, are encouraged.

- Password and encryption settings for a SQL Server Mobile database cannot be changed unless the database is compacted. Note that compacting a database is usually done to compress unused space and check data inconsistency. In addition, the password for the original database is required.

To create a password-protected SQL Server database, you need to specify the password property of the connection string. The following connection string would be used to create a `MyDB007` database with the password MyPassword:

```
Data Source=\My Documents\MyDB007.sdf; password= MyPassword
```

In addition to requiring a password, you can set the data encryption option, which will encrypt data using the symmetric RC4 algorithm with the key value generated from the MD5 hash value of the password. The encryption option is available to a SQL Server Mobile database only when it is password-protected. To programmatically set this option, simply turn on the `encrypt database` option in the connection string, as follows:

```
Data Source=\My Documents\SecDB.sdf; password= MyPassword; encrypt database = TRUE
```

A sample application is provided here to demonstrate how it works. In the UI design page, a `textbox` object is added to the form to accept a user's password. Options are then added to the menu to create and access a secure database. Figure 13-4 illustrates the UI of this simple application. Note that you need to add a reference to `System.Data.SqlServerCe`. In Visual Studio 2005, choose Project➪Add Reference, and then select System.Data.SqlServerCe.

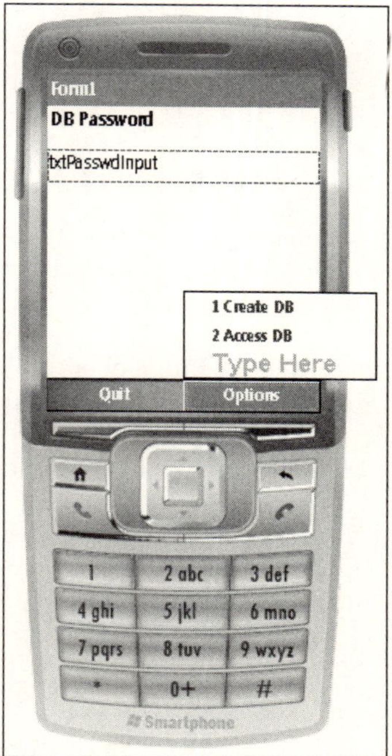

Figure 13-4

Because `txtPasswdInput` is used as a password input box, you should set the `PasswordChar` property so that whatever the user types in will not be printed on the screen:

```
txtPasswdInput.PasswordChar='*';
```

To create the secure database, you first define a connection string `secConn`, which contains the password and enables the `encrypt database` option. You then create a new data engine, `ceEngine`, with the security-enabled connection string `secConn`. A secured database can then be created by calling the `CreateDatabase()` method of `ceEngine`:

```
string secConn = @"Data Source=\My Documents\SecDB.sdf;password="
                + txtPasswdInput.Text+"; encrypt database = TRUE";
SqlCeEngine ceEngine = new SqlCeEngine (secConn);

ceEngine.CreateDatabase();
```

Similarly, when accessing a password-protected database with encryptions on, the connection string should include both the password and the encryption option. A secured SQL Server Mobile connection `SqlCeConnection ceConn` can then be created with the connection string, as shown here:

```
string secConn = @"Data Source=\My Documents\SecDB.sdf;password="
                 +txtPasswdInput.Text+"; encrypt database = TRUE";

SqlCeConnection ceConn = new SqlCeConnection(secConn);
ceConn.Open();
```

In the sample application, we would like to determine whether the connection to a secured SQL Server Mobile database is indeed established. This can be tested by performing a SQL query. Because there is no data in the sample database, a SQL statement to create a table is used. If the user inputs the right password, the table will be created. Otherwise, error information will be shown on the screen.

Following is the full code listing:

```
using System;
using System.Data;
using System.Drawing;
using System.Text;
using System.Windows.Forms;

//Add reference first then...
using System.Data.SqlServerCe;

namespace SecDB
{
    public partial class Form1 : Form
    {
        private string StrPath = @"\My Documents\SecDB.sdf";

        public Form1()
        {
            InitializeComponent();
```

```
            txtPasswdInput.PasswordChar='*';
            txtPasswdInput.Text = "";

        }
```

```
        //Create a secure DB
        private void mnuCreate_Click(object sender, EventArgs e)        {
            string secConn = @"Data Source=\My Documents\SecDB.sdf;password="
                + txtPasswdInput.Text+"; encrypt database = TRUE";
            SqlCeEngine ceEngine = new SqlCeEngine (secConn);

            try {
                ceEngine.CreateDatabase();
                MessageBox.Show("Password protected 'SecDB' is created with
Encryption on");

                mnuCreate.Enabled = false;
                txtPasswdInput.Text = "";

                ceEngine.Dispose();
            }
            catch (SqlCeException seqErr)
            {
                MessageBox.Show(seqErr.ToString());
            }

        }

        //Access the secure DB
        private void mnuAccess_Click(object sender, EventArgs e)        {
            string secConn = @"Data Source=\My Documents\SecDB.sdf;password="
                            +txtPasswdInput.Text+"; encrypt database = TRUE";

            try {
                SqlCeConnection ceConn = new SqlCeConnection(secConn);
                ceConn.Open();

                SqlCeCommand ceCmd = new SqlCeCommand();
                ceCmd.Connection = ceConn;
                string sqlCmd = @"CREATE TABLE StuGrades ( StudentID bigint, Grade
smallint)";

                ceCmd.CommandText = sqlCmd;
                ceCmd.ExecuteNonQuery();

                ceConn.Close();
                MessageBox.Show("Table StuGrades Created");

            }
            catch (SqlCeException sqlEx)
            {
                MessageBox.Show(sqlEx.ToString());
```

```
                }
            }

            private void mnuQuit_Click(object sender, EventArgs e)
            {
                Application.Exit();
            }
        }
    }
```

Because accessing a database is normally error prone, it is always a good practice to put the database-related operations in the `try` block and handle possible exceptions in the `catch` block.

Figures 13-5 and 13-6 show the running results of the application.

The preceding example demonstrates how to protect a database with a password and encryption. Note that the application is expected to run only once because you don't want to create the same database or table repeatedly.

Figure 13-5

Figure 13-6

Securing Communication Channels

The thin-client nature of Smartphone devices means that they are constantly talking to various servers to access information. When passing sensitive data over a network, especially a wireless network, it is extremely important to secure the communication at both ends.

Network Authentication

Network authentication should be applied before a user can access any information. The .NET Compact Framework 2.0 supports both the Microsoft NTLM and Kerberos authentication protocols.

Microsoft NTLM, or Windows NT LAN Manager, is a network authentication protocol based on challenge and response. The server stores a user's password in an encrypted format. The client machine initiates the authentication process with a negotiate message, and then the server sends a challenge message that contains the message type, the security signature, and the negotiation type. The client then encrypts the challenge message with the user's password, typically using the DES algorithm. The encrypted message is then sent out to the server as the response message. The same algorithm is applied on the server side. The result is then compared to the response from the client. If they match, the authentication is successful. An advantage of NTLM is that the user's password is not sent over the network during the authentication process.

Kerberos involves a much more complicated process, but it offers a more secured communication channel. Because of its better security, it will be picked during the authentication negotiation process when both NTLM and Kerberos are supported. NTLM is mainly used in Microsoft's early products, such as Windows NT 4.0.

The `System.Net.NetworkCredential` class can create credentials to authenticate the user with the most secure method supported by the server. The code to generate the network credential is as follows:

```
using System.Net;
...

NetworkCredential myCredentials =
          new NetworkCredential("myUserName", "MyPasswd", "myDomainName");
```

You can then use the generated credentials to secure network applications that require user authentication, such as web services.

Note that the `NetworkCredential` class is not available on the .NET Compact Framework 1.0 and 1.1. You need to install the .NET Compact Framework 2.0 to leverage this service for network authentication.

Secure Sockets Layer (SSL)

Secure Sockets Layer (SSL) is the industry standard for secure web communications. It is used widely in today's e-communication to transmit sensitive information, such as credit card numbers and online banking, over the Internet.

Essentially, SSL is the technology that involves encrypting and decrypting messages between a web browser and the web server. SSL does not prevent hackers from eavesdropping, but the encrypted message makes it much harder for hackers to access the message.

Typically, message communication in SSL proceeds as follows:

1. The client initiates SSL communication, such as an HTTPS request.
2. The client starts an SSL session with a unique public key that is created for the client's web browser.
3. The message is encrypted with the server's public key and sent to the web server.
4. The web server decrypts the information with its private key, which is kept secret; no one else has the knowledge of this private key.

This process is considered fairly secure because each SSL session generates a unique public key that a hacker is unlikely to guess. The message is encrypted with the server's public key, which (in theory), only the web server itself is able to decrypt.

The length of the key in SSL communication is either 40-bit or 128-bit. The longer the key length, the harder it is to crack the message.

To enable SSL communication, you need to install certificates on the server side. Certificates can be obtained from different certificate authorities, such as VeriSign and GeoTrust. The certificate application process is normally pretty straightforward and requires a certificate signing request (CSR) file. On an IIS Server, the steps to generate the CSR file are as follows:

1. Select Start⇨Run, and type **compmgmt.msc** to launch the Computer Management console.
2. On the left panel, expand the Services and Applications and then expand Internet Information Services.
3. Expand Web Sites, right-click Default Web Site, and then choose Properties (see Figure 13-7).
4. Select the Directory Security tab in the Default Web Site Properties window and click the Server Certificate button (see Figure 13-8).
5. Choose "Create a new certificate" from the IIS Certificate Wizard (see Figure 13-9) and input the required information, such as country, website name, and the name of the certificate file.

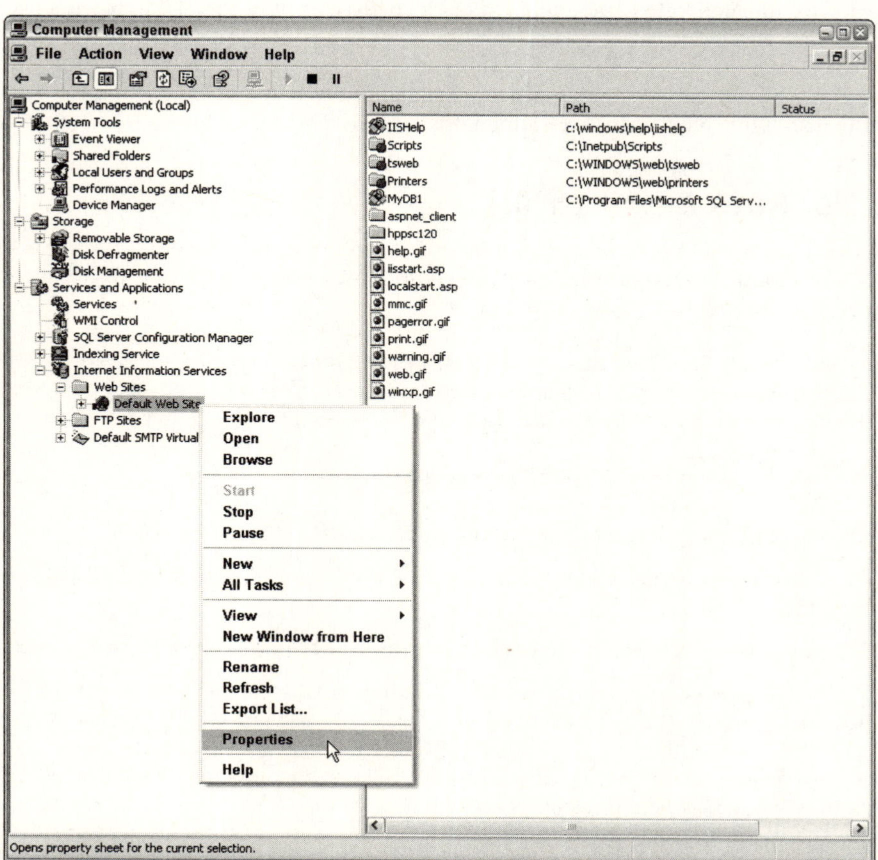

Figure 13-7

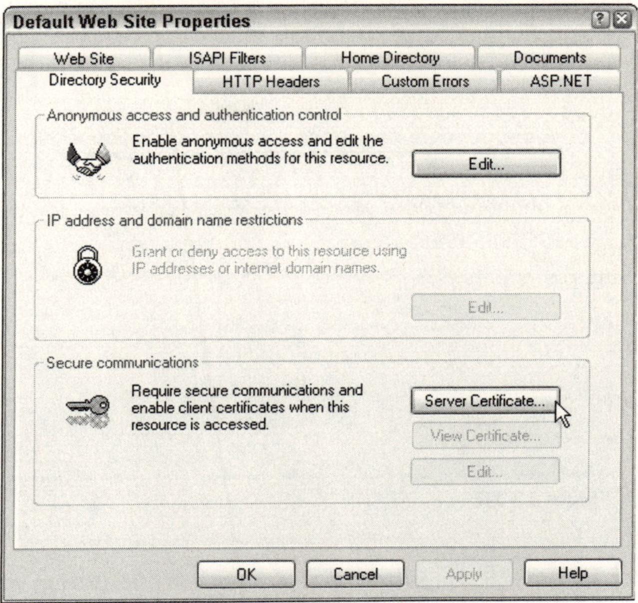

Figure 13-8

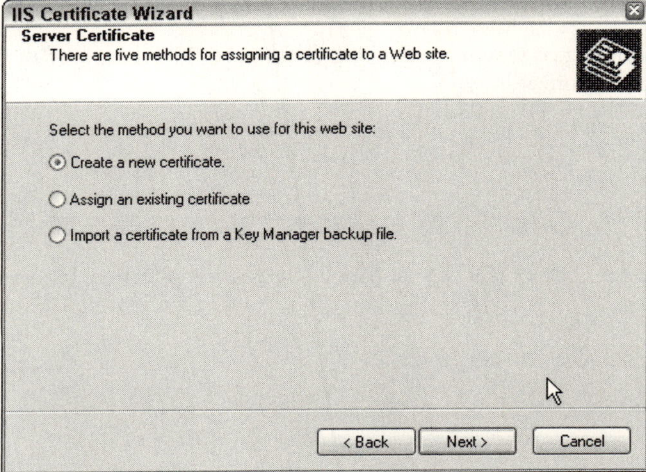

Figure 13-9

When going through the wizard, you need to provide organizational information, geography information, and so on. Remember that the common name has to be the DNS name or NetBIOS name (normally the computer name) of the web server, as illustrated in Figure 13-10.

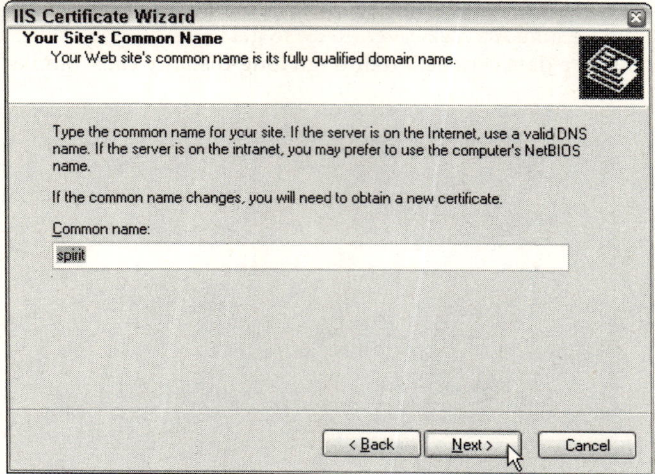

Figure 13-10

You can use the CSR file created by the wizard to request the certificate from your CA. Once the request is granted, follow the certificate authority's instructions to install the certificate on the web server.

Most of the work to set up and enable SSL communication is on the server side. If the web server is correctly installed and protected with SSL, you simply specify an HTTPS request on the client side.

For example, the following code will start an SSL-enabled HTTPS request with a SSL-enabled web server named mySecureServer.Com:

```
using System.Net;
...

WebRequest webReq = WebRequest.Create("https://mySecureServer.com");
```

Likewise, if you want to protect SQL Server Mobile database replication, use `https` in the `InternetURL` property:

```
using System.Data.SqlServerCe;
...
repl = new SqlCeReplication();
repl.InternetUrl = "https://spirit/sqlmobile/sqlcesa30.dll";
...
```

Virtual Private Networks

A virtual private network (VPN) is a common network mechanism to provide secure end-to-end network connections. The idea is to first negotiate and set up a network tunnel between the two communication nodes. Usually, a VPN server also connects to a RADIUS server, allowing only authorized users to have the permissions to establish such tunnels. The data is then encrypted before it is transmitted over the network. Then it will be decrypted on the receiver side. Compared to a dedicated private leased line, a VPN is preferred by many companies because of its low cost.

Consider a corporate network without the support VPN. As shown in Figure 13-11, a remote user needs to dial up the corporate Remote Access Server (RAS) to get access the servers. This could be very pricey if users are trying to make a data connection through long-distance calls. Similar problems also exist on the remote offices and mobile users.

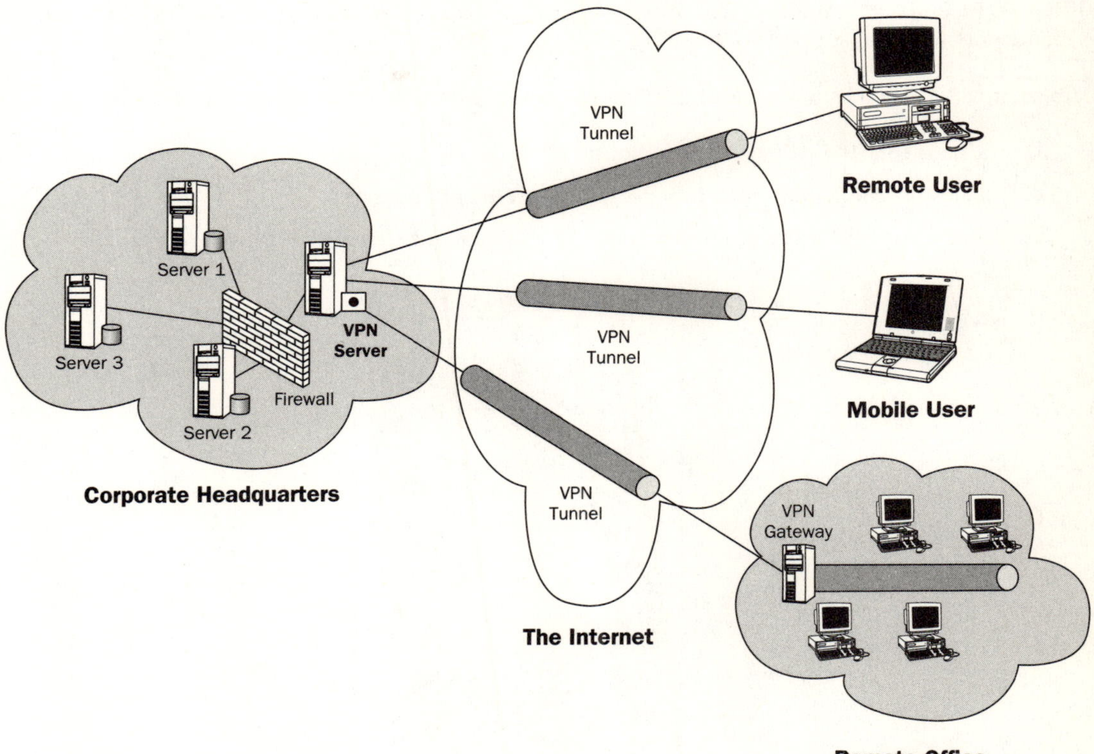

Figure 13-11

With a VPN, however, users do not need to directly dial to the corporate headquarters. A remote user can simply dial into the local access server and rely on the local ISPs to package the data and route the data through a "tunnel" to the remote servers. Of course, you need to pay for the tunnel services offered by the ISPs, but it is typically less than half of what you pay for leased lines or long-distance phone calls. Three major tunneling protocols are supported via the Internet:

- **IP Security (IPSec)** — Developed by the Internet Engineering Task Force (IETF), this protocol operates at the network layer and can be implemented independent of application layer.
- **Point-to-point Tunneling Protocol (PPTP)** — This is the protocol developed by Microsoft, 3Come, and Ascent Communications. It works at the data link layer and is preferred for Microsoft Windows–based network traffic.
- **Layer 2 Tunneling Protocol (L2TP)** — This is the implementation of Cisco, which combines their previously proposed Layer 2 Forwarding with PPTP. It offers more flexibility than PPTP, but need supports from the underlying network devices, such as routers and switches.

The advantages of a VPN include reduced cost, effective use of bandwidth, enhanced scalability, and enhanced connectivity. With added-on services, it also offers better security than conventional Internet protocols. The drawback of a VPN is also obvious: It is highly dependent on the Internet and lacks interoperability of devices and protocols.

The .NET Compact Framework 2.0 supports PPTP, L2TP, and IPSec (as opposed to the .NET Compact Framework 1, which supports only PPTP).

To set up a VPN connection on a Smartphone device, perform the following steps:

1. Choose Start⇨Settings⇨Connections⇨VPN.
2. Click the Menu button and choose Add from the popup menu.
3. In the Add VPN screen, shown in Figure 13-12, enter a description for your VPN.

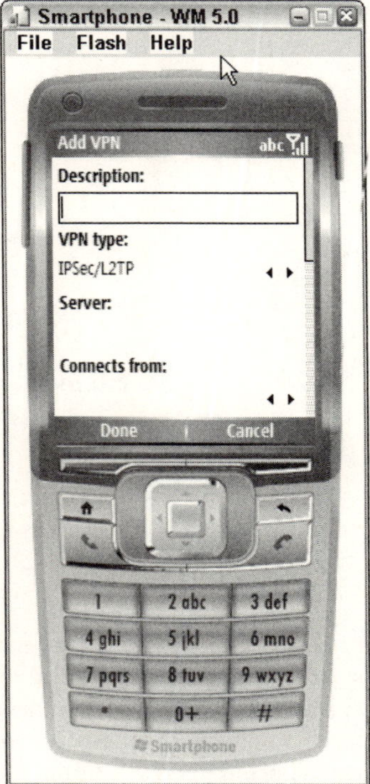

Figure 13-12

Besides setting up the VPN connection correctly at the Smartphone end, you should also make sure that the VPN server is running properly at the other end and that the VPN technology is supported by the underlying networks. Data communication using a VPN is highly recommended for Smartphone applications to enhance communication security.

Wi-Fi

The IEEE 802.11 wireless network, also known as Wi-Fi, is operating at the unlicensed 2.4 GHz band. The nature of radio waves enables it to travel through walls, and therefore makes the indoor and outdoor wireless communication fairly convenient. However, it also imposes a severe network security threat because messages are broadcasted over the air between client and base station. Anyone with a proper device can intercept and tamper with the message.

In 1999, Wired Equivalent Privacy (WEP) was proposed and implemented on Wi-Fi networks. It uses a shared secret key and RC4 algorithm with a key length of 40 bits. Researchers later identified the security flaws of WEP and it is no longer considered secure. The extension of WEP — namely, WEP2 — addresses some of the early concerns and increases the key size to 128 bits. However, for network experts, WEP2 is also flawed and can only be categorized as a weak security protocol.

If you have a choice, don't use WEP. Instead, you should use protocols defined in IEEE802.11i, such as Wi-Fi Protected Access (WPA), Extensible Authentication Protocol (EAP), or Protected Extensible Authentication Protocol (PEAP).

Securing Web Services with SOAP Headers

When deploying web services, one way to implement security is to enforce user authentication from the web server. The Microsoft Internet Information Services (IIS) server provides the following four options for user authentication:

- **Anonymous access** — No authentication is required to access the web resources.
- **Basic authentication** — A user is authenticated by sending his or her username and password to the web server over the network in clear text. This is not a recommended authentication method. It exists simply as a fallback authentication protocol if a more secured authentication protocol is not supported.
- **Digest authentication for Windows domain servers** — This option is enabled if the IIS server is a member server of a Windows domain. A user's password is not sent over the network in this authentication process; rather, the MD5 hash value of the password, termed a *digest*, is transmitted through the network and compared with the digest stored in the domain controller.
- **Integrated Windows authentication** — This uses the NTLM authentication protocol and requires a Windows user account on the machine on which the IIS server is installed.

To change the authentication method for an IIS server, launch the IIS service management console by clicking Start⇨Settings⇨Control Panel⇨Administrative Settings⇨Internet Information Services. On the left panel, expand the name of the IIS server, followed by Web Sites. Right-click Default Web Site and choose Properties. In the Directory Security tab, click the Edit button in Anonymous access and user authentication control. An Authentication Methods window will appear, as shown in Figure 3-13.

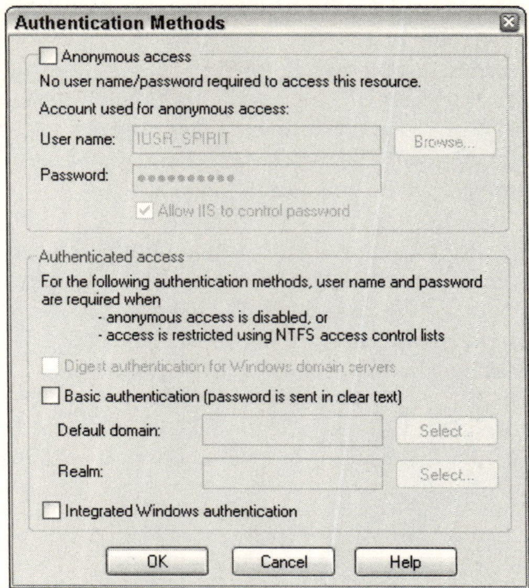

Figure 13-13

Because different web servers have different ways to handle user authentication, it makes sense to authenticate a user without relying on a particular platform or version of the web server. One approach is to customize authentication by using SOAP headers in a XML web service. Note that you should set the user authentication to allow anonymous user logins.

Using a custom SOAP header alone is not a secured solution because usernames and passwords are transmitted over the Internet as clear text. You should apply this approach only in a secured communication channel, such as SSL or a VPN.

In the following two subsections, you are going to learn how to build and publish an ASP.NET web service on the server side and how to connect to the web service from the client side.

Server Side

Following are the major steps to perform to create a web service with an authentication SOAP header on the server side:

1. Create a new website from Visual Studio 2005 with ASP.NET web services.
2. Write a simple web service function that can authenticate a user using the SOAP header.
3. Build and publish the website.
4. Test the web service.

To create a new website from Visual Studio 2005, click File⇨New⇨Web Site, as shown in Figure 13-14.

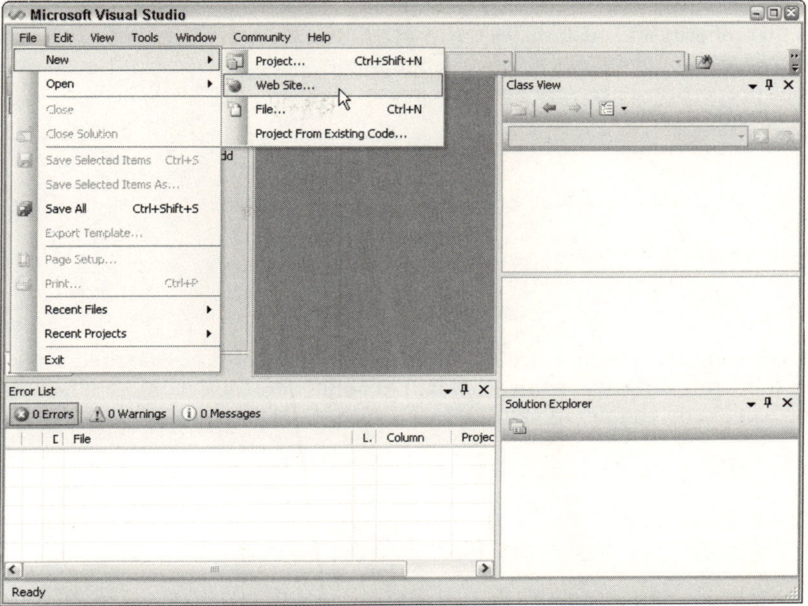

Figure 13-14

The New Web Site dialog will appear, as shown in Figure 13-15. Select the ASP.NET Web Service template.

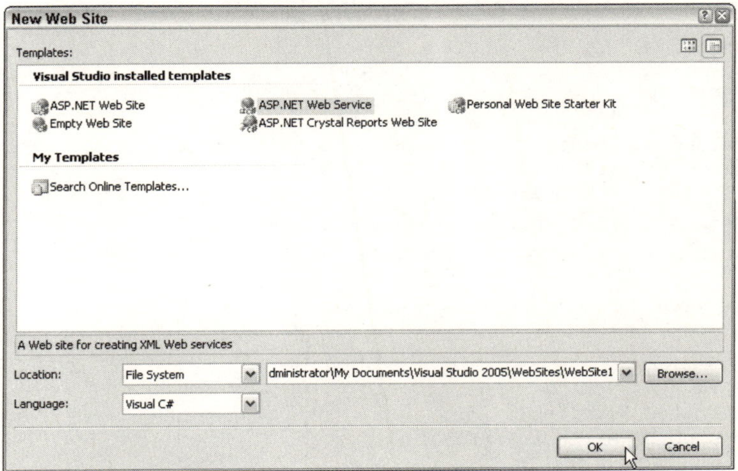

Figure 13-15

To write an ASP.NET web services application that requires a SOAP header, you need to create a SOAP header class that inherits and customizes the `System.Web.Services.Protocols.SoapHeader` class. The following code snippet defines a `SoapAuthenHeader` class that inherits the `SoapHeader` class and has two public properties: `username` and `password`:

```
using System.Web.Services.Protocols;
...
public class SoapAuthenHeader : SoapHeader
{
    public string username;
    public string password;
}
```

Then in the web method, you must declare that a `SoapHeader` object is required using directives. For example, the following directive requires a `SoapAuthenHeader` object `userAuthen`. Setting the `SoapHeaderDirection` enumeration to `In` means the header is sent from the client to the server:

```
public SoapAuthenHeader UserAuthen;

[WebMethod (Description="A simple web service authentication via Soap Header")]
[SoapHeader("UserAuthen",Direction=SoapHeaderDirection.In)]
```

A simple user authentication method can then be written. In the following snippet, the `SimpleAuthen()` method will return a string indicating that authentication is successful if the `username` property is `Test` and the `password` property is `Yes`:

```
public string SimpleAuthen() {
    if (UserAuthen == null)
        return "Sorry, you need to provide username and password to access this web
service";

    if (UserAuthen.username == "Test" && UserAuthen.password == "Yes")
    {
        return "Congratulations, you are authenticated via Soap Header extension!";
    }
    else
        return "Sorry, your username and password are incorrect. Please try again";
    }
```

The code for the server-side web service is as follows:

```
using System;
using System.Web;
using System.Web.Services;
using System.Web.Services.Protocols;

public class SoapAuthenHeader : SoapHeader
{
    public string username;
    public string password;
}
```

```
[WebService(Namespace = "http://192.168.0.88/webServices")]
[WebServiceBinding(ConformsTo = WsiProfiles.BasicProfile1_1)]


public class Service : System.Web.Services.WebService
{

    public SoapAuthenHeader UserAuthen;
    public Service () {

        //Uncomment the following line if using designed components
        //InitializeComponent();
    }

    [WebMethod (Description="A simple web service authentication via Soap Header")]
    [SoapHeader("UserAuthen",Direction=SoapHeaderDirection.In)]
    public string SimpleAuthen() {
        if (UserAuthen == null)
            return "Sorry, you need to provide username and password to access this
web service";

        if (UserAuthen.username == "Test" && UserAuthen.password == "Yes")
        {
            return "Congratulations, you are authenticated via Soap Header
extension!";
        }
        else
            return "Sorry, your username and password are incorrect. Please try
again";
    }

}
```

You can then build the website by choosing Build⇨Build Web Site. After the website is built, it is ready to be published. Choose Build⇨Publish Web Site, as shown in Figure 13-16. You will then be asked where to publish the web service. Choose the WebServices folder in Local IIS, as indicated in Figure 13-17.

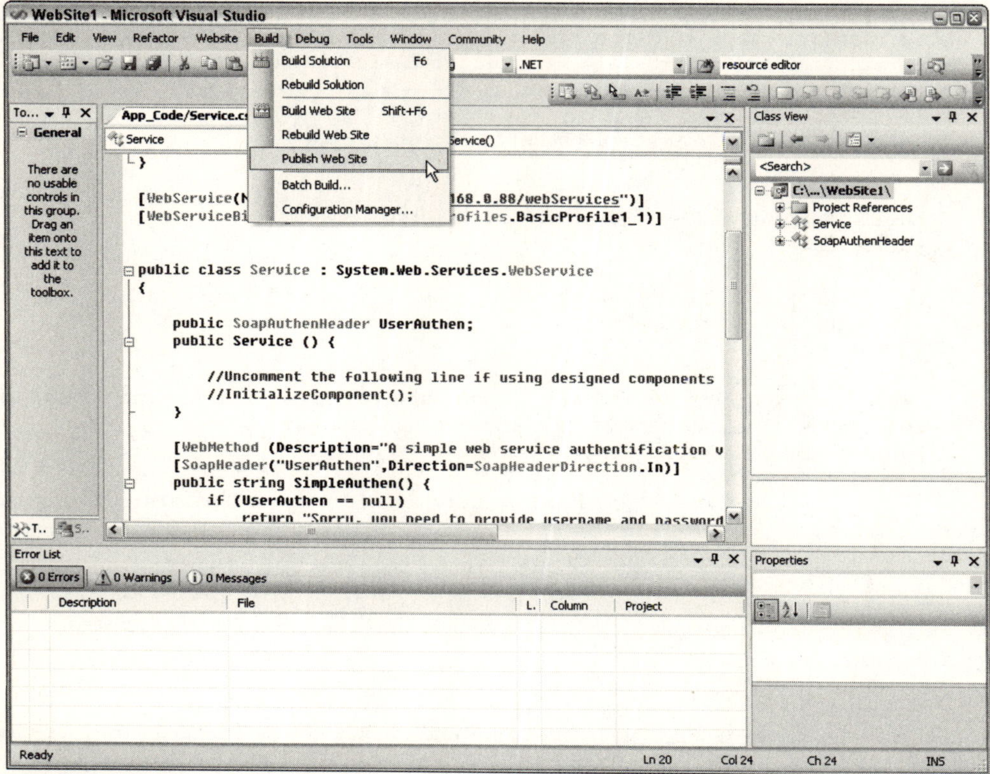

Figure 13-16

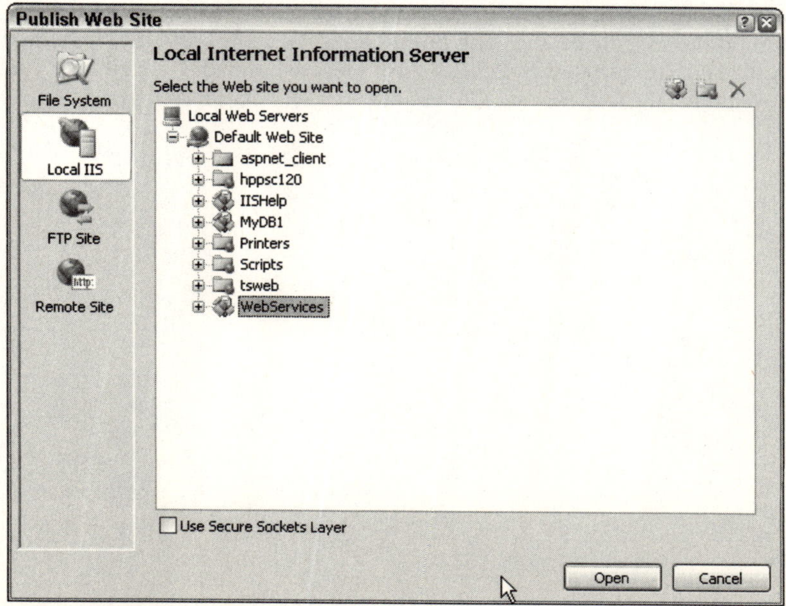

Figure 13-17

The next window, Publish Web Site (shown in Figure 13-18), enables you to configure a number of features. In our example, you can simply use the default setting and click OK.

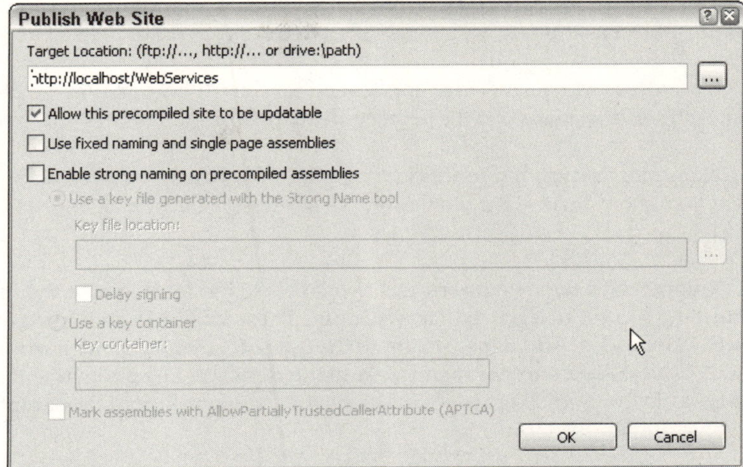

Figure 13-18

As shown in Figure 13-19, the web service can be launched from the web browser by entering the correct URL. In our example, the IP address of the server is 192.168.0.88. The virtual directory of the web service is WebServices. Therefore, the URL is `http://192.168.0.88/webServices/service.asmx`.

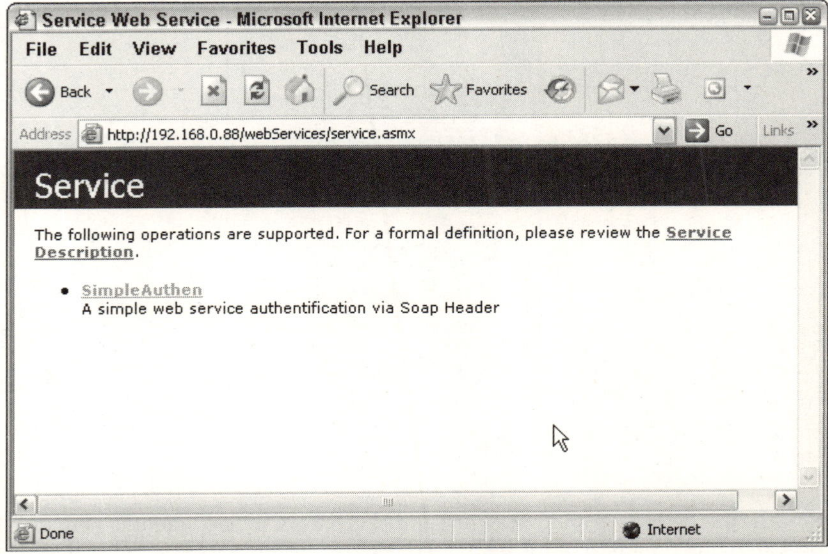

Figure 13-19

Client Side

On the client side, you need to create a new instance of the `SoapAuthenHeader` class defined in the web service and pass the username and password to this `SoapAuthenHeader` object. Then, pass this `SoapAuthenHeader` object as the `SoapAuthenHeaderValue` property of the `Service` class.

First, create a new Windows Mobile 5 device application for Smartphone, name the project **wsAuthClient**, and then add the web service created on the server side by choosing Project➪Add Web Reference.

In the preceding example, the web reference is named `spirit`, which happens to be the name of the IIS server. Obviously, you should name the web reference anything you like, as long as it is meaningful to your code.

From the Form Designer, add two text boxes and two labels to the form. Name the first text box **txtNameInput** and the second text box **txtPasswdInput**. These will be used to pass the username and password to the SOAP header. Add a menu item to the left soft key and name it **mnuQuit** with the `Text` property set to `Quit`. Next, add another menu item to the right soft key and name it **mnuConnect** with the `Text` property set to `Connect`. Figure 13-20 shows the user interface of the sample application.

Figure 13-20

After a user inputs the username and password and clicks the Connect menu, a new web proxy is created by calling the `Service()` method of the web reference:

```
        //Create a new web proxy
        spirit.Service ws = new spirit.Service();
```

Now create a new SOAP header, `AuthHeader`, and pass the `Text` properties of both text boxes to the SOAP header, as follows:

```
spirit.SoapAuthenHeader AuthHeader = new spirit.SoapAuthenHeader();

AuthHeader.username = txtNameInput.Text;
AuthHeader.password = txtPasswdInput.Text;
```

Next, pass the SOAP header to the web proxy `ws` and call the `SimpleAuthen()` method of `ws`. This will call the web services stored on the IIS server `spirit`:

```
ws.SoapAuthenHeaderValue = AuthHeader;
string resp = ws.SimpleAuthen();
```

Following is the full listing of the code:

```
using System;
using System.Collections.Generic;
using System.ComponentModel;
using System.Data;
using System.Drawing;
using System.Text;
using System.Windows.Forms;


namespace wsAuthClient
{
    public partial class Form1 : Form
    {
        public Form1()
        {
            InitializeComponent();

        }

        private void mnuQuit_Click(object sender, EventArgs e)
        {
            Application.Exit();
        }

        //Connect to the web service
        private void mnuConnect_Click(object sender, EventArgs e)
        {
            //Create a new web proxy
            spirit.Service ws = new spirit.Service();
            spirit.SoapAuthenHeader AuthHeader = new spirit.SoapAuthenHeader();

            AuthHeader.username = txtNameInput.Text;
            AuthHeader.password = txtPasswdInput.Text;

            ws.SoapAuthenHeaderValue = AuthHeader;

            string resp = ws.SimpleAuthen();
```

```
                MessageBox.Show(resp);

            }
        }
    }
```

Note that the preceding code works only if the client side can connect to the IIS server through the network. From our experiences, we have found that using the IP address of the server that provides the web service has a better chance of success than using the computer name or DNS name of the web server.

Summary

In this chapter you learned how to encrypt data and files with the classes in the `System.Security.Cryptography` namespace, and how to enable password protection and data encryption in a SQL Server Mobile database. For mobile devices with limited computer power and RAM, you should cautiously pick the encryption algorithms without sacrificing too much performance.

Various techniques to secure the communication channel were discussed in this chapter. We recommend that you apply SSL and a VPN in your application to prevent potential data leaks. The bottom line is, you do not want intruders to hack into your communication with ease and steal the confidential information, whether it is corporate or personal.

User authentication certainly can help you enhance security, but most of the authentication methods are highly correlated with the hardware or software platform used. The custom SOAP header is a hardware- and software-independent approach to implementing user authentication. You are encouraged to apply this approach only in a secured communication channel to avoid the theft of usernames and passwords.

14

Globalization and Localization

Advances in information technologies increasingly enable the international community to become connected. As a result, you are likely to need to develop an application that can be used by people around the world, who speak different languages and come from different cultural backgrounds. Indeed, many U.S.-based Fortune 500 corporations have offices not only in North America but also in Europe, Asia, and beyond. Instead of developing one application for one specific region, it makes more sense to design and implement a "world-ready" application that can be easily localized. The advantages are obvious: a fast and cost-effective development cycle, because you don't need to start from scratch when transforming English-based user interfaces to German- or Chinese-based interfaces.

In this chapter, you will learn about the fundamental techniques available in Windows Mobile 5.0 to support world-ready applications, including the following:

- An overview of globalization and localization concepts
- Developing a culture-aware application
- Localizing data
- Best practices

Globalization and Localization Support

In this book, the term *globalization* describes an application that is not bound to one particular culture and can be used in multiple geographic locations. In the .NET Compact Framework 2.0, globalization is achieved by saving information in neutral data formats and storing a separate set of resource files for each targeted culture or region.

The term *localization* refers to the process that customizes the user interface, text, and date format of an application to a targeted culture. Typical tasks of localization include translating strings into different natural languages, resizing UI elements, and redrawing images.

To gain a better appreciation of the .NET Compact Framework's support for globalization and localization, you need a solid understanding of three concepts: cultures, satellite assemblies, and localized data.

Culture

When talking about cultures in the context of software development and engineering, the focus is on language, calendar, and data formats. In RFC 1766, a hierarchical protocol is defined to distinguish each different culture. In the .NET Compact Framework, a culture name has two parts: a neutral culture that defines the language being used and an optional subculture that defines a geographic location. For instance, "es" refers to Spanish, and whereas "es-PE" represents Spanish in Peru, "es-MX" represents Spanish in Mexico, and "es-ES" means Spanish in Spain. Having a subculture in the culture name is necessary because language itself is not sufficient to distinguish different cultures. Another obvious example is currency. People from different countries or regions may speak the same language but use a different currency. For example, Table 14-1 lists several culture names, locale IDs (LCIDs), and language code pages for both English and Chinese. (For a full list of the table, please refer to the home page of Microsoft's Global Developer Center at `www.microsoft.com/globaldev/default.mspx`.)

Table 14-1 Information of CulturesLocalized Cultural Codes

LCID (HEX)	Culture Name	Locale	ANSI Code Page	OEM Code Page
0x0004	zh-CHS	Chinese (Simplified)	936	936
0x0404	zh-TW	Chinese (Taiwan)	950	950
0x0804	zh-CN	Chinese (People's Republic of China)	936	936
0x0C04	zh-HK	Chinese (Hong Kong)	950	950
0x1004	zh-SG	Chinese (Singapore)	936	936
0x1404	zh-MO	Chinese (Macao)	950	950
0x7c04	zh-CHT	Chinese (Traditional)	950	950
0x0009	en	English	1252	850
0x0309	en-ZW	English (Zimbabwe)	1252	437
0x0409	en-US	English (United States)	1252	437
0x0809	en-GB	English (United Kingdom)	1252	850
0x0c09	en-AU	English (Australia)	1252	850
0x1009	en-CA	English (Canada)	1252	850
0x1409	en-NZ	English (New Zealand)	1252	850

You can see that the cultures are organized first by the language and then by the geographical regions. In addition to its name, each culture has a unique locale ID (LCID). Microsoft assigned LCIDs so that cultures with the same language always have the same endings. For example, the LCIDs of Chinese-speaking cultures end with 0x04, whereas the LCIDs of English-speaking cultures end with 0x09. Two sets of code pages are used in the .NET Compact Framework. The ANSI code page is used for applications using a GUI, and the OEM code page is used for applications using a character-based interface.

On a Smartphone device with Windows Mobile support, users can change cultures by selecting Start⇨ Settings⇨More⇨Regional Settings. If the operating system has additional language support, users can further change their default language on the device.

The CultureInfo Class

To retrieve and set cultures on Windows Mobile devices, the .NET Framework provides the `CultureInfo` class, which is defined in the `System.Globalization` namespace. Two static properties, `CurrentCulture` and `CurrentUICulture`, are available to get and set the current culture information. In most cases, `CurrentCulture` and `CurrentUICulture` are identical. The two methods return different results only when users have installed Multilingual Language Interface (MUI) and chosen a language that is different from the locale. For example, a British software developer assigned to the Germany branch for systems integration will probably need to install the German MUI on the PC. In this situation, the `CurrentCulture` property is still English while the `CurrentUICulture` property is set to Germany.

The .NET Compact Framework supports only a subset of the `CultureInfo` class, and the `CurrentCulture` property is set to read-only, implying that you are not supposed to change the culture for each different process or thread. The philosophy behind this is simple: Mobile devices are normally used by a single user, who probably prefers to stick to one common default setting.

The following code snippet shows how to use the `CultureInfo` class to retrieve the current culture settings:

```
Using System.Globalization;

CultureInfo curCulture = CultureInfo.CurrentCulture;

string cultureName = "Name: "+curCulture.Name+"\t Locale"+curCulture.EnglishName;
MessageBox.show(cultureName);
```

Because the `CurrentCulture` property is read-only in the .NET Compact Framework, you cannot simply assign a value to `CurrentCulture` and hope that you can change the culture settings. Instead, you should create a new instance of the `CultureInfo` class with the desired culture.

The `CultureInfo` class constructor supports four overloads:

- `Public CultureInfo (int culture)`
- `Public CultureInfo (string name)`
- `Public CultureInfo (int culture , bool useUserOverride)`
- `Public CultureInfo (string name, bool useUserOverride)`

The first two constructors will change the culture settings once a new `CultureInfo` object is instantiated. Sometimes, however, you do not want a newly created `CultureInfo` object conflicting with your system's default culture settings. If that is the case, use either of the last two constructors and set the Boolean parameters to `false`. Note that if you create an instance of `CultureInfo` that represents a culture unsupported by the .NET Compact Framework or the device's operating system, then an `ArgumentException` will be thrown.

The following sample code will create a new `CultureInfo` instance and change the culture settings to English (United Kingdom), regardless of the previous culture setting:

```
Using System.Globalization;

CultureInfo newCulture = new CultureInfo("en-GB");
```

Developing a World-Ready Application

In this section you will learn how to develop a world-ready application by using C#'s resource editors and managers.

When developing a world-ready application, you want to construct it so that it can be easily adapted to different cultures without reinventing and redesigning the whole thing. In the .NET Compact Framework, this is achieved by having a single globalized code base to deal with the logical flow and other common tasks. Then, each culture-specific dynamic linked library (DLL) will be loaded during runtime so that the application can display localized, culture-specific content.

Figure 14-1 illustrates the approach. An application named MuiApp has a single code base `MuiApp.exe` and common resources libraries, all saved in an application folder named `MuiAppDir`. Three satellite assemblies are created and saved into three different folders, each corresponding to one culture setting.

Intuitively, you can identify the modules on the left as the globalized modules, whereas the modules on the right are the localized modules.

Creating Localized Resources

Culture-specific satellite assemblies can be created as embedded resources so that the operating system runtime can determine which resources file to load based on the culture setting. Three formats are accepted when you create a resource file: You can write a text file or an XML file, or you can compile a binary resource file directly. The text file has a `.txt` extension, the XML file has an extension of `.resx`, and the binary file has an extension of `.resource`. Note that only binary resource files can be embedded during runtime. However, because binary files are too hard to read, maintain, and debug, it is highly recommended that you edit the resource file using the XML format.

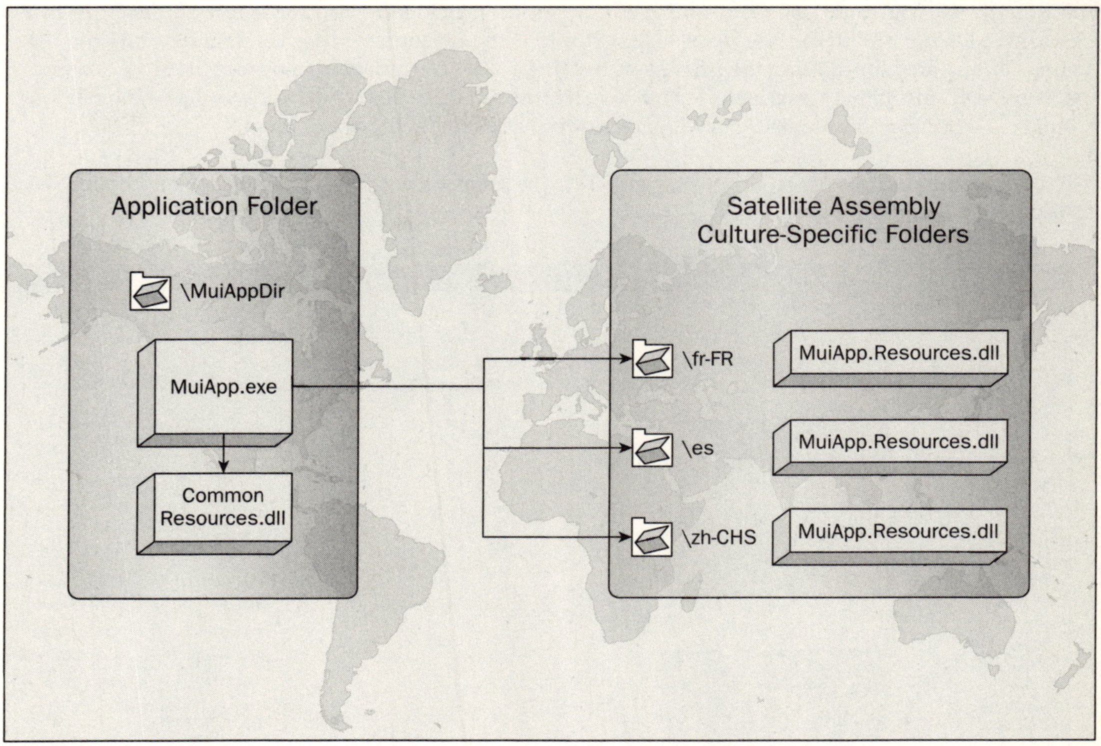

Figure 14-1

To create a culture-specific satellite assembly, an XML-formatted resource file needs to be compiled to the binary format and then linked to the application. Prior to Visual Studio 2005, you needed command-line tools to compile and link a resource file. The .NET-based utility to compile an XML-formatted resource file is `resgen.exe`, which converts `.txt` files and `.resx` files to binary `.resources` files. For example, the following command converts the XML-formatted resource file `App1.Resources.resx` to the binary-formatted resource file `App1.Resources.resources`:

```
RESGEN.EXE App1.Resources.resx App1.Resources.resources
```

After compiling a resource file to binary format, you need to link the resource file to the assembly. Prior to Visual Studio 2005, a .NET-based tool assembly linker, `Al.exe`, was typically used to achieve this. A sample use of this command is as follows:

```
AL.EXE /t: library
 /out:Resources.resources.dll
 /link:Resources.resources
```

Alternatively, you could use the resource editors for C# provided to you as sample applications in Visual Studio .NET 2003. With the release of Visual Studio 2005, a resource editor is integrated into the IDE, streamlining and simplifying the process of creating an XML-formatted resources file: You can visually edit an XML-formatted resources file and Visual Studio 2005 will automatically compile the file to a binary `.resx` file and link the binary-formatted resources file to the application.

To add a new resource file to a project, right-click the project from Solution Explorer, and then choose Add➪New Item (see Figure 14-2).

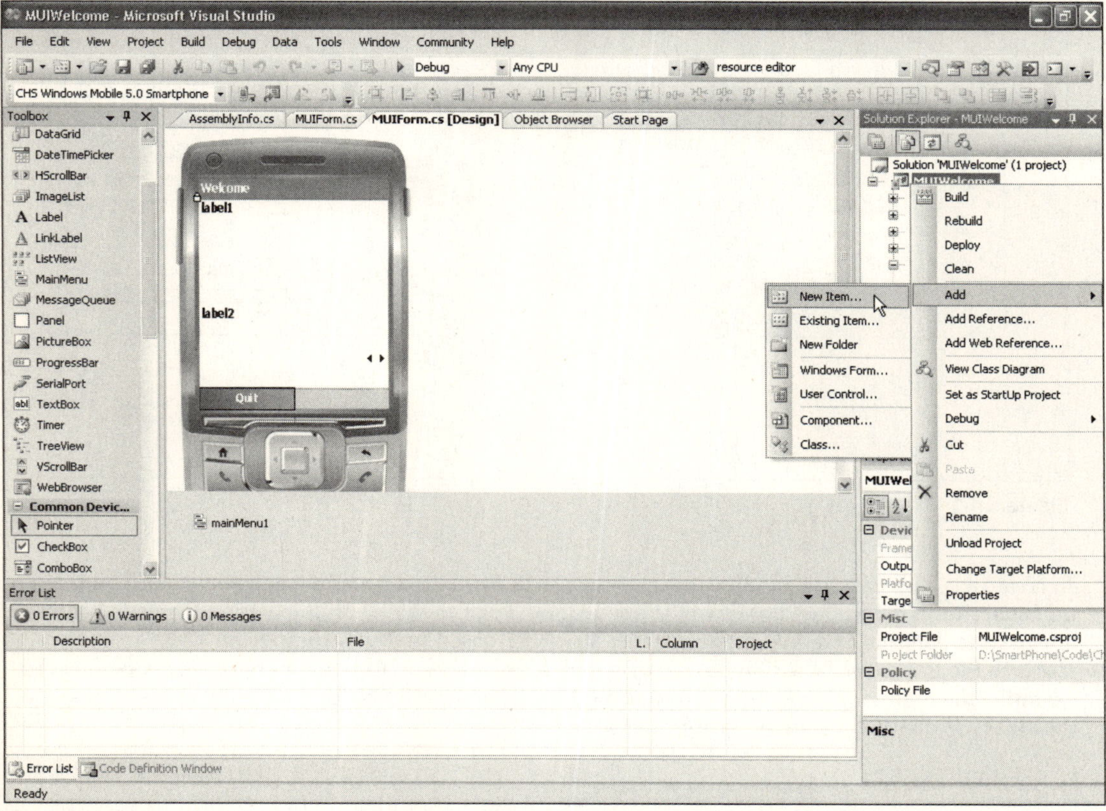

Figure 14-2

When the Add New Item wizard appears, choose Resources File and name the `.resx` resources file to whatever you feel is appropriate — for example, **Resource1.resx**, as shown in Figure 14-3. After the resources file is added to the project, you can double-click the file to edit it. The resources editor in Visual Studio 2005 currently enables you to edit strings and to add or remove strings, images, or other objects. Figure 14-4 shows a resources file named `Resources.resx` that contains two strings. The string value of `Caption` is `Regional Settings` and the string value of `WelcomStr` is `Welcome to Windows Mobile 5.0!`.

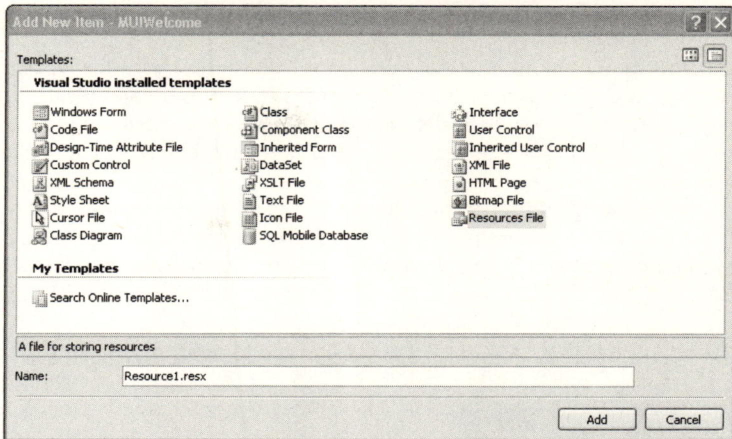

Figure 14-3

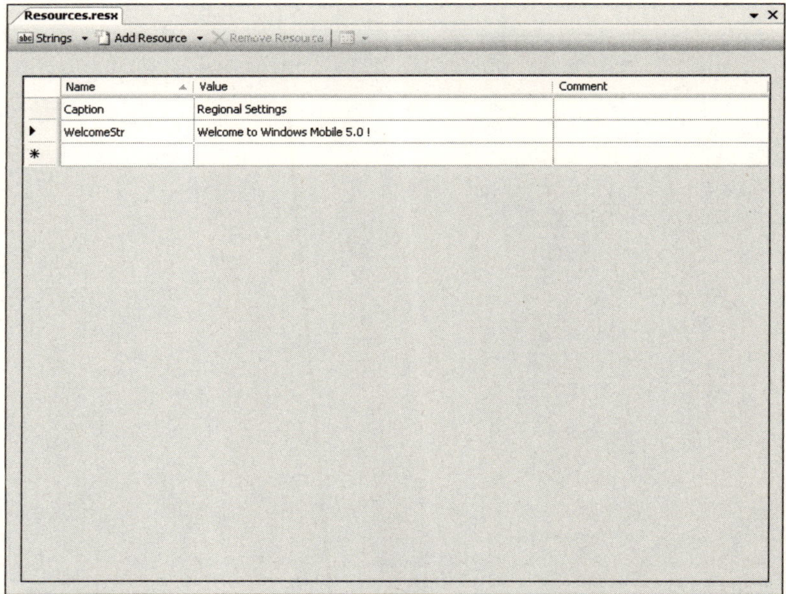

Figure 14-4

Now that you know how to add a resources file to a C# project, the next section describes how to use this resources file for localization purposes.

A Sample Application with a Localized Satellite Assembly

In this section you will create a sample world-ready application that uses the localized satellite assemblies. The function of this application is straightforward: When a user changes the culture preferences, the UI will display a welcome message in the language the user specified. To highlight the programming techniques, this sample application will enable the user to switch between English and Chinese.

Start a new Windows Mobile 5.0 Smartphone device application from Visual Studio 2005 and name the project **MUIWelcome**. Rename the default Form1 to **MUIForm** and change the caption of the form to **Welcome**. Add two Label controls, **label1** and **label2**, to MUIForm. Then add to MUIForm a ComboBox control `comboBox1` and a menu item `menuiItem1` with the caption Quit (see Figure 14-5).

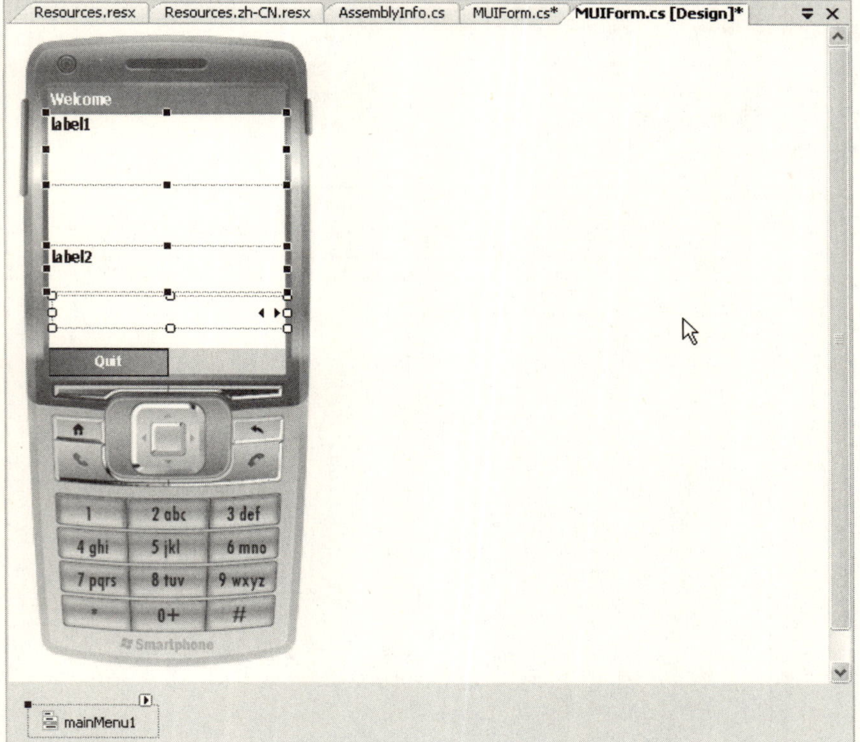

Figure 14-5

`label1` is used to display the welcome message, and `label2` serves as the caption to the `comboBox1`. From the Properties window of `comboBox1`, add `en-US` and `en-CN` to the `Items` collection so that at runtime a user can use `comboBox1` to choose the language settings for the user interface of the sample application. To make the `Text` values of both `label1` and `label2` culture specific, this application separates Chinese culture content from the main code and compiles it to a separate resources file.

When creating a culture-specific resources file, be sure to use the following naming convention:

```
<resourcename>.<culturename>.resx
```

For example, in the sample code, the Chinese resources file is named as `Resources.zh-CN.resx`. The English resources file, which is the fallback resources file that uses the default '`en-US`' culture in this application, is named as `Resources.resx`. By following the resources file-naming convention, the .NET Compact Framework runtime can successfully link the compiled resources binary files; otherwise, the runtime will complain that the resources file cannot be located.

Two strings are created in the project resources file `Resources.resx`. `Caption` is the English-language representation of the `label2` text, and `WelcomeStr` is the English-language welcome message. You then need to create the corresponding Chinese presentation in `Resources.zh-CN.resx`, as shown in Figure 14-6.

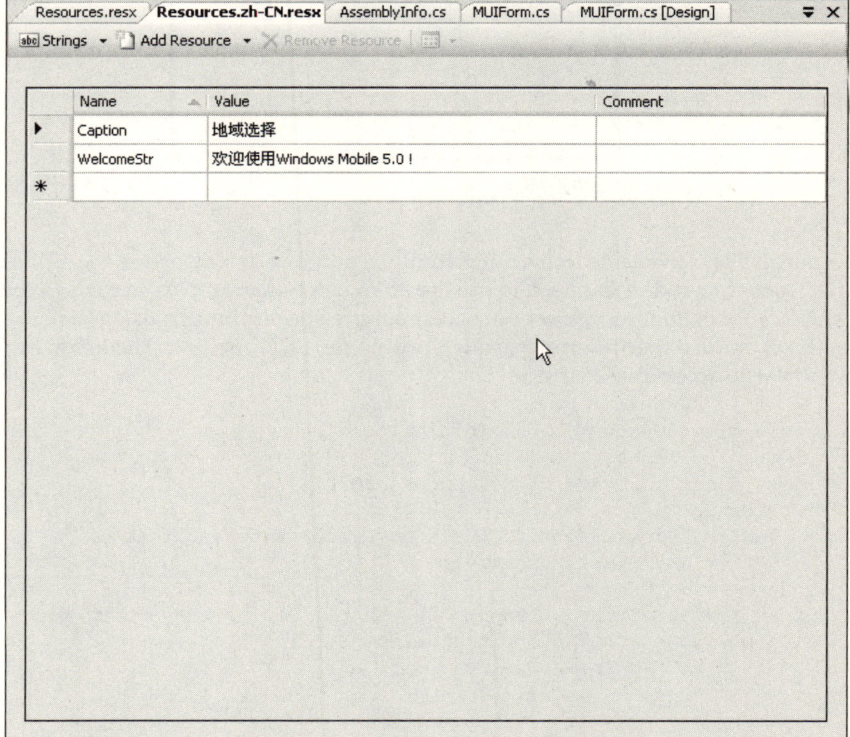

Figure 14-6

Note that the names of the two strings are identical in both resources files. If your application has a good deal of culture-specific content, it is error prone and inefficient to create a new resources file from scratch. A better approach is to work on one resources file first, and then copy and paste it as another resources file. This shortcut is particularly useful if you are building several culture-specific satellite assemblies.

As shown in Figure 14-7, once the resources files are created and compiled, a corresponding culture-specific `MUIWelcome.resources.dll` file is created in the `zh-CN` folder.

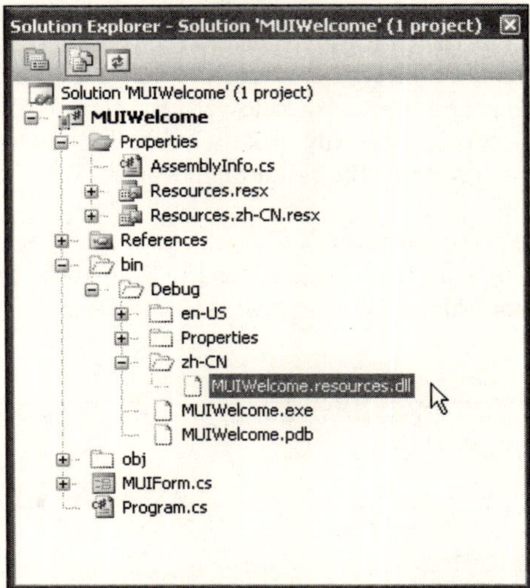

Figure 14-7

Now the resources file is available to the .NET runtime, but how do you access the strings defined in the resources file from your code? You need to use the `ResourceManager` class in the `System.Resources` namespace. A `ResourceManager` object can access culture-specific images using the `GetObject()` method, or access culture-specific strings using the `GetString()` method. The following code example demonstrates how to access those strings:

```
private void displayLabel(CultureInfo culture)
{
    //Create a new instance of ResourceManager
    ResourceManager Rm =
          new ResourceManager("MUIWelcome.Properties.Resources",
            this.GetType().Assembly);

    //Set the value of the string to the entries in corresponding resources file
    this.label1.Text = Rm.GetString("WelcomeStr", culture);
    this.label2.Text = Rm.GetString("Caption", culture);
}
```

The `displayLabel()` function changes the `Text` property of both `label1` and `label2` to culture-related content. To begin, instantiate a `ResourceManager` object. You need to specify the location of the resources file. Referring back to Figure 14-7, the `Resources.zh-CN.resx` file is located under the `Properties` of the `MUIWelcome` project, which is why the location of the resources file should be specified as `MUIWelcome.Properties.Resources`.

When a user selects a different language setting from `comboBox1`, a `SelectedValueChanged` event is raised. In the event handler, you can create a new `CultureInfo` object based on user's selection and pass this `CultureInfo` object as the `displayLabel()` function, as shown in the following code:

```
        //Display the screen using the language a user specifies
        private void comboBox1_SelectedValueChanged(object sender, EventArgs e)
        {
            //Declare a CultureInfo variable
            CultureInfo selectedCulture;

            switch (this.comboBox1.SelectedIndex)
            {
                //Change the language to "en-US" if it is selected
                case 0:
                    selectedCulture = new CultureInfo("en-US");
                    break;

                //Change the language to "zh-CN" if it is selected
                case 1:
                    selectedCulture = new CultureInfo("zh-CN");
                    break;

                //Always use "en-US" as the fallback language
                default:
                    selectedCulture = new CultureInfo("en-US");
                    break;
            }

            //Change labels accordingly
            displayLabel(selectedCulture);
        }
```

Following is the full code listing:

```
using System;
using System.Collections.Generic;
using System.ComponentModel;
using System.Data;
using System.Drawing;
using System.Text;
using System.Windows.Forms;

//Need resources and globalization
using System.Resources;
using System.Globalization;


namespace MUIWelcome
{
    public partial class MUIForm : Form
    {
        public MUIForm()
        {
            InitializeComponent();

            //Get current culture info
            CultureInfo curCulture = CultureInfo.CurrentCulture;

            //Display label in current language (culture)
```

```
            displayLabel(curCulture);

        }

        private void displayLabel(CultureInfo culture)
        {
            //Create a new instance of ResourceManager
            ResourceManager Rm =
                  new ResourceManager("MUIWelcome.Properties.Resources",
                    this.GetType().Assembly);

            //Set the value of the string to the entries in corresponding resource
file
            this.label1.Text = Rm.GetString("WelcomeStr", culture);
            this.label2.Text = Rm.GetString("Caption", culture);

        }

        //Quit application
        private void menuItem1_Click(object sender, EventArgs e)
        {
            if ( MessageBox.Show ( "Really want to Quit?","Confirmation",
                  MessageBoxButtons.OKCancel, MessageBoxIcon.Question,
                  MessageBoxDefaultButton.Button2 ) == DialogResult.OK)
              this.Close();
        }
        //Display the screen using the language a user specifies
        private void comboBox1_SelectedValueChanged(object sender, EventArgs e)
        {
            //Declare a CultureInfo variable
            CultureInfo selectedCulture;

            switch (this.comboBox1.SelectedIndex)
            {
                //Change the language to "en-US" if it is selected
                case 0:
                    selectedCulture = new CultureInfo("en-US");
                    break;

                //Change the language to "zh-CN" if it is selected
                case 1:
                    selectedCulture = new CultureInfo("zh-CN");
                    break;

                //Always use "en-US" as the fallback language
                default:
                    selectedCulture = new CultureInfo("en-US");
                    break;
            }
```

```
            //Change labels accordingly
            displayLabel(selectedCulture);

        }
    }
}
```

Figure 14-8a illustrates the execution results of this application when `zh-CHS` is selected from the regional setting. And Figure 14-8b shows the language changes to English when `en-US` is selected.

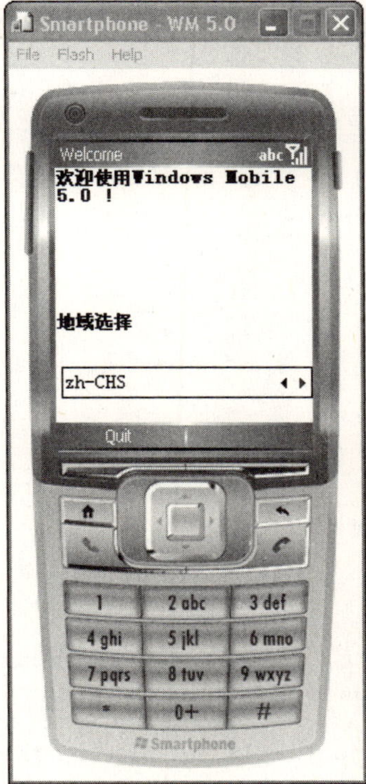

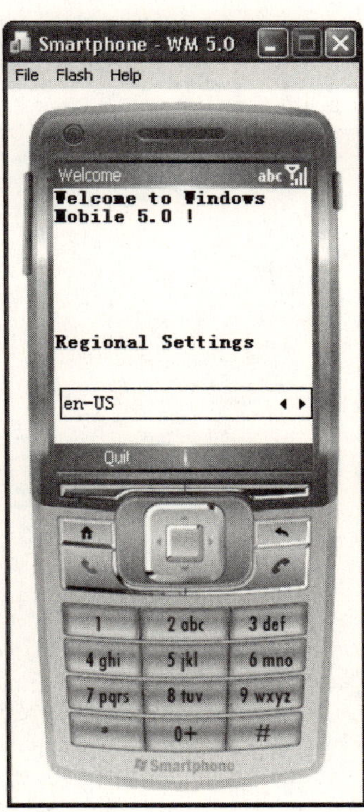

Figure 14-8

Some additional issues are worth mentioning. You do not need to add culture-specific resources for the texts of the buttons on a MessageBox control. For example, in the sample application, when a user presses the left soft key to quit, a message box will appear and ask for confirmation. If the user set the language to English, the texts of the two buttons are shown in English (see Figure 14-9a); otherwise, the texts are automatically displayed in Chinese (see Figure 14-9b).

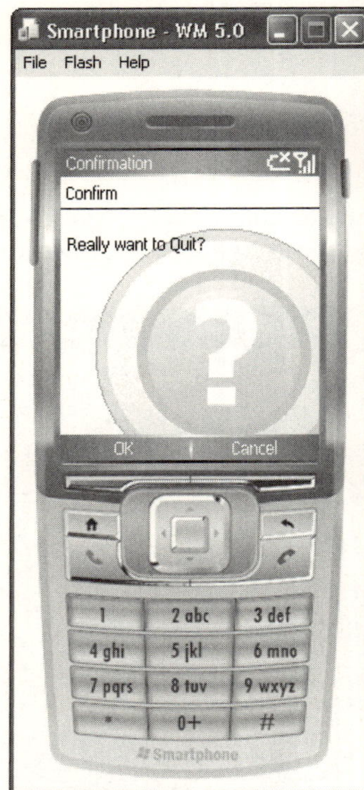

Figure 14-9

The second issue is regarding the fallback process of resources file retrieval. As you can imagine, if you need to add support for more than one culture, you need to create one resources file for each culture. Problems may occur, however, if the resource loader is expecting a resources file but it cannot be found anywhere. Suppose, for example, that you have created a resources file for a neutral culture such as `es` (Spanish). In your program, however, the `CurrentCulture` property is set to `es-MX`, which does not have a corresponding resources file in your project. How does the .NET Compact Framework handle this situation? Fortunately, it is smart enough to first fall back to the neutral culture name. In this case, when `es-MX` is not found, it will try to load a resources file marked `es`. If it cannot locate the `es` resources file, the .NET Compact Framework will use the default culture resources specified in the project assembly file as the last resort. In Visual Studio 2005, the fallback culture is not specified by default. Rather, it uses an empty string, which indicates a LCID of 0x007F and is associated with English. In C#, the name of the assembly file is `AssemblyInfo.cs`, which can be found in the `Properties` folder of the project in the Solution Explorer. However, it is safer to specify your default resources language in the assembly information file. For example, adding the following line in the `AssemblyInfo.cs` file will guarantee that the .NET Compact Framework falls back to using general English:

```
[assembly: System.Resources.NeutralResourcesLanguage("en")]
```

Finally, what if you are trying to retrieve an image from the resources rather than a string? You can achieve this by calling the `GetObject()` method of the `ResourceManager` class. For instance, if you have created a .bmp image file named myIcon in your resources file, then you can pass this image file to a PictureBox object aPicbox, as follows:

```
Windows.Forms.PictureBox aPicbox;
aPicbox.image = (image) Rm.GetObject("myIcon",culture);
```

Localizing Data

Another very important step in globalizing your application concerns with how data is stored and represented to the end users. In most cases, this can be taken care of by setting the CurrentCulture property of the CultureInfo class. You should be aware, however, of the pitfalls when you localize data, as described in the following subsections.

Dates, Time, and Calendars

For a globalized application, it is important to keep in mind that the world has many different time zones. Caution should be taken to avoid time discrepancies when synchronizing the data. For example, suppose user A in the U.S. Pacific time zone just uploaded data to the server. A few minutes later, user B in the U.S. Eastern time (3 hours later) tries to synchronize with the server. If the date and time are not handled properly, user B may either override the data entry with outdated entries or retrieve the wrong timestamps. Of course, a solution could be to synchronize the data based on the server's time, rather than the client's local time. However, what if you have more than one server located in more than one time zone? The best practice is always to save the value of the date and time to Greenwich Mean Time (GMT), also termed as Coordinated Universal Time (UTC), on the server side and translate them to local time when the mobile devices retrieve the data.

In the .NET Framework, the DateTime class provides a pair of functions to help you solve the problem. The ToUniversalTime() method will translate the local time to the GMT time; conversely, the ToLocalTime() method will translate the GMT time to your local time.

In addition to the time zone issue, different cultures normally have different formats for dates. For example, July 4, 2006, is expressed as 04/07/06 in British culture, while it is 07/04/06 in American culture.

To help eliminate the confusion, the DateTime structure in the .NET Framework offers a number of different formats to display the date and time. The most commonly used methods are as follows:

- ToString()
- ToShortDateString()
- ToShortTimeString()
- ToLongDateString()
- ToLongTimeString()

You can simply call one of these methods to display the date in the format specified by CurrentCulture. If you want to display the date in a format other than CurrentCulture, you can use the overloaded

`ToString()` method while indicating your desired date format. For example, the following code snippet will display the long date pattern in the `fr-CA` format:

```
DateTime today = DateTime.Now;
CultureInfo FrenchCA = new CultureInfo("fr-CA");
string msg = today.ToString(FrenchCA.DateTimeFormat.LongDatePattern);
MessageBox.Show(msg);
```

Similarly, you can use the `Parse()` method of the `DateTime` class to create a `DateTime` object from a string.

The following sample code demonstrates how date and time can be stored and retrieved in a persistent manner without being subjected to time zone and cultural differences:

```
//Save date and time to a culture invariant string with Universal time
private void SaveTime ( ) {

    //Convert local time to GMT
    DateTime uniTime = DateTime.Now.ToUniversalTime();

    //Save the value with culture invariant format
    string TimeStr = uniTime.ToString(CultureInfo.InvariantCulture);

    //Write data to a file
    StreamWriter aWriter = new StreamWriter("myData.txt");
    aWriter.WriteLine(TimeStr);
    aWriter.Close();
}

//Load culture invariant Universal date and time and translate them to local time
with local format
private DateTime LoadTime ( ) {

    //Read data from the file
    streamReader aReader = new StreamReader("myData.txt");
    string TimeStr = aReader.ReadLine();
    aReader.Close();

    //Create DateTime object by parsing the string
    DateTime uniTime = DateTime.Parse(TimeStr,CultureInfo.InvariantCluture);

    //Convert time from universal to GMT and return
    return uniTime.ToLocalTime();

}
```

You may also want to familiarize yourself with the `Calendar` class in the `System.Globalization` namespace. A `Calendar` object represents time in divisions, such as weeks, months and years. Even though the .NET Framework offers a variety of culture-specific calendars, only the following five are available on Windows Mobile for Smartphone:

- `GregorianCalendar`
- `JapaneseCalendar`

- ThaiBuddhistCalendar
- KoreanCalendar
- TaiwanCalendar

As an aside, it's a pity that the `EastAsianLunisolarCalendar` abstract class is not implemented in the .NET Compact Framework because many older people in Asia are still celebrating their birthdays based on the East Asian lunar calendar, rather than the Gregorian calendar.

To use a culture-specific calendar, you need to first create a new instance of the calendar. A variety of methods are available to display the culture-specific calendars. The following code snippet demonstrates how to interpret current date and time in the Korean calendar:

```
using System;
using System.Globalization;
...

//Create an instance of the KoreanCalendar class
KoreanCalender Kcal = new KoreanCalendar();

//Get current date and time
DateTime DtNow = DateTime.Now();

//Display current Korean era
Console.WriteLine("Era: {0}", Kcal.GetEra(DtNow) );

//Display current Korean year
Console.WriteLine("Year: {0}", Kcal.GetYear(DtNow) );

//Display current Korean month
Console.WriteLine("Month: {0}", Kcal.GetMonth(DtNow) );

...
```

Numbers and Currency

The ways to represent numeric data also differ across cultures. For example, in the United States "one million" is written as 1,000,000, whereas in China it is denoted as 100,0000.

To support different number formats, the `CultureInfo` class exposes a `NumberFormat` property that enables you to represent data consistent with a particular culture. By default, when you call the `ToString` method of number types, such as `System.Int32`, `System.Double` and `System.Decimal`, the numeric format is determined by the `NumberFormat` property of the current culture. If you want to display data in a different culture, you can simply create a new object of the `CultureInfo` class and call the overloaded `ToString()` method. The following code snippet shows how to format data for an Italian (Switzerland) culture (Italian-speaking Switzerland):

```
Cultureinfo sw = new CultureInfo ("it-CH");
double aNumber = 1234567.89;
string swFormat = aNumber.ToString(sw);
```

The signature of the `ToString()` method in this example is as follows:

```
System.Double.ToString(System.IFormatProvider)
```

The `IFormatProvider` interface provides developers with a means to control the formatting of an object, and the `CultureInfo` class is one of the derived types of this `IFormatProvider` interface.

If you want to fully customize the numeric format for any reason, you can also feed the format string code to the `ToString()` method. Table 14-2 summarizes the major format strings in the .NET Compact Framework.

For a more detailed description of format strings, please refer to the MSDN website at `http://msdn.microsoft.com/library/default.asp?url=/library/en-us/cpguide/html/cpconNumericFormatStrings.asp`.

Table 14-2 Main Format Strings

Format Specifier	Name	Description
C or c	Currency	Converts the number to a string that represents a currency amount. By default, use the currency of the current culture.
D or d	Decimal	Converts an integer to a string of decimal digits (0–9).
E or e	Scientific	Converts a number to a string of digits in exponential format.
F or f	Fixed-point	Converts a number to a string of digits with a fixed number of decimal places.
G or g	General	Converts a number to a string of digits in general format, which is the most compact of either fixed-point format or scientific format.
N or n	Number	Converts a number to a string of digits with thousand separators inserted between each group of three digits.
X or x	Hexadecimal	Converts a number to a string of hexadecimal digits (0–9, a–f).

The following code snippet illustrates how to use format strings to present a double type number in a different format:

```
using System;
...

int MyDouble = 98765.4321;

Console.WriteLine(MyDouble.ToString("C"));
Console.WriteLine(MyDouble.ToString("E"));
Console.WriteLine(MyDouble.ToString("F"));
Console.WriteLine(MyDouble.ToString("G"));
Console.WriteLine(MyDouble.ToString("N"));

...
```

The preceding code generates the following strings to the console:

```
$98,765.43
9.876543E+004
98765.43
98765.4321
98,765.43
```

Automatic formatting also works for currency value. You can apply the same technique to display a currency value for an individual culture. However, the currency format will not handle the currency exchange rate appropriately. For example, a currency value of 234.18 is represented as "$234.18" in the en-US culture. This same value is shown as £234.18 if you force the string to be formatted for en-GG culture. Therefore, if you want to convert a value of currency from one culture to another, then you need to convert the value in addition to changing the formats. To ensure the correctness of this conversion, of course, you need access to the current exchange rate.

Another thing you should be aware of is the euro currency. The .NET Compact Framework assigns the symbol € (Euro) as the default currency for countries that have adopted the euro as their currency unit. It conflicts with the default currency settings of the operating system that runs the .NET Compact Framework, which is set to the national currency.

Strings

For an application that may contain more than one language, *character encoding* is an important issue you should never overlook. The .NET Compact Framework supports various character encodings, such as Unicode, UTF8, UTF7, and ASCII. By default, the .NET Compact Framework stores data with Unicode encoding. Therefore, if characters are always saved and represented with Unicode, you won't run into any hassles. However, to read external data that is not encoded in Unicode, you have to convert the encoding properly to avoid invalid translations.

The .NET Compact Framework provides the `Encoding` class in the `System.Text` namespace to enable the easy conversion of encoding. In the following example, assume a string is encoded as GB2312, with a code page 936. Your task is to convert this string into Unicode. This is how you can do it:

```
//Assume you get a byte array from method 'readFromSomewhere()'
//Also assume it is encoded as GB2312
byte[] GB2312Ary = readFromSomewhere();

//Get an encoding for GB2312 (code page 936)
Encoding GB2312Enc = Encoding.GetEncoding (936);

//Convert the GB2312Ary to UnicodeAry with Encoding.Convert
byte[] UnicodeAry = Encoding.Convert(GB2312Enc, Encoding.Unicode, GB2312Ary);
```

The key function here is the `Encoding.Convert()` method, which takes three parameters: source encoding (`GB2312Enc`), target encoding (`Encoding.Unicode`), and source byte array (`GB2312Ary`). Although `Unicode` is defined as a property of the `Encoding` class, GB2312 encoding needs to be generated separately; hence, the `GetEncoding (code page)` method is called before the `Convert()` method.

If later you need to retrieve the Unicode text and convert it to a GB2312 string, you can convert the Unicode byte array back to a GB2312 byte array and call the overloaded string constructor with the desired encoding, as follows:

```
String GB2312Str = new string(GB2312Ary, 0, GB2312Ary.length-1, GB2312Enc)
```

In addition to the character encoding issue, comparing string values in a different culture could return quite different results. This difference, however, if it correctly reflects the cultural difference, is considered a feature, not a bug. As a developer, you should preserve this cultural difference so that the string comparison results make more sense to end users.

By default, the `Compare()` method of the `String` class will compare a pair of strings in the context of the current culture. If you need to compare two strings according to a particular culture, you can use the overloaded string `Compare()` method and specify the culture name in the argument lists when you make the function call. The signature of this function is as follows:

```
public static int Compare(string strA, string strB, bool ignoreCase,
System.Globalization.CultureInfo culture)
```

The default culture-aware string comparison is great, but be aware that there is no guarantee that a database such as SQL Server can do the same thing for you. When your application is dealing with database-related data access, the best practice is to sort the data in your .NET Compact Framework application, rather than use a SQL statement to return a sorted list of data. For example, you can retrieve an unsorted list of data from a database to an `ArrayList` object. Then call the `Sort()` method of `ArrayList` to get a culture-aware sorted list.

Best Practices

Following are the best practices we recommend to develop a world-ready Smartphone application:

- **Make your application culture-friendly** — Displaying information for a particular culture correctly is good, but not necessarily good enough. As a visionary developer, you should always try to avoid pitfalls that may cause political or religious difficulties simply because you picked the wrong word.
- **Consider font and screen size when designing the UI** — You should realize that the size of the letters, characters, and symbols in other languages could differ from your own. The same is true for screen sizes of mobile devices. For example, a line of a message that fits the 176-pixels-wide English version of a Smartphone may require more pixels when displayed on a Chinese version of the device. You are encouraged to leave enough space between visual components on a user interface and always remember to test and run the application on the real devices with different cultures.
- **Consider text directions when designing the UI** — Remember that text in some languages goes from right to left or vertically from top to bottom. When designing a culture-aware UI, take text direction into consideration.

- **Separate the content from the UI** — This is particularly true if you are going to patch your world-ready applications with resources files targeted at a number of cultures.
- **Redraw the screen when necessary** — When switching from one culture to another culture, calling the methods in the `System.Drawing` namespace to display the texts on the screen can avoid possible UI distortions.

Summary

When building an application that can be easily adapted to another locality, not only are the development costs saved, but the production cycle process is simplified, which in turn helps you grab the international market quicker than your competitors. Even though you have to spend extra time and plan carefully in advance, developing a culture-aware application is certainly a wise investment that can return tangible profits.

This chapter discussed the concepts and practices of world-ready applications in the .NET Compact Framework, including the following highlights:

- The best practice to develop a world-ready application is to create localized satellite assemblies, one for each culture.
- The `CultureInfo` class in the `System.Globalization` namespace can be used to make an application culture-aware and enable it to present data in a culture-related format.
- The `Calendar` class in the `System.Globalization` namespace can be used to display calendar information other than the general Gregorian calendar.
- The `Encoding` class in the `System.Text` namespace can be used to convert character encoding.

In the next chapter, you will learn how to draw graphics on a Smartphone.

15

Graphics

In computer programming, *graphics* refers to the software platform facility and programming language construct that enable the display and control of objects such as lines, curves, two- and three-dimensional shapes, surfaces, text, and images. Many Smartphone applications use graphics heavily to enrich the user experience. Although controls can be used most of the time, in some cases the application must directly draw the entities.

Managed Smartphone applications can take advantage of GDI+ in the .NET Compact Framework for graphics-related tasks. The `System.Drawing` and `System.Drawing.Text` namespaces consist of a number of classes for vectors, text, images, and a set of common graphics components. This chapter discusses these classes as well as some graphics terminology in Windows GDI+, such as clipping, double-buffering, and embedded resource. Granted, we can't cover every aspect of this broad area, but we identify the most frequently used classes and enumerations, with the intention of making your further exploration easier. Also included in this chapter are numerous examples.

The following topics are discussed in this chapter:

- The `System.Drawing.Graphics` class
- Working with vector graphics
- Drawing text and using fonts
- Manipulating and drawing images

Note that the graphics discussed in this chapter are all two-dimensional. For 3D graphics, Windows Mobile 5.0 provides Mobile Direct 3D — a set of native APIs to control and render 3D graphics on Windows Mobile devices. There is also managed Mobile D3D support in the .NET Compact Framework for Windows Mobile 5.0 and later, Pocket PC, and Smartphone. Because programming 3D graphics requires a solid knowledge of 3D-related terminology and concepts, this topic is beyond the scope of this book. Interested readers should look into the two namespaces for details: `Microsoft.WindowsMobile.DirectX` *and* `Microsoft.WindowsMobile.DirectX.Direct3D`.

.NET Compact Framework Graphics

Developers of desktop graphics applications on Windows are familiar with the functions in the GDI library (`gdi.dll`). Starting with Windows XP and Windows Server 2003, Microsoft introduced GDI+, which is a new 2D graphics environment with advanced features supporting complex 2D graphics composition and manipulation, and intrinsic support for JPEG and PNG formats. In the .NET Framework, the managed GDI+ graphics library is `gdiplus.dll`. On Windows CE (thus Windows Mobile), the core graphics functions for native code are wrapped in a library called `coredl.dll`. The .NET Compact Framework provides a very compact managed library called `system.drawing.dll`, which includes two namespaces: `System.Drawing` and `System.Drawing.Text`.

The `System.Drawing` namespace contains the following four types of classes:

- Graphics classes for outputting on display surfaces such as a screen or an image
- Text and font classes for creating and displaying text
- Vector classes for creating and displaying lines, rectangles, ellipses, etc.
- Image classes for creating and displaying bitmaps and gif images

You should not use `System.Drawing` *classes in a desktop Windows service or an ASP.NET service, as they are primarily designed for graphic user-interface applications, not services.*

Although basic font functionality is implemented in a number of classes in the `System.Drawing` namespace, the `System.Drawing.Text` namespace provides three enhanced font collection classes and an enumeration: The `FontCollection` class provides a base class for installed and private font collections; the `InstalledFontCollection` class represents the fonts installed on the system; the `PrivateFont Collection` class represents a collection of font families provided by applications (private font collections); and the `GenericFontFamilies` enumeration specifies a list of three font families: `Monospace`, `SansSerif`, and `Serif`.

Unless absolutely necessary, you should use controls rather than output directly on the display surface. Graphics can be very complicated to handle and error prone to code. Hence, if there is no special need (such as performance concerns and a unique GUI), we suggest using a control to display graphical components.

The Graphics Class

The `System.Drawing.Graphics` class, the core class for .NET graphics, encapsulates a display surface for graphical output. Any instance of a `Graphics` class has a context that defines the target surface of the output. You can use the `Graphics` class to do the following:

- Create color, pens, and brushes
- Draw lines, rectangles, ellipses, polygons, etc.
- Draw text

Note that the `Graphics` *class cannot be inherited.*

Before discussing each category, we first introduce the methods of obtaining a `Graphics` object.

Creating a Graphics Object

One way to create a `Graphics` object is to use the `CreateGraphics()` method of a control or form. This will return a reference to the `Graphics` instance of the drawing surface associated with the control or the form. For example, in your main form class (a class derived from `System.Windows.Forms.Form` class), generated automatically by Visual Studio when you create a Smartphone project, you can do the following to obtain a `Graphics` reference:

```
Graphics g = this.CreateGraphics();
// Do the drawing here
....
g.Dispose();
```

The `Graphics::Dispose()` method is used to release all resources used by the object.

Another way to create a `Graphics` object is to use the `PaintEventArgs` parameter of a control or a form's `Paint` event. A `Paint` event is generated whenever the control or the form is redrawn. For example, when a form pops up another form, and that second form is later closed, the original form (and its controls) will be redrawn and the `Paint` event will be fired. When a form is redrawn, the `Paint` event handler of every control on the form will also be called to redraw the controls automatically. As a result, you don't need to redraw the controls. However, everything that needs to be done in addition to redrawing controls for proper window drawing should be put into the `Paint` event handler. To trigger the `Paint` event in a method, use the `Control::Invalidate()` method. You can invalidate the entire control or a specific rectangle. Either way, a `Paint` event will be sent to the control. In the following example, whenever the form is redrawn, the event handler `Form_Paint()` will be called:

```
private void Form1_Paint(object sender, PaintEventArgs e)
{
     Graphics g = e.Graphics;
     // Do the drawing here
     ...
     g.Dispose();
}
```

The third way to create a `Graphics` object is to use the `Graphics.FromImage()` method. Simply pass an instance of a class that is derived from the `Image` class, and you will obtain a `Graphics` object representing that image. The following is an example:

```
Bitmap newImage = new Bitmap(16, 16);
Graphics gBitmap = Graphics.FromImage(newImage);
```

Creating a `Graphics` object is expensive in terms of performance. A general guideline is to try to reduce the number of `Graphics` objects created in your application. If possible, you may create a single `Graphics` object and use it whenever you need to draw something on the screen or on an image.

The Color, Pen, and Brush Objects

A `Color` structure represents an ARGB (alpha, red, green, and blue) color, with alpha as the transparency value. It has a number of static properties that define a set of named colors. For example,

System.Drawing.Color.LightBlue returns a Color object for the system-defined light-blue color. There are more than 140 named colors defined in the Color structure.

ARGB is sometimes referred to as RGBA; the last datum is the alpha value.

You can also create your own Color object by specifying ARGB values in the static Color.FromArgb() method. The alpha value and each of the RGB values are 8 bits each. The following two lines create two Color objects that share the same ARGB value (A=FF, Red=16, Green=32, Blue=48):

```
Color colorA = Color.FromArgb(0xFF102030);
Color colorB = Color.FromArgb(16, 32, 48);
```

Note that in the first line of the preceding example, the default alpha value is 0xFF, which specifies fully opaque. Even if you use 32-bit integers for each of the RGB values, only the lowest 8 bits will be used to construct the Color object. You can use Color::ToArgb() to get the 32-bit integer value of ARGB:

```
int blue = Color.Blue.ToArgb();
```

Another way to obtain a Color object is by using a SystemColors enumeration such as SystemColors .WindowText or SystemColors.Menu. These are Color objects that the system is currently using to display windows elements.

A Pen object represents the attributes of a pen that will be used for drawing. You can specify the color of a Pen, its line width (a float value), and the line style (either Solid or Dash defined in the System .Drawing.Drawing2D.DashStyle enumeration of the .NET Compact Framework). There are other properties in the Pen class of the .NET Framework, but these three are the only properties supported in the .NET Compact Framework.

The following example creates a Pen object:

```
Pen penBlue = new Pen(Color.Blue);
penBlue.DashStyle = System.Drawing.Drawing2D.DashStyle.Dash;
penBlue.Width = 2.0F;
```

A Pen object is used to create a Brush object (as shown below) and in common drawing methods of the Graphics class.

A Brush object is used to fill a shape with a selected color. In the .NET Framework, you can also specify a pattern when creating a Brush object, but this is not available in the .NET Compact Framework. To create a Brush object with a specific color, use the SolidBrush class, as follows:

```
Brush bshYellow = new SolidBrush(Color.Yellow);
```

Of course, you can specify a system color defined in the SystemColors enumeration, or create an ARGB color on-the-fly and use it for the SolidBrush.

Once a Brush object is created, you can use it in "fill" methods of the Graphics class, including FillRectangle(), FillEllipse(), FillPolygon(), and FillRegion().

Vector Graphics

Compared with the full .NET Framework, the .NET Compact Framework supports only a limited number of vector graphics methods in the `System.Drawing.Graphics` class. For `DrawXXX` methods (where `XXX` represents the shape), a `Pen` object is needed; for those `FillXXX()` methods, a `Brush` object is required.

Many classes and enumerations in the .NET Framework `System.Drawing.Drawing2D` *namespace are not available in the .NET Compact Framework. In fact, in the .NET Compact Framework, there are only two enumerations,* `DashStyle` *(for line style) and* `CombineMode` *(for region clipping mode) in the* `System.Drawing.Drawing2D` *namespace.*

Before we talk about the vector graphics methods in the `Graphics` class, we need to introduce two related terms: *region* and *clip*. Other graphics-related terms, such as rectangle, polygon, and ellipse, are self-explanatory.

Region

A *region* describes an area of the display surface. In the .NET Framework, a region can be any shape that is defined by a series of points; it can also be a region obtained from a rectangle. In the .NET Compact Framework, however, you can create a region only from a rectangle, meaning you don't have the flexibility to create customized shape regions.

The `Graphics` class has a method to fill a region with a specified `Brush` object. When a `Region` object is not needed anymore, the `Region::Dispose()` method should be called to release the resource of the `Region object.`

Another useful method is `Region::IsVisible()`, which accepts a `Point` structure, a `Rectangle` structure, or a coordinates pair. You can use this method to determine whether a point or a rectangle is contained in the region. In the following example, the `IsVisible()` method is used to determine whether a rectangle is in the region:

```
// Create a region
Rectangle regionRect = new Rectangle(0, 0, 100, 100);
Region myRegion = new Region(regionRect);

RectangleF myRect = new RectangleF(10, 10, 20, 20);

// Determine the rectangle is contained in the region.
bool contained = myRegion.IsVisible(myRect);
```

The `Region` class provides five different ways to perform operations with two regions. The parameters passed to these methods are either a `Region` object or a `Rectangle` object.

- `Intersect` operations update the region with the intersection with the specified rectangle or region.
- `Union` operations combine two regions or rectangles.
- `Xor` operations update the region with a union of two regions minus the intersection.

- `Complement` operations simply update the region to the portion of the specified rectangle or region.
- `Exclude` operations update the region to the portion of its interior that does not intersect with the specified region or rectangle.

The following code snippet demonstrates how to perform region operations: Two rectangles, `recA` and `recB`, which partially intersect, are created and drawn on the screen. A `Region` object `reg` is created from `recA`. Then we perform one of the five region operations:

```
private void Form1_Paint(object sender, PaintEventArgs e)
{
    Graphics g = e.Graphics;

    // Rectangle A
    Rectangle recA = new Rectangle(40, 40, 100, 100);
    g.DrawRectangle(new Pen(Color.Red), recA);

    // Rectangle B
    Rectangle recB = new Rectangle(60, 60, 140, 140);
    g.DrawRectangle(new Pen(Color.Black), recB);

    // Region reg is created from Rectangle A
    Region reg = new Region(recA);

    // Demo the region operations; Here five region operations can be used
    reg.Exclude(recB);

    // Fill the region
    g.FillRegion(new SolidBrush(Color.LightSkyBlue), reg);

    // Release resource
    reg.Dispose();
    g.Dispose();
}
```

Figure 15-1 illustrates the region operations with five screenshots of the preceding code on a drawing pane, each corresponding to a region operation method. A simple logical expression is used to explain the operations: A and B are the two entities (representing Rectangle A and Rectangle B, respectively). The operations are `A.Intersect(B)`, `A.Union(B)`, `A.Xor(B)`, `A.Complement(B)`, and `A.Exclude(B)`. The expressions in the parentheses of the screenshot description define the set of points in the result region. AND generates points in both A and B (intersect); `OR` generates points in either A or B, or both (union); and – excludes some specified set of points.

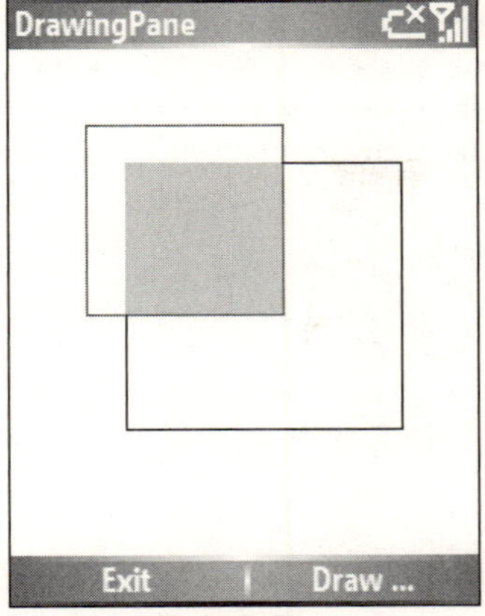

Intersect (A AND B)

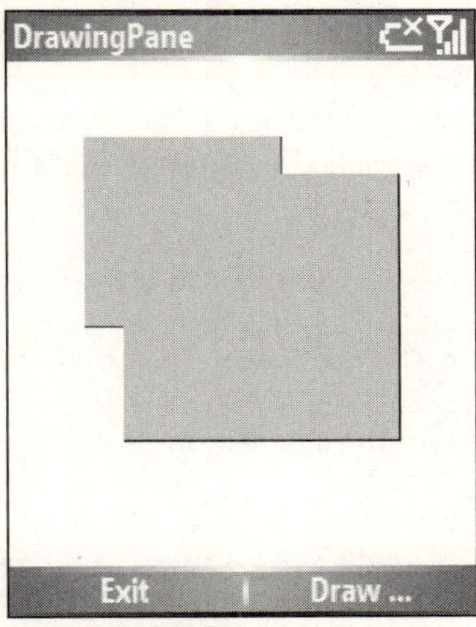

Union (A OR B)

Xor (A OR B) – (A AND B)

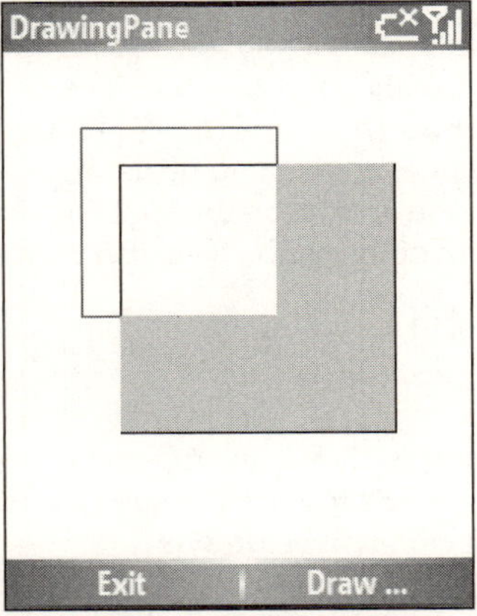

Complement (B – (A AND B)

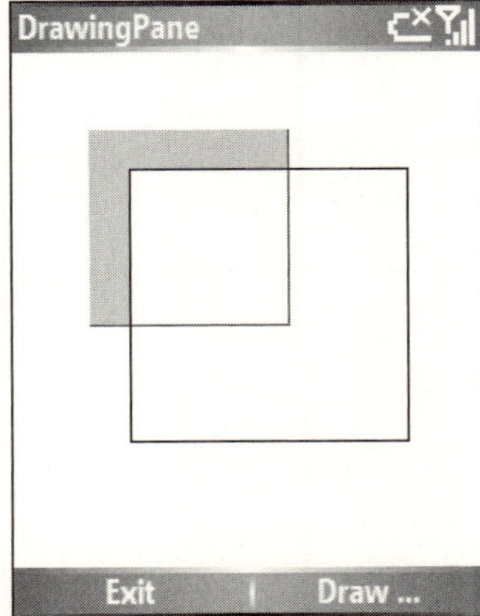

Exclude (A – (A AND B)

Figure 15-1

Clipping

The *clip region* is the area within which you can draw. You can set the clip region so that all subsequent drawing will be restricted within it. To do this, set the `Clip` property of the `Graphics` object. For example, the following code sets the clip region of the current form or control to a rectangle defined by the `Region` object `reg`:

```
Graphics g = CreateGraphics();
Region reg = new Region(new Rectangle(50,50,180,180));
g.Clip = reg;
```

You can always obtain the clip region using the same property:

```
Region reg2 = g.Clip;
RectangleF clipRect = reg2.GetBounds(g);
// Then we can obtain the region's location and size
int x = clipRect.Left;
...
```

The preceding code creates a `Region` object `reg2` and associates it with the clip region of the `Graphics` object `g`. Then we use the `Region::GetBounds()` method to get the detailed coordinates and size data of the region. The `RectangleF` structure is almost the same as the `Rectangle` structure, except that the coordinates and size values are float numbers. You can also use the `Region` object (in this example, `reg2`) to further clip the region area using the region operation methods discussed earlier. If you just want to obtain the rectangle bound of the region area, there is a quicker way to do that: use the `Graphics::ClipBounds` property:

```
RectangleF rf = g.ClipBounds;
```

The following example shows an ellipse on a clipped region in a form. Because the clip region is set to a rectangle and the circle's top-left portion is outside the rectangle, only the portion of the circle falling into the clipped region will be drawn:

```
private void Clipping()
{
    Graphics g = CreateGraphics();
    RectangleF rf0 = g.ClipBounds;

    Region reg = new Region(new Rectangle(50,50,150,150));
    g.FillRegion(new SolidBrush(Color.DarkGray), reg);

    g.Clip = reg;
    g.FillEllipse(new SolidBrush(Color.Gold),20,20,100,100);
    /*RectangleF rf = g.ClipBounds;
    Region reg2 = g.Clip;
    RectangleF clipRect = reg2.GetBounds(g);*/

    reg.Dispose();
    g.Dispose();
}
```

Figure 15-2 shows the screenshot of the clip region of the form. The gray rectangle is the clipped region that does not cover the whole circle. As a result, only a portion of the circle is drawn.

Figure 15-2

Vector Methods

Common drawing tasks are implemented as a set of drawing methods in the `Graphics` class. Table 15-1 lists the vector–related methods in the `System.Drawing.Graphics` class.

Table 15-1 System.Drawing.Graphics Vector-Related Methods

Method	Parameters	Description
`DrawLine()`	A `Pen` object and a coordinates pair	Using a pen, draws a line between the two points
`DrawLines()`	A `Pen` object and an array of `Point` structures	Using a pen, draws a series of lines that connect those points in the array
`DrawEllipse()`	A `Pen` object and a bounding `Rectangle` object of the ellipse, or A `Pen` object and a bounding rectangle specified by a coordinates pair for the top-left corner vertex, width, and height	Using a pen, draws an ellipse bounded by the rectangle
`DrawPolygon()`	A `Pen` object and an array of `Point` structures	Using a pen, draws a polygon defined by the point array
`DrawRectangle()`	A `Pen` object and a `Rectangle` structure, or A `Pen` object, a coordinates pair, width, and height	Using a pen, draws a rectangle

Table continued on following page

Method	Parameters	Description
`FillEllipse()`	A `Brush` object and a bounding `Rectangle` object of the ellipse, or A `Brush` object and a bounding rectangle specified by a coordinates pair for the top-left corner vertex, width, and height	Using a brush, fills the interior of an ellipse defined by the rectangle
`FillPolygon()`	A `Brush` object and an array of `Point` structures	Using a brush, fills the interior of a polygon defined by the point array
`FillRectangle()`	A `Brush` object and a `Rectangle` structure, or A `Brush` object, a coordinates pair, width, and height	Using a brush, fills the interior of a rectangle
`FillRegion()`	A `Brush` object and a `Region` object	Using a brush, fills the region

The coordinates origin used in a display screen is at the top-left corner. The *X* axis runs from left to right, and the *Y* axis runs from top to bottom. To get the size of the drawing area, use the `ClientSize` property of the form or control, which is a `Size` structure containing the width and height of the drawing area of the form or control. If a form has a scroll bar, borders, or a menu bar, the `ClientSize` will be the bounds of the form/control minus these nondrawing areas.

```
   //In an instance method of a Form
       Size controlSize = this.ClientSize();
```

The following example draws a grid on the screen by calling the `DrawLine()`and `FillRegion()` methods. The grid looks like the board used in the popular game Go, with 19 × 19 lines. We also draw two small circles, one black and one white, at two special locations on the board.

The main `Form` class has three fields to define some variables used in the `DrawGrid()` method (listed below): a `Point` structure of the location of the grid, the line spacing value, and the number of lines. The grid has 19 horizontal lines and 19 vertical lines, all evenly separated by 10 points:

```
           private Point pntOrigin;
           private int lineSpacing = 10;
           private int lineCnt = 19;
```

The following code is the `DrawGrid()` method used to draw the grid in a form. The location of the grid was determined in the `DrawGrid()` method based on the size of the `ClientSize` of the form — we want to place the grid in the middle of the screen.

The `Pen` object used to draw the lines is a blue path with a width of 2 and a dash style of `System.Drawing.Drawing2D.DashStyle.Solid`. The brush used to paint the background of the grid is solid yellow:

```
private void DrawGrid()
{
    Graphics g = this.CreateGraphics();

    // Get the client area size of the form
    Size szeClient = this.ClientSize;

    // Calculate the origin of the lines
    pntOrigin = new Point((szeClient.Width - 18 * lineSpacing) / 2,
                          (szeClient.Width - 18 * lineSpacing) / 2);

    // First, we paint the background
    Brush bshYellow = new SolidBrush(Color.Yellow);
    Rectangle recDrawArea = new Rectangle(pntOrigin.X - 10,
                                          pntOrigin.Y - 10,
                                          (lineCnt - 1)* lineSpacing + 20,
                                          (lineCnt - 1)* lineSpacing + 20);

    Region rgnDraw = new Region(recDrawArea);

    g.FillRegion(bshYellow, rgnDraw);

    int x1, y1;  // Line's starting point
    Pen penBlue = new Pen(Color.Blue);
    penBlue.DashStyle = System.Drawing.Drawing2D.DashStyle.Solid;
    penBlue.Width = 2.0F;

    // Draw a grid on the screen
    int i = 0;

    // Draw horizontal lines
    int lineLength = (lineCnt - 1) * lineSpacing;
    for (i = 0, x1 = pntOrigin.X, y1 = pntOrigin.Y; i < lineCnt; i++)
    {
        g.DrawLine(penBlue, x1, y1, x1 + lineLength, y1);
        y1 += lineSpacing;
    }

    // Draw vertical lines
    for ( i = 0, x1 = pntOrigin.X, y1 = pntOrigin.Y; i < lineCnt; i++)
    {
        g.DrawLine(penBlue, x1, y1, x1, y1 + lineLength);
        x1 += lineSpacing;
    }
    rgnDraw.Dispose();
    g.Dispose();
}
```

The two small circles are drawn in the following method. Here we use one of the overloaded `Graphics::DrawEllipse()` methods that requires a `Pen` object, a coordinates pair identifying the top-left vertex of the bounded rectangle of the ellipse, and the width and height of the rectangle. The size of the rectangle (actually a square) is two-thirds of the `lineSpacing`. Then we call the `Graphics::FillEllipse()`

method to paint the interior of the circle. Note that the bounding rectangle of each circle (whose diameter is 2/3 of the line spacing) must be placed right above the crossing of the two grid lines. Thus, the rectangle's top-left vertex should be `dotSize/2` to the left and top of the grid crossing, as shown in Figure 15-3.

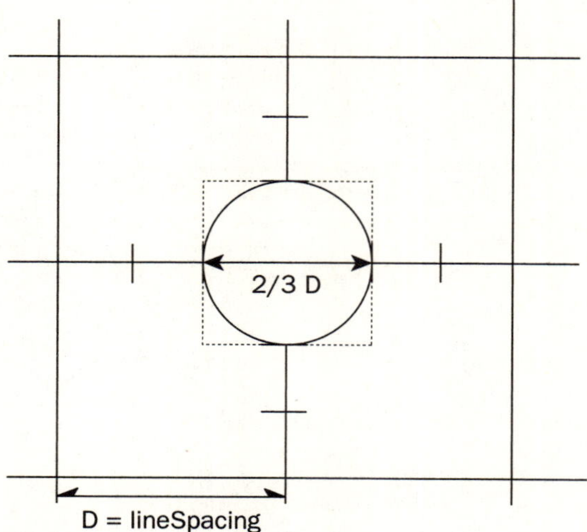

Figure 15-3

```
private void DrawElipse()
{
    Graphics g = CreateGraphics();
    Pen penBlack = new Pen(Color.Black);
    int dotSize = lineSpacing * 2 / 3;

    // Draw the first 'black' dot
    g.DrawEllipse(penBlack,
        pntOrigin.X + 4 * lineSpacing - dotSize / 2,
        pntOrigin.Y + 4 * lineSpacing - dotSize / 2,
        lineSpacing / 2,
        lineSpacing / 2);
    Brush bshBlack = new SolidBrush(Color.Black);
    g.FillEllipse(bshBlack,
        pntOrigin.X + 4 * lineSpacing - dotSize / 2,
        pntOrigin.Y + 4 * lineSpacing - dotSize / 2,
        lineSpacing / 2,
        lineSpacing / 2);

    // Draw the second 'white' dot
    g.DrawEllipse(penBlack,
        pntOrigin.X + 15 * lineSpacing - dotSize / 2,
        pntOrigin.Y + 15 * lineSpacing - dotSize / 2,
        lineSpacing / 2,
        lineSpacing / 2);
    Brush bshWhite = new SolidBrush(Color.White);
```

```
            g.FillEllipse(bshWhite,
                pntOrigin.X + 15 * lineSpacing - dotSize / 2,
                pntOrigin.Y + 15 * lineSpacing - dotSize / 2,
                lineSpacing / 2,
                lineSpacing / 2);

            g.Dispose();
        }
```

Figure 15-4 shows the grid on the Smartphone emulator.

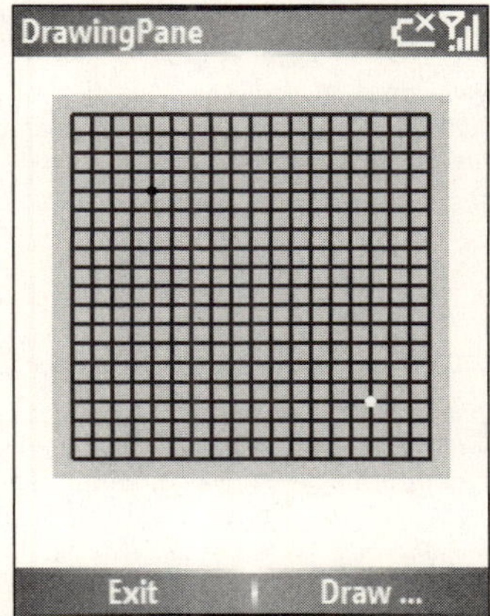

Figure 15-4

Drawing Text

You can draw text on a form or a control by using one of the overloaded `Graphics::DrawString()` methods. However, you might not achieve the best visual result if the string stretches too long or is badly aligned. Regardless of which overloaded `DrawString()` method you use, you always need to pass a `String`, a `Font`, and a `Brush`. In addition, you may pass a `RectangleF` structure as the bounding rectangle for the string box. Another setting you can use to draw text is a `StringFormat` object that specifies line alignment on the horizontal plane, text alignment on the vertical plane, and some format flags controlling text clipping and wrapping against the bounding rectangle.

Graphics::DrawString()

Every Window form and control has a default font. Unless you want to use a special font for the text, you may simply use the form or control's `Font` property. The following code snippet is an example of using the `Font` property of the form to draw a text string at a specific place:

```
            g.DrawString(".NET Compact Framework rocks!",
                Font,
                new SolidBrush(Color.Blue),
                60,
                60);
```

Normally, before calling the `DrawString()` method, you should first call `Graphics::MeasureString()` to obtain a `RectangleF` structure of the bounding text box that can be passed to `DrawString()`. The underlying reason is that when a string in a font is measured, some extra spacing and overhanging glyphs will be added to the result. With this piece of information, you can use the `DrawString()` method to finely control the string display by placing the text at the appropriate location or adjusting the alignment. The following is an example of using the `MeasureString()` method to obtain the size of the text:

```
// Measure string size first
SizeF szeText = g.MeasureString(".NET Compact Framework rocks!", Font);
g.DrawRectangle(new Pen(Color.Black),
    0,
    0,
    (int)szeText.Width,
    (int)szeText.Height);
g.DrawString(".NET Compact Framework rocks!",
    Font,
    new SolidBrush(Color.Blue),
    0,
    0);
```

To place your text string into a rectangular box, you must pass a `Rectangle` object to `DrawString()`:

```
RectangleF rct = new RectangleF(30F, 30F, 200F, 100F);
g.DrawString("Smartphone Application rocks!",
    Font,
    new SolidBrush(Color.Red),
    rct);
```

If the rectangle is not wide enough to hold the string, then the text will wrap around to the next line if there is sufficient space. If no space is available, the text string will be truncated. The text is left-aligned and top-aligned by default. If you want to change the formatting, use a `StringFormat` object and pass it to `DrawString()`:

```
RectangleF rct = new RectangleF(0, 30F, 200, 60);
StringFormat strFmt = new StringFormat();
// Text will be put at the bottom of the Rectangle
strFmt.LineAlignment = StringAlignment.Far;
// Text will be put in the middle on a horizontal line
strFmt.Alignment = StringAlignment.Center;
g.DrawString(" Mobile ",
    Font,
    new SolidBrush(Color.Black),
    rct,
    strFmt);
```

The `StringFormat::Alignment` property determines how the text is aligned horizontally, and the `StringFormat::LineAlignment` property specifies the vertical position of the text in a rectangle. Both of these properties get the value from the `StringAlignment` enumeration, where `StringAlignment::Center` means "in the middle," `StringAlignment::Near` means "close to the beginning of a text," and `StringAlignment::Far` means "to the other side of the beginning." In a left-to-right writing system such as English, "near" refers to left-aligned and top-aligned.

Working with Fonts

The preceding examples use a form's `Font` property to write text. Alternatively, you can create a `System.Drawing.Font` object by specifying a font name string, font size, and font style, chosen from the `FontStyle` enumeration, as shown in the following example:

```
Font fntArial = new Font("Arial", 14, FontStyle.Bold | FontStyle.Underline);
```

Another way to create a `Font` object is by specifying a `System.Drawing.FontFamily`, which represents a generic font type from which several fonts may be derived, as shown in the following example. The code also uses `StringFormat` objects to format the text within a `RectangleF` box. Note that the second `StringFormat` object, `strFmt2`, has a `NoClip` setting with its `FormatFlags` property. This specifies that if the text can't be displayed within the rectangle (`recF2` in the example), then it will continue outside the bounding box. To clearly illustrate these settings in `StringFormat`, the bounding boxes for the two strings are displayed as well.

```
            Font fntMono = new
Font(FontFamily.GenericMonospace,10,FontStyle.Italic);

            RectangleF rctF = new RectangleF(0F, 100F, 200F, 40F);
            RectangleF rctF2 = new RectangleF(0F, 150F, 40F, 60F);
            g.DrawRectangle(new Pen(Color.Black), (int)rctF.Left,
(int)rctF.Top,(int)rctF.Width,(int)rctF.Height);
            g.DrawRectangle(new Pen(Color.Red), (int)rctF2.Left, (int)rctF2.Top,
(int)rctF2.Width, (int)rctF2.Height);

            g.DrawString("Smartphone Application rocks!",
                fntMono,
                new SolidBrush(Color.Red),
                rctF);

            // Create a StringFormat object for text display settings
            StringFormat strFmt2 = new StringFormat();
            // Align center
            strFmt2.Alignment = StringAlignment.Near;
            // Align bottom
            strFmt2.LineAlignment = StringAlignment.Center;
            // Set "No clipping" for the text
            strFmt2.FormatFlags = StringFormatFlags.NoClip;

            // Draw the string with the StringFormat object
            g.DrawString("Mobile rocks!",
                fntArial,
                new SolidBrush(Color.Red),
                rctF2,
                strFmt2
                );
```

Figure 15-5 shows the display of these two text strings with different formatting. The first is top- and left-aligned by default because no `StringFormat` is applied, and the other is center- and left-aligned without clipping.

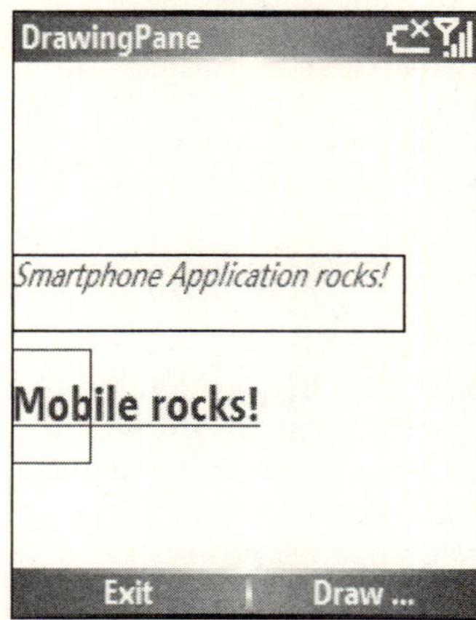

Figure 15-5

As you can see from the example, drawing text in the .NET Compact Framework for Windows Smartphone typically involves creating a number of graphics objects, such as `Pen`, `Brush`, `Font`, `RectangleF`, `Region`, and so on. Generally, to get better performance while a user interacts with the GUI, you need to create these objects before the user interaction starts, rather than create them on-the-fly while handling control events. For example, if a black pen will be used many times in your application, you can just create it once during application startup and use it whenever needed. In addition, using the form or control's `Font` property to draw text may lead to slower drawing performance, especially when many text strings are drawn. In this case, a `Font` object should be created before the GUI is loaded to the screen.

Drawing Images

You can draw images directly on the display surface or onto an `Image` object. If you only want to display an image with a format of bitmap (bmp), jpeg (jpg), png, gif, or icon (ico), simply use an image control (for example, a `PictureBox`). In addition, you can use an `ImageList` control to display images in a `ListView` control or a `TreeView` control. You can add images manually at design time on the `ImageList` control's properties page; you can also do this programmatically in the code (by adding a dynamically created image to the `ImageList`).

The following example shows how to use an `ImageList` control in a `ListView` control. We drag and drop these two controls to a Windows form, and manually add three 16 ×16 images (png files) to the `ImageList`. Then, when the form is shown, we generate the fourth image and add it to the `ImageList`. The `ListView` control is then populated with four text strings and is associated with the `ImageList`. After setting the `ListView`'s `View` property to `SmallIcon`, we show the `ListView`.

```
private void Form3_Paint(object sender, PaintEventArgs e)
{
    Bitmap newImage = new Bitmap(16, 16);
    Graphics gBitmap = Graphics.FromImage(newImage);
    gBitmap.Clear(Color.Red);
    gBitmap.DrawLine(new Pen(Color.White, 2.0F), 8, 0, 8, 16);
```

```
            imgList.Images.Add(newImage);

            // Draw the four images first
            Graphics g = CreateGraphics();
            g.DrawImage(imgList.Images[0], 0, 0);
            g.DrawImage(imgList.Images[1], 20, 0);
            g.DrawImage(imgList.Images[2], 40, 0);
            g.DrawImage(imgList.Images[3], 60, 0);

            // Another way to use the ImageList: associate each image with an item
in ListView
            for (int i = 0; i < 4; i++)
            {
                ListViewItem lstItem = new ListViewItem("Test " + i);
                lstItem.ImageIndex = i;
                listView1.Items.Add(lstItem);
            }

            listView1.SmallImageList = imgList;
            listView1.View = View.SmallIcon;
            listView1.Show();
            g.Dispose();
            gBitmap.Dispose();
        }
```

While reading the preceding code, you can safely skip the beginning part of it, as we discuss the bitmap creation and drawing functions shortly. Pay attention to the lines that relate to the `ListView` control `listView1`. To add items onto the `ListView` control, use a `ListViewItem` object, which is constructed with a text string and is associated with an index to the `ImageList` associated with the `ListView`. Note that a `ListView` has four types of views — `SmallIcon`, `LargeIcon`, `List`, and `Details` — each corresponding to a standard Windows list method. The images in our `ImageList` are used as small icons in the `ListView` accompanying the list items when `View.SmallIcon` is enabled. Figure 15-6 shows the `ListView` control with four list items and four small images.

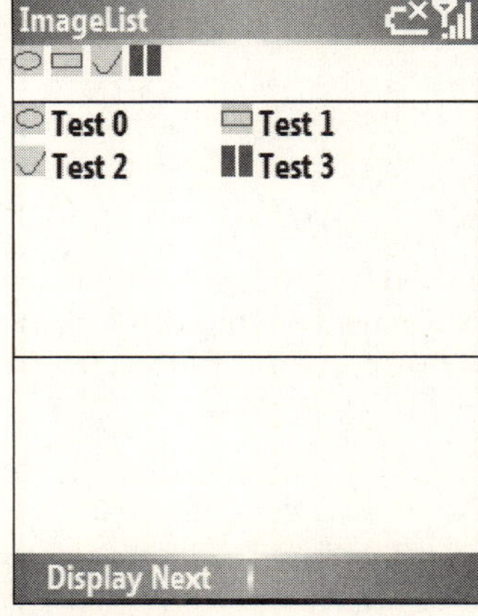

Figure 15-6

There might be other scenarios in which you want to have direct control over what is displayed in order to create some special visual effects. In these cases, you need to use the `Bitmap` class and the `Graphics::DrawImage()` method to display the image.

Drawing Bitmaps

The `System.Drawing.Bitmap` class is derived from the `System.Drawing.Image` class. To create a `Bitmap` class, use one of the five constructors with different parameters. You can create a `Bitmap` object based on an existing `Image` instance, a data stream (for an embedded resource; explained later), or an image filename. Or, you can create an empty bitmap with a specific size, and a pixel format as enumerated in `System.Drawing.Imaging.PixelFormat`. `PixFormat` defines the number of bits used for RGB, respectively. Following are the available `PixFormat` values for the .NET Compact Framework; there are more for the .Net Framework:

- `Format16bppRgb555`: 16 bits per pixel, with 5 bits each for red, green, and blue; the remaining one bit is not used
- `Format16bppRgb565`: 16 bits per pixel, with 5 bits for red, 6 bits for green, and 5 bits for blue
- `Format24bppRgb`: 24 bits per pixel, with 8 bits each for red, green, and blue
- `Format32bppRgb`: 32 bits per pixel, with 8 bits each for red, green, and blue; the remaining 8 bits are not used

The following are some examples of creating `Bitmap` objects:

```
            Bitmap b1 = new Bitmap(@"\Storage Card\wind.jpg");
            Bitmap b2 = new Bitmap(100, 100,
System.Drawing.Imaging.PixelFormat.Format16bppRgb555);
            Bitmap b3 = new Bitmap(b1);

            Graphics g = CreateGraphics();

            // Draw an image
            g.DrawImage(b3,0, 0);

            // Dispose images to release resource
            b1.Dispose();
            b2.Dispose();
            b3.Dispose();
```

In these examples, `Bitmap b3` is created from `Bitmap b1`, which in turn is loaded from a jpeg file at the specified path. Then `b3` will be displayed on the screen using one of the overloaded methods, `Graphics::DrawImage()`. The parameters in this call are the `Image` instance (`b3`) and a coordinates pair specifying where the image will be displayed.

Drawing on a Bitmap

As mentioned earlier in this chapter, the `Graphics` object can be created with an `Image` object by calling the `Graphic::FromImage()` method. For example, after creating an empty `Bitmap` object, you can obtain a `Graphics` object from the `Bitmap` and use it to draw vectors and text onto the `Bitmap`, just as you would draw on a display surface:

```
        // Create a bitmap image
        int size = 60;
        Bitmap bmp = new Bitmap(size, size,
System.Drawing.Imaging.PixelFormat.Format16bppRgb555);

        Graphics g = Graphics.FromImage(bmp);

        // Draw onto the bitmap
        Pen penWhite = new Pen(Color.White, 2.0F);
        g.Clear(Color.DarkBlue);
        g.DrawLine(penWhite,0, 5, size, 5);
        g.DrawLine(penWhite, 0, 10, size, 10);
        g.DrawLine(penWhite, 5, 0, 5, size);
        g.DrawLine(penWhite, 10, 0, 10, size);
        g.DrawEllipse(new Pen(Color.Yellow, 2.0F), 15, 15, 40, 40);

        // Draw the bitmap on screen
        Graphics gScreen = CreateGraphics();

        // Display the image at its original size
        gScreen.DrawImage(bmp, 0, 0);
        // Rectangle of the original bitmap
        Rectangle rct0 = new Rectangle(
            0,
            0,
            bmp.Width,
            bmp.Height);
        // Enlarge rectangle for the bitmap to fit in
        Rectangle rct1 = new Rectangle(
            bmp.Width + 5,
            bmp.Height + 5,
            bmp.Width * 2,
            bmp.Height * 2);
        // Display the bitmap, but enlarged
        gScreen.DrawImage(bmp, rct1, rct0, GraphicsUnit.Pixel);
        g.Dispose();
```

The preceding example demonstrates a technique called *double buffering*, which draws the image in memory on an off-screen `Graphics` object, and then displays the prepared image from memory to the screen (commonly known as *blitting*). This can significantly reduce display flickering and is widely using in desktop graphics applications. In the code, a `Graphics` object is created from the `Bitmap` instance. The `g.Clear(Color.DarkBlue)` call clears the background of the bitmap with a selected color. Four lines are drawn on it, two horizontal and two vertical, and then a circle is drawn. To draw the bitmap on the screen, a different `Graphics` object is created. The bitmap is drawn twice on the screen: the first time at the origin (`0, 0`) at its original size (`bmp.Width, bmpHeight`), and the second time at another place (`bmp.Width + 5, bmp.Height - 5`) at doubled its original size.

The first `Graphics::DrawImage` call

```
        gScreen.DrawImage(bmp, 0, 0);
```

passes three parameters:

- The Bitmap instance (`bmp`)
- A coordinates pair specifying the place to draw the image (`0, 0`)

Thus, this method is used to draw the entire image at a specified location.

The second `Graphics::DrawImage()` call

```
gScreen.DrawImage(bmp, rct1, rct0, GraphicsUnit.Pixel);
```

passes four parameters:

- The image instance (`bmp`).
- The bounding rectangle in which the image or a portion of the image will be drawn (`rct1`).
- A rectangle representing the portion of the original image to be drawn (`rct0`).
- The `GraphicsUnit` value (`GraphicsUnit.Pixel`) indicating the unit of measure. Only `GraphicsUnit.Pixel is supported in the .NET Compact Framework.`

Thus, this call draws the specified portion of the image at the specified location and with the specified size. In our example, we drew the entire bitmap and doubled its size. Figure 15-7 shows the screen display of the emulator.

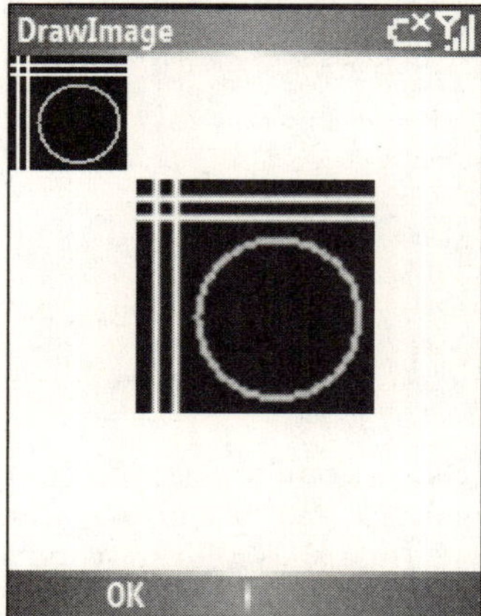

Figure 15-7

There are similar overloaded methods of `Graphics::DrawImage()` that require a different signature but essentially do the same thing. One of the overloaded `DrawImage()` methods accepts a parameter of a `System.Drawing.Imaging.ImageAttributes` object that controls color transparency. You need to define a color key that consists of a high key value and a low key value; any color that has each of its three components (red, green, blue) falling into the corresponding range between the low key and high key is made transparent. You can use the `SetColorKey()` method of the `ImageAttributes` object to define a color key.

In the .NET Compact Framework, the `ImageAttributes::SetColorKey(Color lowKey, Color highKey)` *method can accept only a single color key, meaning that both the low key and the high key must be the same. However, in the .NET Framework, the two keys can be different, specifying a color key range.*

If the image has a matching color, that color will not be drawn. The overloaded `DrawImage()` method using `ImageAttributes` has seven parameters: the `Image` object, a `Rectangle` object specifying the target rectangle for the image, *X* and *Y* coordinates, width and height that collectively specify a rectangle relative to the source image that will be drawn, `GraphicsUnit`, and an `ImageAttributes` object. The following is an example of using this method. The bitmap is drawn in the rectangle `destRct`. The bitmap's dark-blue color will not be drawn, as it matches the color key defined in the `ImageAttributes imgAttr object imgAttr`:

```
    Rectangle destRct = new Rectangle(bmp.Width, 0, bmp.Width, bmp.Height);
    System.Drawing.Imaging.ImageAttributes imgAttr = new
System.Drawing.Imaging.ImageAttributes();

    // Set the transparent color key; both high key and low key must be the same
    imgAttr.SetColorKey(Color.DarkBlue, Color.DarkBlue);
    gScreen.DrawImage(bmp,
        destRct,
        0,
        0,
        bmp.Width ,
        bmp.Height,
        System.Drawing.GraphicsUnit.Pixel,
        imgAttr);
```

Manipulating a Bitmap

A bitmap is a series of pixels. Graphics entities drawn on the image are eventually saved pixel by pixel, each taking a color value that can be 16 bits or 32 bits, as defined in `System.Drawing.Imaging.PixelFormat`. The `Bitmap` class provides the `GetPixel()` and `SetPixelI()` methods to control each pixel in the bitmap. `GetPixel(int x, int y)` returns a `Color` object at `Pixel (x, y)` on the bitmap, whereas the `SetPixel(int x, int y, Color c)` method sets the color of `Pixel (x, y)` to `Color c`.

The following sample program demonstrates using `GetPixel()`, `SetPixel()`, and a timer to create some animation on an image file. The idea is to create a bitmap with a background color (`Color.SeaGreen`), and then randomly draw some pixels onto the bitmap. It the pixel selected has a color of the background, draw a different color for it; otherwise, draw the pixel with the background color. We use a timer that fires every 500 milliseconds; each time, a `DrawPixels()` method will be called and 100 randomly selected pixels on the bitmap will be drawn. The Enter button on the navigation pad can enable and disable the timer, thus pausing and restarting the animation. When the program runs, at first you will see more and more red dots show up on the green background. While the number of red dots continues to increase, the color of the display will gradually stabilize and there will be roughly the same number of dots of each color. Overall, the bitmap looks like many blinking dots on a green background.

The following is the complete `form` class:

```
using System;
using System.Collections.Generic;
```

```
using System.ComponentModel;
using System.Data;
using System.Drawing;
using System.Text;
using System.Windows.Forms;

namespace Chap15Graphics
{
    public partial class Form4 : Form
    {
        // Random number generator
        private Random r = null;
        // Timer for drawing
        Timer timer = new Timer();
        bool firstDraw = true;

        // The bitmap object for drawing
        Bitmap bmp = null;

        public Form4()
        {
            InitializeComponent();
            r = new Random();
            // Timer fires every 500 milliseconds
            timer.Interval = 500;
            // Timer's event handler
            timer.Tick += new EventHandler(DrawPixels);

        }
        private void DrawPixels(object sender,  System.EventArgs e)
        {
            for (int i = 0; i < 100; i++)
            {
                // Randomly select a pixel
                // x and y are both with the range of [0,200)
                int x = r.Next(200);
                int y = r.Next(200);
                Color c1 = bmp.GetPixel(x, y);

                // Toggle the color of the pixel
                if (c1 == Color.SeaGreen)
                    bmp.SetPixel(x, y, Color.Red);
                else
                    bmp.SetPixel(x, y, Color.SeaGreen);
            }
            // Draw the image. Is there a light-weight way to do this?
            Graphics g = CreateGraphics();
            g.DrawImage(bmp, 0, 0);
            g.Dispose();

        }
        private void Form4_Paint(object sender, PaintEventArgs e)
        {
            if (firstDraw)
```

```
        {
            // Create a Bitmap
            bmp = new Bitmap(200, 200,
System.Drawing.Imaging.PixelFormat.Format24bppRgb);

            // Draw a background for this Bitmap
            Graphics g = Graphics.FromImage(bmp);
            g.Clear(Color.SeaGreen);

            Graphics gScreen = e.Graphics;
            gScreen.DrawImage(bmp, 0, 0);
            gScreen.Dispose();
            g.Dispose();

            firstDraw = false;
            // Enable the timer
            timer.Enabled = true;

        }

    }

    private void Form4_KeyDown(object sender, KeyEventArgs e)
    {
        if ((e.KeyCode == System.Windows.Forms.Keys.Left))
        {
            // Left
            Application.Exit();
        }
        if ((e.KeyCode == System.Windows.Forms.Keys.Enter))
        {
            // Enter
            if (timer.Enabled)
                timer.Enabled = false;
            else
                timer.Enabled = true;
        }
    }
  }
}
```

The class has a private member of the `Random` object, which is used to obtain random coordinates within the range of the bitmap size. The `Random` object is initialized in the constructor of the class. Together with the `Random` object, a `System.Windows.Forms.Timer` object (not to be confused with `System.Timers.Timer` or `System.Threading.Timer`) is initialized in the class constructor. The timer's interval (`timer.Interval` property) is set to `500` milliseconds. In fact, due to the runtime overhead, a timer will need to take many milliseconds to fire, even if you can set this `Interval` property to 1 millisecond. An `EventHandler` delegate is created to pass the timer procedure, `DrawPixels()`, to the `Timer` object via the `Timer::Tick` event. The form's `OnPaint` event handler draws the bitmap with a background color and starts the timer. When the timer expires, the `DrawPixels()` method is called by the main UI thread and 100 randomly pixels are drawn. Note that we don't need to reset the timer's interval; the timer will keep going until it is disabled.

Be cautious when you use a timer in graphics applications. If the timer fires too frequently and the timer procedure takes quite long to complete (for example, the timer interval in the preceding example is changed to 50 milliseconds) , the system's performance may be significantly degraded, as the time procedure may take many CPU cycles, resulting in higher CPU usage. Therefore, a short thread procedure is always preferred. In addition, if possible, use asynchronous methods in the timer procedure so that some tasks are being done in another thread and the main UI thread is unblocked.

Saving a Bitmap to a File

You can save an in-memory `Bitmap` object into a file or a stream using the `Bitmap::Save()` method. You need to specify a filename or a `Stream` object, and a `System.Drawing.Imaging.ImageFormat` instance. The `ImageFormat` enumeration consists of four values — `Bmp`, `Jpeg`, `Gif`, and `Png` — each corresponding to an image format. Here is an example of using `Bitmap::Save()`:

```
            bmp.Save(@"Storage Card\MyImage.png",
   System.Drawing.Imaging.ImageFormat.Png);
```

Embedded Resources

You can make an image file available to your application by including it in your Visual Studio 2005 project. A general guideline for this practice is to make such an image an "embedded resource" that is built into the assembly as a binary stream. Doing so will eliminate the risk of users accidentally deleting the separated image files that come with the assembly.

To add an image file to the project, select Project⇨Add Existing Items, and then select the file you want to add. Once the file appears in the Solution Explorer, open its properties page and change the Build Action option from Content to Embedded Resource, as shown in Figure 15-8. The Content setting means that the image file, like HTML files in a web application project, will be part of the output group downloaded to the client during deployment.

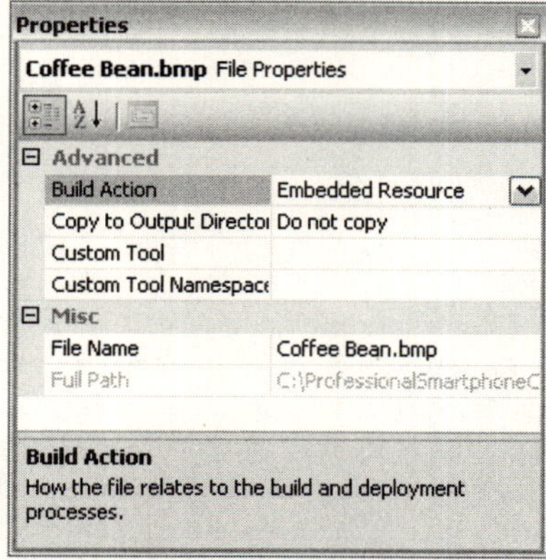

Figure 15-8

If the image file is named `a.bmp` and the project's namespace is `MyApp`, the embedded resource will be named `MyApp.a.bmp` (case-sensitive). This information is stored in the assembly's manifest. You can use the `GetManifestResourceStream()` method in the `System.Reflection.Assembly` class to retrieve the embedded resource as a binary stream object, as follows:

```
            // Display a bitmap resource
            System.Reflection.Assembly asse =
System.Reflection.Assembly.GetExecutingAssembly();
            Stream stream = null;
            try
            {
                stream = asse.GetManifestResourceStream("Chap15Graphics.Coffee
Bean.bmp");
                Bitmap resBmp = new Bitmap(stream);
                gScreen.DrawImage(resBmp, 10, 10);
            }
            catch (System.IO.FileNotFoundException e)
            {
                MessageBox.Show(e.Message);
            }
```

In this example, the application's namespace is `Chap15Graphics` and the image file is named `Coffee Bean.bmp`. Thus, the embedded stream is being retrieved as `Chap15Graphics.Coffee Bean.bmp` by the `GetManifestResourceStream()`. A `Bitmap` object is created with the returned `Stream` object.

Summary

If your Smartphone application needs to have direct control over the display surface for drawing, you need to use graphics in the .NET Compact Framework. You can obtain a `Graphics` object from the current form or the control, or from an `Image` object. With this `Graphics` object, you can start to draw graphical entities using specified pens and brushes, or draw formatted text. Usually, you can choose one of the multiple overloaded methods for one task, as demonstrated by the examples in this chapter. For example, to display an image, you can specify the source image, the target location, and optionally the bounded rectangle of the destination image, a portion of the source image, and the color key setting. This enables you to select the most appropriate method to use, rather than use a big, complex method that does everything.

We want to emphasize that while developing graphics applications for Smartphone, you must keep one thing in mind: *performance*. This is extremely critical for managed applications, which tend to be slower than native code. A rule of thumb is to make use of off-screen buffers to minimize direct screen drawing and reduce the number of `Graphics` objects in your application.

16
Performance

A user's experience with a Smartphone application is determined largely by the application's performance. Consider an on-the-go salesperson launching a Smartphone application to access an enterprise database via a wireless data network. If the application runs very slowly or the UI is virtually unresponsive during a transaction, the salesperson would look for a new application. A Smartphone application must be able to run fast enough to be adopted by users.

Improving the performance of a Smartphone application is sometimes quite challenging. You need to have a basic understanding of the .NET Compact Framework, the CLR, garbage collection, and so on. Choosing which class or method to use for a specific topic is also vital in many cases. The GUI's performance is probably the most important aspect of application performance. As a developer, you must ask yourself: What are the general guidelines to improving application performance? What are the best practices for specific tasks? (Note that you have already read about some performance-related issues in previous chapters.) This chapter answers those questions. It presents a collective overview of various performance topics, including the following:

- General performance principles
- .NET Compact Framework performance counters
- CLR factors
- Class library guidelines
- XML data
- Windows forms

General Principles

As discussed in previous chapters, programming for Smartphone means you have to work with resource-constrained devices: memory, processor, form factor, power, network bandwidth, and so on. Smartphone application performance tuning is somewhat more challenging than on desktop computers, as developers have to explore quite a few factors. For managed applications, the highly optimized CLR relieves application code from directly interacting with the operating system. Still, developers must carefully select appropriate .NET Compact Framework classes and apply correct coding techniques.

The following are some general guidelines for Smartphone application developers:

- **Design with performance in mind at the very beginning of the development process.** Don't follow the pattern of "work first, performance second," because for Smartphone application development, as well as all device application development, performance is often a sophisticated issue with many factors (device capability, power, form factor, operating system, runtime, the application, etc.). Usually, there is no handy tool for performance investigation as there is on desktop platforms. Be aware of the overhead of method calls and object memory consumption when you write code, and make design trade-offs for better performance.
- **Use fewer managed objects, GUI objects, less code, and short code path.** Objects take time to create and initiate, and consume memory, which, in turn, leads to more frequent garbage collection. On mobile devices, a small code footprint is already preferred, so try to reduce your code size as much as possible.
- **Multithread your code to achieve better GUI performance.** Users perceive performance in the GUI's responsiveness. Arrange your code with multiple threads to unblock the UI update so that the UI reacts to user input in a timely manner. Use threads that work in the background to perform data access and computation, and use asynchronous I/O whenever possible. Finally, use the double-buffering technique (explained in Chapter 15) to manipulate and display bitmaps.

Using .NET Compact Framework Performance Counters

To enable performance counters on a Smartphone device, use the Windows CE remote registry editor (a Visual Studio 2005 tool) to add the following key:

```
HKEY_LOCAL_MACHINE\SOFTWARE\Microsoft\.NETCompactFramework\PerfMonitor
```

Under this key, add a DWORD entry named `Counters` with a value of `1`.

After adding the `Counters` key, you can run your program on the device or the emulator. A performance statistics file will be created under the root directory of the device. The file's name is the program you deployed onto the device, with a `.stat` extension. It is a plaintext file in a simple table format, each row describing a specific performance counter. Table 16-1 shows an example of this `.stat` file.

Table 16-1 .NET Compact Framework Performance Counters

Counter	Total	Last Datum	N	Mean	Min	Max
Total Program Run Time (ms)	385252	–	–	–	–	–
App Domains Created	1	–	–	–	–	–
App Domains Unloaded	1	–	–	–	–	–
Assemblies Loaded	5	–	–	–	–	–
Classes Loaded	321	–	–	–	–	–
Methods Loaded	716	–	–	–	–	–
Closed Types Loaded	0	–	–	–	–	–
Closed Types Loaded per Definition	0	0	0	0	0	0
Open Types Loaded	1	–	–	–	–	–
Closed Methods Loaded	0	–	–	–	–	–
Closed Methods Loaded per Definition	0	0	0	0	0	0
Open Methods Loaded	0	–	–	–	–	–
Threads in Thread Pool	–	0	2	0	0	1
Pending Timers	–	0	2	0	0	1
Scheduled Timers	1	–	–	–	–	–
Timers Delayed by Thread Pool Limit	0	–	–	–	–	–
Work Items Queued	1	–	–	–	–	–
Uncontested Monitor.Enter Calls	9	–	–	–	–	–
Contested Monitor.Enter Calls	0	–	–	–	–	–
Peak Bytes Allocated (native + managed)	2862248	–	–	–	–	–
Managed Objects Allocated	150733	–	–	–	–	–
Managed Bytes Allocated	5005609844	28	150733	33208	8	200064
Managed String Objects Allocated	50158	–	–	–	–	–
Bytes of String Objects Allocated	5002566140	–	–	–	–	–

Table continued on following page

Counter	Total	Last Datum	N	Mean	Min	Max
Garbage Collections (GC)	4636	–	–	–	–	–
Bytes Collected by GC	5019204248	486468	4636	1082658	486468	1222824
Managed Bytes in Use After GC	–	6208	4636	135325	6208	214084
Total Bytes in Use After GC	–	296056	4636	1577820	296056	1656704
GC Compactions	0	–	–	–	–	–
Code Pitchings	1	–	–	–	–	–
Calls to GC.Collect	0	–	–	–	–	–
GC Latency Time (ms)	142496	24	4636	30	0	66
Pinned Objects	0	–	–	–	–	–
Objects Moved by Compactor	81	–	–	–	–	–
Objects Not Moved by Compactor	55	–	–	–	–	–
Objects Finalized	8	–	–	–	–	–
Boxed Value Types	14	–	–	–	–	–
Process Heap	–	360	617	7559	72	16240
Short Term Heap	–	0	1286	965	0	24944
JIT Heap	–	0	1572	50813	0	119912
App Domain Heap	–	0	4122	102829	0	156080
GC Heap	–	0	67326	1880365	0	2588672
Native Bytes JITed	134220	168	440	305	80	5540
Methods JITed	440	–	–	–	–	–
Bytes Pitched	110104	112	409	269	0	5540
Methods Pitched	409	–	–	–	–	–
Method Pitch Latency Time (ms)	11	11	1	11	11	11
Exceptions Thrown	0	–	–	–	–	–
Platform Invoke Calls	0	–	–	–	–	–
COM Calls Using a vtable	0	–	–	–	–	–
COM Calls Using IDispatch	0	–	–	–	–	–
Complex Marshaling	0	–	–	–	–	–

Counter	Total	Last Datum	N	Mean	Min	Max
Runtime Callable Wrappers	0	–	–	–	–	–
Socket Bytes Sent	0	–	–	–	–	–
Socket Bytes Received	0	–	–	–	–	–
Controls Created	3	–	–	–	–	–
Brushes Created	2	–	–	–	–	–
Pens Created	0	–	–	–	–	–
Bitmaps Created	0	–	–	–	–	–
Regions Created	0	–	–	–	–	–
Fonts Created	1	–	–	–	–	–
Graphics Created (FromImage)	0	–	–	–	–	–
Graphics Created (CreateGraphics)	0	–	–			

Each performance counter has several fields, as follows:

Counter Field	Description
Total	A summation of all data items of this counter
Last datum	The last data item
N	The total number of data items
Mean	The mean of the data items
Min	The minimum of data items
Max	The maximum of data items

Not all of the performance counters have all the six field values; some fields may not be meaningful for them. For example, the Total Bytes in Use counter does not have the Total value:

Counter	Total	Last Datum	N	Mean	Min	Max
Total Bytes In Use After GC	–	296056	4636	1577820	296056	1656704

You can see that for this performance counter, the last piece of data is 296056 bytes, which is the amount of bytes after last garbage collection. There are 4636 data items available for this counter. The mean value of this counter is 1577820, and the minimum and maximum are 296056 and 1656704, respectively. The total value of this performance is not listed because it is apparently not useful for performance diagnostics.

The performance counters can be divided into the following four categories, according to the order in which they appear in the stat file:

- **Loader counters** — A set of counters collected in the CLR's assembly loader, including the total program runtime in milliseconds, the number of application domain created, the number of applications unloaded, the number of assemblies loaded, the number of classes loaded, and so on. These counters give a summary of the program.
- **Generics counters** — A set of counters for generic types, including the number of closed types and the number of open types loaded across all application domains, the maximum number of unique generic types created for a given definition across all application domains, the number of closed methods loaded and maximum number of unique generic methods for a given definition, and so on. If your program uses generics, these counters can tell you the summation of each type of generic type.
- **Threads counters** — A set of counters for threads and timers, including the number of threads in the thread pool, the number of running timers, the number of scheduled timers, the number of work item queued, and so on. These counters are mainly useful for programs using multiple threads and the thread pool.
- **Garbage collection** — A set of counters for garbage collection operation, including peak bytes allocated, managed objects allocated, unused managed objects allocated, managed bytes and unused managed bytes allocated, the number of managed string objects allocated, the number of bytes of string objects allocated, the number of times the garbage collector has run, the number of bytes the garbage collector collected, the number of times the garbage collector compacted the heap, the number of boxed valued types, and so on. To make use of these counters, you need to understand how garbage collection works in the .NET Compact Framework.
- **Memory usage** — A set of counters for memory usage, including heap size for the CLR, the JIT compiler, the CLR's application domain, and the garbage collector. Smartphones generally are memory-constrained mobile devices. These counters, along with the garbage collection counters, are major performance counters you should look into to improve program performance. We discuss garbage collection and memory management in more detail in the next section.
- **JIT compilation** — A set of counters for JIT operations, including the number of native bytes and native methods generated by the compiler and the number of bytes and methods generated by the JIT compiler that have been discarded (also as known as *pitched*).
- **Exceptions** — The number of managed exceptions thrown while the program is running. This counter can be used to verify that managed exceptions have been thrown and handled.
- **Interop** — A set of counters for P/Invoke and COM Interop operations, including the number of P/Invoke classes, the number of COM calls, and so on. These counters are mainly for debugging performance issues of native code interoperability from within a managed Smartphone application.
- **Networking** — A set of counters for networking performance: the number of bytes sent via sockets and the number of bytes received via sockets.

- **GUI-related** — A set of counters for Windows Forms controls and various graphics objects, including the number of controls created and the number of graphics objects created (graphics handles, pens, brushed, bitmaps, fonts, etc.). For GUI-rich applications, these counters can give you some insight into the performance of graphics and Windows Forms.

CLR Performance

Understanding how the CLR operates with respect to system and application performance is vital to developing fast Smartphone applications. Although the CLR itself can be very complex, here we focus on those issues pertaining to a developer's concerns regarding application performance. This section introduces some best practices corresponding to the CLR:

- Garbage collection
- Call overhead
- Math
- Reflection
- Generics

Garbage Collection

A managed application in the CLR has a per-process 32MB virtual address space. Each application has a garbage collection (GC) heap located in the virtual address space for managed objects. The GC heap consists of a number of 64KB segments that are allocated from system memory. As more managed objects are allocated, the GC heap will grow on a segment basis but will not go beyond 1MB. According to the .NET Compact Framework team of Microsoft, garbage collection is triggered when one of the following conditions is met:

- 1MB of managed objects have been allocated in the per-process GC heap
- An application is moved to the background
- An application receives the WM_HIBERNATE message from the operating system
- A failure of memory allocation occurs
- An application explicitly calls `GC.Collect()`

Whenever one of these conditions is satisfied, the running thread at that exact point in time will run the garbage collection code.

The first thing it will do is bring all threads in the process to a state in which they are not running any managed code or unmanaged portion of the CLR. All managed objects allocated in the GC heap are examined: Those still being referenced (they are descended from what are called *live roots*, meaning that they are still in use in current scope) are marked.

Next, unmarked objects are freed (but that segment of the GC heap may not be returned to the operating system because the GC heap itself may cache those segments as long as the 1MB limit is not reached) except those that have a finalizer defined, which will be put into a finalization queue. Two optional steps may happen: If the GC heap is quite fragmented, it may be compacted at this point. The CLR caches some JITed code in memory such that it does not need to JIT compile it every time. When memory consumption is in pressure and some allocation has failed, the JITed code cache will be freed, or *pitched*.

Then, all threads in the process resume while a dedicated thread is running in the background to drain the finalization queue.

In addition to the managed object the application created explicitly using the `new` keyword, the CLR may implicitly create many managed objects for the application. Remember that boxing of value types (encapsulating value types within a object) will create a managed object. Various `String` object operations may end up creating a new `String` object.

It is generally not a good idea to force garbage collection by calling `GC.Collect()`, as the CLR will trigger that automatically at the appropriate time. Depending on the number of objects the application and the CLR allocated, garbage collection may slow down the performance of the application and the system. Technically, because garbage collection is a dynamic procedure closely related to memory usage over time, there is very little chance that calling `GC.Collect()` will absolutely improve performance by freeing up memory.

Call Overhead

Method calls are very common in any application. Considering the fact that Smartphones are resource-constrained, you need to make smart choices when it comes to choosing what types of methods to implement for a specific task. Table 16-2 lists the five types of function calls in managed application development on Smartphone.

Table 16-2 Managed Application Function Calls

Call Type	Description	Performance (Ratio to Native Calls)
Native calls	Windows API calls	1
Instance method calls and static calls	Managed code that calls an object's instance method or static method, which are both bound at compilation time	2–3
Virtual calls	Managed code that calls a virtual method that requires dynamic binding	3–5
P/Invoke calls	Managed code that calls native functions exported in a windows DLL	10–15
COM calls	Managed code that calls an interface method of a COM object	10–15

P/Invoke Calls and COM Calls

As you can see from Table 16-2, calling from managed code into native code is quite expensive due to the overhead of marshaling. If possible, avoid using Platform Invoke calls and COM calls (collectively known as *interop calls*), as they are dramatically slower than managed instance calls. If you do need to use them, you may want to maximize the amount of work in each such call. In addition, use those types that can be easily marshaled as parameters. For example, data types such as integer, byte, and string do not require any transformation across the managed and unmanaged boundary, as their representation is the same in both domains. Arrays, structs that contain only these "blittable" data types (except strings), and pointers or references (ref) to them, are also marshaled quickly. Table 16-3 lists the blittable data types. Note that even if `System.String` is a blittable type, it is not blittable within a struct or a class object.

Table 16-3 Blittable Data Types

.NET Compact Framework Type	C# Type
`System.Byte`	`sbyte`
`System.Int16`	`short`
`System.Uint16`	`ushort`
`System.Int32`	`int`
`System.Int64`	`long` (only for "ref")
`System.Uint64`	`ulong`
`System.IntPtr`	* (unsafe)
`System.Char`	`char`
`System.String`	`string`
`System.Boolean`	`bool`

You may also consider using `Marshall.Prelink()` and `Marshall.PrelinkAll()` to force JIT compilation of the interop stubs in your managed code so that when they are called the application is directly executing native code. The following is an example of using `Marshall.Prelink()` with the P/Invoke call `GetDiskFreeSpaceEx()`:

```
public static void TestPInvoke()
    {
        string dir = @"\Windows";
        UInt64 freeByteAvail, totalBytes, totalFreeBytes;
        int retVal;

        int start = Environment.TickCount;
        int end;
        //
        // Call the Win32 API with specified directory name
        //
        retVal = GetDiskFreeSpaceEx(
            dir,
            out freeByteAvail,
            out totalBytes,
            out totalFreeBytes
        );
        end = Environment.TickCount;
        Debug.WriteLine("P/Invoke time (ms): " + (end - start));
```

```
            if (retVal != 0)
            {
                //
                // Print P/Invoke call result
                //
                Debug.WriteLine("Free Bytes Available: " + freeByteAvail
                                + " Total Bytes: " + totalBytes
                                + " Total FreeBytes: " + totalFreeBytes
                );
            }
        }
```

In the calling method, we pre-link the `TestPInvoke()`. You need to add `System.Runtime.InteropServices` and `System.Reflection` to the imported namespace list. The `Type.GetType()` call specifies the fully qualified name `BCLTest.Program`, where `BCLTest` is the namespace of this assembly and `Program` is the class name:

```
            Type t = Type.GetType("BCLTest.Program");
            MethodInfo m = t.GetMethod("TestPInvoke");
            Marshal.Prelink(m);
            TestPInvoke();
```

A test run shows that with `Marshall.Prelink()`, the P/Invoke call can be as much as two times faster than the same call without `Prelink`. On the other hand, `Prelink` introduces some overhead when the program is loaded — this is just another example of balancing design preferences and end users' perceived performance.

Virtual Calls

Virtual calls are comparatively slower than other managed calls. Virtual calls are interpreted by the .NET Compact Framework; no *vtable* (a table of method pointers in the class) is built during JIT compilation for virtual calls (in an effort to reduce the working set). This means when calling a virtual call, the compact CLR walks the class hierarchy to locate the requested method. Conversely, for instance method calls, the compact CLR builds a fixed vtable for all instance method calls in the class. In addition, whereas the CLR may optimize instance method calls by inlining some of them, virtual calls are never inlined.

Consider the following classes:

```
    class Fruit
    {
        internal virtual void Peel()
        {
            Type t = this.GetType();
            Debug.WriteLine(t.Name + " peels ...");
        }
    }
    class Apple : Fruit
    {
        internal override void Peel()
        {
```

```
            Type t = this.GetType();
            Debug.WriteLine(t.Name + " peels ...");
        }
    }
    class BigApple : Apple
    {
        internal override void Peel()
        {
            Type t = this.GetType();
            Debug.WriteLine(t.Name + " peels ...");
        }
        internal void Peel2()
        {
            Type t = this.GetType();
            Debug.WriteLine(t.Name + " peels (instance method)...");
        }
    }
```

This is a simple class hierarchy: Fruit⇨Apple⇨BigApple. Both the `Apple` and `BigApple` classes override the virtual method `Peel()` derived from its superclass. The `BigApple` class also has an instance method `Peel()`. We use these classes to explore the cost of virtual calls. Here is the test code:

```
public static void PeelAFruit(Fruit f)
        {
            f.Peel();
        }
public static void TestVirtualCall()
        {
            BigApple bigApple = new BigApple();
            int start = Environment.TickCount;
            int end = 0;
            //
            // Virtual call from a BigApple object
            //
            PeelAFruit(bigApple);
            end = Environment.TickCount;
            Debug.WriteLine("Virtual call 1 (ms): " + (end - start));

            //
            // Instance method call from a BigApple object
            //
            start = Environment.TickCount;
            bigApple.Peel2();
            end = Environment.TickCount;
            Debug.WriteLine("Instance call (ms): " + (end - start));

            //
            // Virtual call from an Apple object
            //
            Apple apple = new Apple();
            start = Environment.TickCount;
            PeelAFruit(apple);
            end = Environment.TickCount;
```

```
            Debug.WriteLine("Virtual call 2 (ms): " + (end - start));

            //
            // Virtual call from a Fruit object
            //
            Fruit fruit = new Fruit();
            start = Environment.TickCount;
            PeelAFruit(fruit);
            end = Environment.TickCount;
            Debug.WriteLine("Virtual call 3 (ms): " + (end - start));
        }
```

The test code creates three objects: `BigApple`, `Apple`, and `Fruit`. For each of them, the static method `PeelAFruit()` is called, which essentially makes a virtual call to the virtual `Peel()` method of each individual class. The following shows the IL (Intermediate Language) code of the `PeelAFruit()` method:

```
.method public hidebysig static void  PeelAFruit(class BCLTest.Fruit f) cil managed
{
  // Code size       9 (0x9)
  .maxstack  8
  IL_0000:  nop
  IL_0001:  ldarg.0
  IL_0002:  callvirt   instance void BCLTest.Fruit::Peel()
  IL_0007:  nop
  IL_0008:  ret
} // End of method Program::PeelAFruit
```

Notice the `callvirt` on line `IL_0002`. The first time the code is JIT compiled, this instruction tells the CLR to work through the class hierarchy to find the proper method depending on runtime information of the object passed to the `PeelAFruit` method. Without applying any optimization to the code, we can run this simple test to get a general idea of virtual performance. Here is a sample output of the test code:

```
BigApple peels ...
Virtual call 1 (ms): 543
BigApple peels (instance method)...
Instance call (ms): 94
Apple peels ...
Virtual call 2 (ms): 137
Fruit peels ...
Virtual call 3 (ms): 98
```

As shown in the result, `BigApple`'s virtual call is the slowest (543 milliseconds), whereas `Apple`'s is the second slowest (137 milliseconds). You can make two interesting observations from this result. First, the more steps the `callvirt` takes through the class hierarchy from the base class, the longer time it takes to find and run that corresponding virtual method. Second, the instance method call is notably faster than the virtual call (refer to the `BigApple` calls). In addition, the virtual call to the base class virtual method performs as fast as an instance method call.

The "callvirt" instruction does not necessarily mean that it is a virtual call during runtime. In some cases, the JIT compiler can determine the proper method during compilation. For example, when a virtual method is sealed, it will not be bound during runtime anymore.

In short, virtual calls should generally be avoided whenever possible. If you do need to use virtual calls in a class design, try to reduce the number of virtual calls and perform more tasks in a single virtual call to amortize the overhead.

Math

Math operations in the .NET Compact Framework perform well with 32-bit integers and floating-point values — almost the same as native math operations. But 64-bit integer operations are 5 to 10 times slower than native 64-bit integer operations, mainly due to the code importing and morphing during JIT compilation.

Reflection

Reflection operations can be very expensive. You want to use reflection only when you really need to. The cost of reflection comes from type information retrieval from manifest, type comparison, and type member access. Because reflection also enables you to instantiate types and load assemblies at runtime, those operations can be significantly slower (10 to 100 times) than in JIT-compiled regular assembly. This is clearly a design trade-off you must consider. For performance-sensitive applications or time-critical code paths, reflection should not be used.

Generics

Generics are the support for the runtime parameter polymorphism supported in the .NET Compact Framework 2.0 and later. Using generic collections, you can avoid the overhead of boxing/unboxing value types in a collection object. Thus, using generic collections turns out to be faster than using non-generic collections. The following code snippet shows a quick benchmarking of running `sort()` with a generic collection object. For comparison, the code uses an `ArrayList` to store the same number of integers and does a `sort()` with the `ArrayList` object:

```
static void TestGenerics()
{
    int start, end;
    List<int> lStr = new List<int>();
    for (int i = 0; i < 100000; i++)
    {
        lStr.Add(i);
    }
    start = Environment.TickCount;
    lStr.Sort();
    end = Environment.TickCount;
    Debug.WriteLine("Generic List sort (ms): " + (end - start));
}
static void TestArrayList()
{
    int start, end;
    ArrayList aList = new ArrayList();
    for (int i = 0; i < 100000; i++)
    {
        aList.Add(i);
    }
    start = Environment.TickCount;
```

```
            aList.Sort();
            end = Environment.TickCount;
            Debug.WriteLine("ArrayList sort (ms): " + (end - start));
        }
```

The results show that on average, using a generic list is about six to seven times faster than using an `ArrayList` object. The disadvantage of using generics is code size, as the .NET Compact Framework implements generics in a way (known as *JIT code specialization*) that may result in a large JIT code size.

Class Library Guidelines

We have already covered some of these guidelines in previous chapters. Here we want to discuss some general rules and guidelines using the BCL (Base Class Library).

BCL Collections

`System.Collections` classes — such as `ArrayList`, `Hashtable`, `Queue`, and `Stack` — always require value types to be boxed before they can be added to the collection. Conversely, unboxing will be performed while retrieving those objects from the collection. If you have a large number of value types to be managed in a collection object, performance of this procedure may suffer due to the boxing/unboxing overhead. To see how this overhead looks, here's a simple test program. It compares the time it takes for an `ArrayList` to box and store 10000 `int` types with the time it takes a raw array:

```
    public static void TestCollection()
        {
            int item_cnt = 10000;
            int start = Environment.TickCount;
            int end = 0;

            ArrayList aList = new ArrayList(item_cnt);
            for (int i = 0; i < item_cnt; i++)
            {
                // Boxing i into an object
                aList.Add(i);
            }
            end = Environment.TickCount;
            Debug.WriteLine("ArrayList delay (ms): " + (end - start));

            //
            // For comparison, we use a raw int array
            //
            start = Environment.TickCount;
            int[] aRawArray = new int[item_cnt];
            for (int i = 0; i < item_cnt; i++)
            {
                aRawArray[i] = i;
            }
            end = Environment.TickCount;
            Debug.WriteLine("Raw array delay (ms): " + (end - start));

        }
```

For this simple test, using a raw int array takes only less than half of the time of using ArrayList — every integer to be added to the ArrayList will be boxed, which means a managed object has to be created. Therefore, if you want to store a fixed number of value types and do not intend to take advantage of the functionality provided by a System.Collections class, such as searching, sorting, and so on, you do not use to have a Collection object.

Pre-sizing a Collection object is another good practice when you know the number of items that will be placed into the Collection object. If you don't specify an initial capacity, the default capacity will be used. While you are adding items into the collection, once the Collection object's capacity is full, it will be resized (usually doubled), and mostly likely a new Collection object will be created, with every existing item copied to the new Collection object. This is a significant overhead that should be avoided if possible.

foreach()

foreach() offers an easy way to iterate through a Collection object, but keep in mind it is compiled into a number of virtual calls (callvirt in IL language) to System.Collections.IEnumerator::get_Enumerator, get_Current() and move_Next(). Because virtual calls are heavier than common instance method calls, you may not want to use them when the collection is huge. The following code snippet demonstrates foreach() overhead:

```
public static void TestForeach()
{
    ArrayList aList = new ArrayList();
    for (int i = 0; i < 1000; i++)
    {
        aList.Add("String " + i);
    }
    int start = Environment.TickCount;
    int end = 0;
    foreach (String s in aList)
    {
        s.ToLower();
    }
    end = Environment.TickCount;
    Debug.WriteLine("foreach delay (ms): " + (end - start));

    //
    // For comparison, we use indexer to access each item
    //
    start = Environment.TickCount;
    for (int i = 0; i < 1000; i++)
    {
        String s = (String)aList[i];
        s.ToLower();
    }
    end = Environment.TickCount;
    Debug.WriteLine("indexer delay (ms): " + (end - start));
}
```

The code uses both foreach() and an indexer to access each of the 1000 String objects in the ArrayList. Test runs shows that using an indexer takes only about half the time it takes to use foreach() — roughly 210 milliseconds for foreach() and 100 milliseconds for the indexer.

StringBuilder versus String

Both String and StringBuilder can be used to handle strings. If the program uses only a few string objects, it probably doesn't matter which class you use. (Actually, using String would be better because the String object initializes faster than StringBuilder). However, if you are creating many String objects, you need to think about it.

Use the String class for strings that do not change; otherwise, use StringBuilder.

Specifically, if you are performing multiple string concatenation operations, definitely use StringBuilder rather than String. If you use String, a new object will be created each time string addition (+) is performed. On the other hand, StringBulider will use a single object for concatenation. The following code snippet illustrates the difference. Two label objects, label1 and label2, are used to display the elapse times of using String and StringBuilder.

```
static void TestString()
        {
            DateTime start, end;
            //
            // Using String class
            //
            start = DateTime.Now;
            String strTest = "This is a test string";
            Debug.WriteLine("Start time : " + start);
            for (int i = 0; i < 5000; i++)
            {
                strTest += " i";
                // Another way to concat strings
                //strTest = String.Concat(strTest, i);
            }
            end = DateTime.Now;
            Debug.WriteLine("End time : " + end);
            Debug.WriteLine("String elapse time (ms): " +
                (end - start).TotalMilliseconds);
            Debug.WriteLine("elapse time (ms)" + (end - start).TotalMilliseconds);

            //
            // Using StringBuilder class
            //
            start = DateTime.Now;
            Debug.WriteLine("Start time : " + start);
            StringBuilder strbTest = new StringBuilder("This is a test string");
            for (int j = 0; j < 5000; j++)
            {
                strbTest.Append("j");
            }
            end = DateTime.Now;
            Debug.WriteLine("End time : " + end);
            Debug.WriteLine("StringBuilder elapse time (ms): " +
                (end - start).TotalMilliseconds);
    }
```

This code uses `String` and `StringBuilder` for the same task: concatenating strings 5000 times. Both strings are initialized with the same string literal: "This is a test string". We measure the elapsed time in both cases. Here is the result of one run on a Smartphone emulator:

- `String` elapse time: 4 milliseconds
- `StringBuilder` elapse time: 2 milliseconds

As you can see, using `StringBuilder` for a large number of string concatenations is much faster than using `String`. Using `String.Concat()` instead of the addition operator of `String` yields similar results. By turning on performance counters (see below), you can obtain more detailed information about managed string objects. The following example shows the result of using the selected performance counters introduced early in this chapter to compare `String` and `StringBuilder` performance.

Using only `String` for the concatenation:

Managed String Objects Allocated	5125
Bytes of String Objects Allocated	50232832
Garbage Collections (GC)	53

Using only `StringBuilder` for the concatenation:

Managed String Objects Allocated	134
Bytes of String Objects Allocated	35580
Garbage Collections (GC)	0

In the first case, there are more than 5000 `String` objects allocated, and the garbage collector has run 53 times. The overhead of creating and garbage-collecting those `String` objects is probably the leading cause of poor performance. When using `StringBuilder`, only 134 `String` objects are created, and because not too much heap space has been taken, the garbage collection does not even run.

> *Note that the preceding sample code uses* `Debug.Writeln()` *to display text strings in the console. The* `Debug` *class is in the* `System.Diagnostics` *namespace. In fact, you can build a console project for Windows Mobile Smartphone devices and use* `Debug.Writeln()` *to output debug information to standard output. (If you debug the program in Visual Studio, standard output can be viewed in the output window.)*

Regular Expression

By default, regular expressions in the .NET Compact Framework are interpreted at runtime. However, the regular expression class `RegEx` provides the option to compile the pattern defined in the `RegEx` object on-the-fly. Internally, during runtime, the regular expression engine first parses the regular expression into some code, which is then transformed into IL code using reflection. This is actually a technique to trade runtime speed for startup speed. Because the regular expression is first compiled into IL code when the program starts, it runs faster than it does when being interpreted. You can specify the compilation mode of `RegEx` by using `RegexOptions.Compiled`, as shown in the following code:

```
Regex r = new Regex("re*",RegexOptions.Compiled);
```

or

```
Regex.Match("test string", @"(d*)abc", RegexOptions.Compiled);
```

Use the compilation mode if you have a small number of RegEx objects that will be used repeatedly in the program. This way you can take advantage of the runtime speed improvement with moderate startup overhead. On the other hand, if you are more concerned with the startup speed of your application or you have many RegEx objects, you should not use the compilation mode.

The .NET Framework features a third operation mode — precompilation — in which regular expression parsing is performed when you compile the source code into IL code. The precompiled RegEx is put into a separate assembly. Other than the initial load time of this assembly, there is no overhead of RegEx interpretation and runtime compilation. However, this feature is not supported in the .NET Compact Framework.

XML and Data Access

As you'll recall from the brief discussion about XML parsing performance in Chapter 9, XMLDocument is a heavyweight object — that is, it builds a tree of untyped objects. Type information is stored as a string object, which is then used for conversion into appropriate managed types. Not surprisingly, in the .NET Compact Framework 1.0, the general guideline is to use XmlTextReader and XmlTextWriter, both of which take less memory than XmlDocument. XmlTextReader does not hold all XML data; rather, it employs a pull model of forward-only data retrieval, meaning data is read only portion by portion, like a sliding window.

XMLReader and XMLWriter

In .NET Compact Framework 2.0 and later, combining the XmlReader class and the XmlReaderSettings class provides a better way of reading XML data. Similarly, the XmlWriter class and the XmlWriterSettings class will do the same thing for writing XML data. You don't need to create XmlTexterReader or XmlTextWriter yourself; as long as you set the desired settings in an XmlReaderSetting or XmlWriterSetting object, and associate them with the XmlReader or XmlWriter object, an optimized XmlReader or XmlWriter object will be generated and ready for use.

XML Schema

A frequently asked question is whether to use XML schema. The answer (as always) is: it depends. Be aware of the validation overhead incurred when you use a schema against an XML document. For example, using XmlReader and XmlReaderSettings with schema validation is two to three times slower than using XmlTextReader. Therefore, in general, whenever possible do not use XML schema for XML parsing. You may consider using a schema document when you want to load an XML document instance into a DataSet or there is no other way except a schema document to validate data in an XML document.

XML Serialization

You can use the `XMLSerializer` class to serialize and deserialize objects to and from an XML document. `XMLSerializer` will serialize the metadata of your object — an expensive operation. As a result, you should create only one instance for a single type.

Choosing XML as the vehicle for data serialization is not suggested for Smartphone application development with respect to performance. Although XML has been widely used as a standard data structuring format on desktop applications, the generation, transferring, parsing, and validation of XML data is sometimes too heavy for mobile devices. Unless you need to provide a standard data structure of your application to other programs (for example, you build a web service that exports data), it would be wise to serialize data into a custom binary stream, and compress it if needed.

Data Access

Depending on the amount of data to be stored locally, you can access local data using either XML or a SQL Server Mobile in-proc database. If the amount of data is small — for example, less than 200–300KB — using an XML file to store it is sufficient; otherwise, use a SQL Server Mobile database. By using a SQL Server Mobile database, you can achieve high performance with a very small footprint. Note that you should query data with `DataReader`. When writing a SQL statement, consider completely bypassing the query processor and using the `TableDirect` command type, which provides a faster index search. In both cases, try to avoid using `DataSet`; it adds too much memory overhead.

In the following example, the SQL comment has been set to `TableDirect` mode, and the `IndexName` property is set to the index name `idxId` on the table. No query process is involved in this procedure. A `SqlDataReader` object is used for the SQL command. You can, of course, use other data reader objects.

```
SqlCeConnection conn = new SqlCeConnection(@"Data Source=\EmployeeInfoDB.sdf");
Conn.Open();
SqlCeCommand cmd = new SqlCeCommand("EmployeeInfo", conn);
cmd.CommandType = CommandType.TableDirect;
cmd.IndexName = "idxId";
SqlCeDataReader sr = cmd.ExecuteReader();
While (sr.Read())
{
   ...
}
```

For remote data access, you generally want to cache the data locally using SQL Server Mobile replication. Frequent network-based data access is still considered quite expensive and should be avoided as much as possible. Whenever possible, reduce the amount of data transferred over the air. For example, always use DiffGrams (see Chapter 9 for details about this setting) to perform incremental data updates to `DataSets`, or apply some compression to the data being transferred. As mentioned earlier, using XML Web services to pass data is suggested only when the amount of transactional data is small or there is no other open interface.

Windows Forms

A basic goal of Windows forms in a Smartphone application is to reduce the loading time. In the form's initialization method (usually `InitializeComponent()` if the form is created using Form Designer in Visual Studio), reduce the number of method calls if possible. In addition, do not populate data within the form's `Load()` and `Show()` methods. Instead, pre-populate data or do it asynchronously.

Form Loading Performance

The first thing to do to improve form loading performance is to minimize the work you have to perform in the `Load` event. Put those heavy tasks into a background thread so that they don't block the loading process of the form. The .NET Framework has a `BackgroundWorker` class to help you achieve this goal. Simply place those heavy jobs into a `BackgroundWorker` component that will be performed in a worker thread, enabling the main thread that is loading the form to continue. In addition, when that worker thread is finished, the main thread can be notified. Unfortunately, the .NET Compact Framework does not support this class, but you can get around this by using a `Threadpool` thread.

In the following example, when the form is being loaded, frequent updating of the text box control on the form occurs. If we perform this update in the form's `Load` event, the form will not be loaded during that time. A test run using the following commented code indicates that it takes more than four seconds to finish the `Load` event. To quickly display the form and enable the user to see the update progress, we use a `Threadpool` thread by calling `ThreadPool.QueueUserWorkItem()`:

```
private void Form2_Load(object sender, EventArgs e)
    {
        //
        // Hide the ListView for now but other controls can be displayed
        //
        int start = Environment.TickCount;
        int end = 0;

        //
        // If we do the 1000-time update here, the form will not be loaded
        // while we are doing this
        //
        /*
        for (int num = 0; num < 1000; num++)
        {
            textBox1.Text = num.ToString();
        }
        */

        //
        // Instead, we can launch another thread to do this
        //
        ThreadPool.QueueUserWorkItem(new WaitCallback(SomeSlowWork));

        end = Environment.TickCount;
        Debug.WriteLine("List-populating time (ms) " + (end - start));

    }
```

The worker thread in the `Threadpool` will pick up a `WaitCallback` method, `SomeSlowWork`, and run it. It will update the `TextBox` control using `Control::BeginInvoke()`, which asynchronously updates the control's `Text` property:

```
private void SomeSlowWork(object o)
{
    int start = Environment.TickCount;
    int end = 0;
    int count = 1000;

    for (int i = 0; i < count; i ++)
    {
        if (textBox1.InvokeRequired)
        {
            object[] args = new object[1];
            args[0] = i;
            textBox1.BeginInvoke(new AddNumHandler(AddNum),args);
        }
        else
        {
            textBox1.Text = i.ToString();
        }
    }
    end = Environment.TickCount;
    Debug.WriteLine("Work-thread list-populating time (ms) "
        + (end - start));
}
private void AddNum(int num)
{
    textBox1.Text = num.ToString();
}
```

Now, with this change, the form's `Load` event is handled much faster. A test runs shows only 138 milliseconds, and the worker thread's update takes about 1500 milliseconds, which is much shorter than four seconds if we do the update in the `Load` event in the same thread.

The example uses a worker thread to do some work and update a control on the form. If some of the data processing has nothing to do with controls on the form, you may also launch a separate thread to do that task, while quickly displaying the form to the user. Although this example uses a `Threadpool` thread, other options include creating your own thread and managing it or using `AsyncCallBack delegate`.

Form Layout

If you have ever taken a look at the following code generated by the Form Designer of Visual Studio, you may have noticed two `System.Windows.Forms.Form` method calls, `SuspendLayout()` and `ResumeLayout()`, in the `Form`'s `InitializeComponent()` method.

When the `Form` object is created, the constructor calls the `InitializeComponent()` method. In this method, right before the controls are set up and positioned on the form, `this.SuspendLayout()` is called to temporarily suspend layout logic. When the positioning is done, `this.ResumeLayout(false)` is called to resume layout logic. This way, the layout engine does not need to deal with the layout event of each control addition. The `false` parameter of the `this.ResumeLayout()` call in `InitializeComponent()` indicates there is no need to execute pending layout requests. This is expected because the controls are already positioned and it is unnecessary to adjust the layout of them at this point.

The following is an example:

```
#region Windows Form Designer generated code

        // <summary>
        // Required method for Designer support - do not modify
        // the contents of this method with the code editor.
        // </summary>
        private void InitializeComponent()
        {
            this.mainMenu1 = new System.Windows.Forms.MainMenu();
            this.menuItem1 = new System.Windows.Forms.MenuItem();
            this.treeView1 = new System.Windows.Forms.TreeView();
            this.SuspendLayout();
            //
            // mainMenu1
            //
            this.mainMenu1.MenuItems.Add(this.menuItem1);
            this.mainMenu1.MenuItems.Add(this.menuItem2);
            //
            // menuItem1
            //
            this.menuItem1.Text = "Exit";
            this.menuItem1.Click += new System.EventHandler(this.menuItem1_Click);
            //
            // treeView1
            //
            this.treeView1.Location = new System.Drawing.Point(0, 16);
            this.treeView1.Name = "treeView1";
            this.treeView1.Size = new System.Drawing.Size(176, 161);
            this.treeView1.TabIndex = 0;

            //
            // Form1
            //
            this.AutoScaleDimensions = new System.Drawing.SizeF(96F, 96F);
            this.AutoScaleMode = System.Windows.Forms.AutoScaleMode.Dpi;
            this.ClientSize = new System.Drawing.Size(176, 180);
            this.Controls.Add(this.treeView1);
            this.KeyPreview = true;
            this.Menu = this.mainMenu1;
            this.Name = "Form1";
            this.Text = "Form1";
            this.KeyDown += new System.Windows.Forms.KeyEventHandler(this
.Form1_KeyDown);
            this.Load += new System.EventHandler(this.Form1_Load);
            this.ResumeLayout(false);

        }

        #endregion
```

Following this auto-generated code, it is suggested that if a form has a number of controls, or the form will adjust the size or position of its controls at runtime, you may use these two methods to suppress those layout events. The more controls that need to be repositioned, the better performance you can get with this trick.

BeginUpdate and EndUpdate

You can utilize the `BeginUpdate()` and `EndUpdate()` methods of the `TreeView` and `ListView` control classes to achieve better performance. The `BeginUpdate()` method stops repainting the control, and the `EndUpdate()` method resumes repainting. You should use the methods around the node update task of the control — for example, if you already have a `TreeView` control with many nodes displayed on the screen and you want to update the `TreeNode` collection or part of it. You should call `TreeView::BeginUpdate()` first, and then do the update. Once the update is complete, call `TreeView::EndUpdate()` to refresh the control. This way the screen will not keep refreshing while you update `TreeNodes` in the `TreeView` control.

If the `TreeView` *control does not have many* `TreeNodes` *displayed, you will not benefit from using this technique because repainting the* `TreeView` *control does not take too much time anyway.*

The following methods demonstrate how to use this trick when reading a directory tree into a `TreeView` control named `treeView1`. A test run of this code (within a Windows form application on the emulator) shows that without using the trick, it takes 1622 milliseconds to clear and reload the entire collection, whereas with the trick it only takes 962 milliseconds to finish the update:

```
void TestTreeViewLoad(bool bTrick)
{
    int start = Environment.TickCount;
    int end = 0;

    //
    // Use the BeginUpdate/EndUpdate trick or not?
    //
    if(bTrick)
        treeView1.BeginUpdate();

    treeView1.Nodes.Clear();

    //
    // Root node
    //
    TreeNode tn = new TreeNode(@"\Windows");
    treeView1.Nodes.Add(tn);
    DirectoryInfo rootDir = new DirectoryInfo(@"\Windows");

    //
    // Update the TreeView control with the directory tree
    //
    UpdateTreeView(tn, rootDir);
    if(bTrick)
        treeView1.EndUpdate();

    end = Environment.TickCount;
```

```
            Debug.WriteLine("Treeview load time (ms): " + (end - start));
        }

        //
        // Recursively update tree view
        //
        private void UpdateTreeView(TreeNode tn, DirectoryInfo dir)
        {
            //
            // Display folder names; Recursively call the method
            //
            foreach (DirectoryInfo di in dir.GetDirectories())
            {
                TreeNode tn2 = new TreeNode(di.Name);
                tn.Nodes.Add(tn2);
                UpdateTreeView(tn2, di);
            }

            //
            // Display file names
            //
            foreach (FileInfo fi in dir.GetFiles())
            {
                TreeNode tn3 = new TreeNode(fi.Name);
                tn.Nodes.Add(tn3);
            }
        }
```

The `bTrick` parameter of the `TestTreeViewLoad()` method dictates whether `BeginUpdate()` and `EndUpdate()` will be used. Before this call, the `TreeView` control already has a long list of the folders and files displayed. Thus, using this trick will make a difference in the amount of time it takes for the `TreeView` control to update the entire node list. The `UpdateTreeView()` is recursively called to enumerate all files and folders under the specified root directory, which is `"\Windows"` in this case. In fact, the `TreeView` control class already uses this trick. When you call the `AddRange()` method with the `Nodes` collection of a `TreeView` control, it internally calls `BeginUpdate()` and `EndUpdate()` before and after an array of `TreeNodes` is added into the `Nodes` collection.

Summary

This chapter is dedicated to performance issues of managed applications on a Smartphone. Although we have talked about application performance along with selected topics in previous chapters, the general guidelines to performance-centric application design and programming can only be covered after individual topics are all covered.

Although this is the last chapter, developers should not consider performance as the last step in developing a Smartphone application. Instead, the word "performance" should be kept in mind throughout the entire application development process.

In the design stage, when choosing a technique or approach to a problem, among all of the options performance should be considered along with other metrics, such as portability, ease of maintenance, and development cost. Due to the intrinsic nature of mobile applications, performance is especially crucial to save power and time, and to improve the user's experience.

Developers should also keep performance in mind while implementing classes, the GUI, or data access. Sometimes, choosing a proper class or a more suitable method can make a big difference. To this end, the chapter has discussed how to choose XML classes, BCL guidelines, suggested data access approaches, and more. To take better advantage of the .NET Compact Framework, developers need to understand how it works internally and the various overhead associated with the CLR.

For developers who intend to dig further into performance issues, Microsoft provides .NET Compact Framework performance counters, along with tools to view these counters. The chapter outlines these counters and describes a sample performance counter file.

To summarize, application performance is indeed a sophisticated topic encompassing operating system issues, runtime issues, programming languages issues, and hardware issues. Understandably, one chapter can only touch on these topics from the standpoint of a Smartphone developer. As you delve into this broad area, you will surely learn more from your practice.

New Features in .NET Compact Framework 2.0

.NET Compact Framework 2.0 expands the support for classes in the full .NET Framework. It also introduces new features and improves a number of existing features. .NET Compact Framework 2.0 also provides significant performance improvements; many operations in the .NET Compact Framework 2.0 runtime are approximately twice as fast as the same operations executed in the .NET Compact Framework 1.0.

This appendix summarizes the key new features you will find in the .NET Compact Framework 2.0.

- **Application domains** — Application domains are supported to provide isolation between applications in runtime environments.
- **Assemblies** — A feature of C#, friend assemblies enable you to access another assembly's internal members.
- **CAB file installation** — CAB files can be installed not only to system ROM, but also to a storage card (from .NET CF 2.0 SP1).
- **Cryptographic support** — The following new cryptographic features are now supported:
    - X.509 certificates
    - MD5 and SHA1 hashing
    - RC2, RC4, DES, and 3DES symmetric key encryption
    - RSA and DSA asymmetric key encryption
- **DirectX and Direct3D support** — Managed DirectX and Direct3D classes are now supported (requires the Windows Mobile 5.0 SDK).
- **Generics** — The generics feature in C# and Visual Basic is supported.
- **Globalization** — Additional encodings are supported.

- **Graphics** — The following new features have been added:
    - Images can be saved using the `System.Drawing.Image.Save()` method.
    - The `LockBits()` method is added to access bitmap data.
    - Bitmap serialization (including JPG images) is supported.
    - ClearType fonts are supported.
    - The `Pen` class is supported to draw lines and curves with different width, color, and style.
    - The QVGA display is supported on Smartphone 2003 and later.
- **Interoperability** — The following features are supported to enhance interoperability:
    - *Native code interoperability* — Additional data types are supported, such as arrays, strings, and structures. In addition, data marshaling in platform invoke is enhanced.
    - *COM interoperability* — The runtime callable wrapper (RCW) enables calling from managed applications into COM objects, and the COM callable wrapper (CCW) supports callbacks from native code to managed code.
- **Keyboard events** — The controls can receive `KeyUp`, `KeyDown`, and `KeyPress` events. In addition, the new `KeyPreview` property can be set to send the keyboard events before it is sent to the focused control.
- **Layout management** — The following new features are now available to simplify the process of creating and managing user interfaces:
    - *Automatic scaling* — Enables a container control to be automatically adjusted to a different screen resolution by setting the `AutoScaleMode` property to `AutoScaleMode.Dpi`
    - *Control anchoring* — Defines the distance from one or more edges of a control to its parent window and automatically resizes the control when the parent window is resized
    - *Control docking* — Positions the control against the edge of the parent control and automatically resizes the control when the parent window is resized
    - *DipX and DipY properties* — Indicates the number of horizontal and vertical dots per inch
    - *Logical layout control* — Enables you to temporarily suppress multiple layout change events while adjusting multiple control attributes by calling the new `SuspendLayout()` and `ResumeLayout()` methods
- **Message queuing** — The core functionality of the Windows CE MSMQ component, which enables an application to communicate with other applications across a network when they are offline, is supported.
- **Networking** — The following new features have been added:
    - Support for the `CredentialCache` class, which stores multiple credentials for multiple Internet resources
    - Support for IPV6, the next-generation IP protocol
    - Support for Kerberos, Negotiate, NTLM, and SOAP 1.2
- **Partial classes** — Partial classes are supported (requires Visual Studio 2005).
- **Registry keys** — The `RegistryKey` class is added to set the registry keys.

- **Remote Performance Monitor** — Remote Performance Monitor enables you to debug and tune the performance of Windows Mobile applications (available from .NET Compact Framework 2.0 SP1).
- **Serial ports** — Up to four serials ports are supported in both Windows Mobile devices and Visual Studio 2005 emulators.
- **SQL Server Mobile** — SQL Server 2005 Mobile Edition replaces SQL Server CE and substantially improves performance and enhances reliability. The enhanced features are as follows:
    - Supports multiple subscriptions and multi-user synchronization
    - Supports column-level tracking
    - Optimizes the storage engine for a mobile architecture with shared memory pool
    - Reuses the empty pages to save storage space
    - Provides an updatable, scrollable cursor to access a SQL Server Mobile database with the new `SqlCeResultSet` class
    - Supports data binding to the `DataGrid` class (new in Smartphone)
- **Threads support** — Threads support is enhanced and is more in line with the `Threads` class in the full .NET Framework. The asynchronous execution of a delegate of a control thread is supported with the new `Control.BeginInvoke()` and `Control.EndInvoke()` methods.
- **WindowsCE Forms controls** — The following controls in the `Mirosoft.WindowsCE.Forms` namespace are supported from .NET Compact Framework 2.0:
    - *DocumentList* — Displays and manages documents in a consistent manner (Pocket PC only)
    - *HardwareButton* — Enables users to override the functionality of hardware buttons (Pocket PC only)
    - *InputModeEditor* — Enables users to specify input methods, such as text mode or numerical mode (requires Smartphone 2003 or later)
    - *LogFont* — Defines the logical characteristics of a font for creating special text effects, such as rotated text
    - *MessageWindow* — Enhances the capability to send and receive windows-based messages with the newly introduced `Text` property
    - *MobileDevice* — Provides access to the `Hibernate` event
    - *Notification* — Displays and responds to user notifications
    - *SystemSettings* — Adds a new `ScreenOrientation` property to support different screen orientations (Pocket PC only)
- **Windows Forms controls** — The following controls in the `System.Windows.Forms` namespace are supported from .NET Compact Framework 2.0:
    - *DataTimePicker* — Enables users to select a date and a time with a specified format
    - *LinkLabel* — Supports and displays hypertext links in a Label control
    - *MonthCalendar* — Enables users to select a date using the graphical monthly calendar

  - *PictureBox* — Displays an image
  - *ProgressBar* — Displays a progress bar
  - *ScrollableControl* — Provides a base class for controls that support the auto-scrolling feature
  - *TabControl* — Manages a set of tabbed pages
  - *UserControl* — Defines an empty control that can be used to create other controls
  - *WebBrowser* — Enables users to navigate web pages inside a form
- **XML support** — XML support is enhanced with the following new classes:
  - The `XmlSerializer` class serializes and de-serializes objects into and from XML documents, which controls how objects are encoded in XML.
  - The classes in the `System.Xml.XPath` namespace support navigating and editing XML information items using the XQuery 1.0 and XPath 2.0 data model.
  - The classes in the `System.Xml.XSchema` namespace support XML schema definition language (XSD) schemas.

B

A Glance at the .NET Compact Framework 2.0 Class Library

The following is a complete list of the namespaces in the .NET Compact Framework 2.0. Key classes discussed in the book are outlined in each namespace. In addition, a number of classes not mentioned in the book but important to Smartphone development are described. The module (DLL file) of each namespace is listed as well.

Microsoft.VisualBasic

- **Module:** microsoft.visualBasic.dll
- **Description:** Contains classes and enums that support the Visual Basic runtime
- **Key types:**
  - `Collection` — Provides an ordered set of items
  - `Conversion` — Provides number and format conversion procedures

Microsoft.VisualBasic.CompilerServices

- **Module:** microsoft.visualbasic.dll
- **Description:** Contains internal use types that support the Visual Basic compiler

- **Key types:**
  - `ProjectData` — Provides helpers for the Visual Basic `Err` object
  - `Operators` — Provides late-bound math operators, such as `AddObject` and `CompareObject`, which the Visual Basic compiler uses internally

Microsoft.WindowsCE.Forms

- **Module:** microsoft.windowsce.forms.dll
- **Description:** Contains classes that are available only for programming applications with the .NET Compact Framework
- **Key types:**
  - `InputPanel` — Provides the soft input panel (SIP) for every Windows Form on Windows CE devices
  - `Message` — Provides a wrapper structure for Windows-based messages
  - `MessageWindow` — Provides the functionality to send and receive Windows-based messages

System

- **Module:** mscorlib.dll
- **Description:** Contains fundamental classes and base classes that define commonly used value and reference data types, events and event handlers, interfaces, attributes, and processing exceptions
- **Key types:**
  - `AppDomain` — Represents an application domain, which is an isolated environment within a process that an application can execute
  - `ApplicationException` — Represents a non-fatal application exception
  - `Array` — Provides general array functionality, including creation, manipulation, and sorting. Limited capability compared with the .NET Framework.
  - `AsyncCallback` — Provides a delegate that represents a callback method to be called in a separate thread when an asynchronous operations is completed
  - `Boolean` — Contains a structure that represents a Boolean value
  - `Byte` — Contains a structure that represents a byte value
  - `Char` — Contains a structure that represents a char value
  - `Console` — Represents the standard input, output, and error streams for console applications. This class can also be used in Smartphone applications for debugging purposes, as the standard output and error streams will be directed to the output window in Visual Studio.

- `Convert`—Contains procedures for data type conversions
- `DateTime`—Contains a structure for a time instance
- `DayOfWeek`—Specifies the day of the week
- `Decimal`—Contains a structure that represents a decimal number
- `Delegate`—Contains the base class for delegate types
- `Double`—Contains a structure that represents a double number
- `Environment`—Provides basic information about the platform
- `EventHandler`—Provides a delegate that handles an event that has no event data. If your event does generate data, you must 1) supply your own custom event data type and 2) either create a delegate whereby the type of the second parameter is your custom type, or use the generic `EventHandler` delegate class and substitute your custom type for the generic type parameter.
- `Exception`—Provides general exceptions
- `GC`—Controls the garbage collector
- `Guid`—Contains a structure that represents a GUID
- `IAsyncResult`—Contains an interface that represents the status of an asynchronous operation
- `Int16`—Contains a 16-bit signed integer
- `Int32`—Contains a 32-bit signed integer
- `Int64`—Contains a 64-bit signed integer
- `IntPtr`—Contains a pointer used to perform P/Invoke or interop
- `Math`—Provides common math functionality
- `Object`—Contains the generic object
- `Random`—Provides a pseudo random number generator
- `Single`—Represents a single-precision floating-point number
- `StackOverflowException`—Provides an exception that is thrown when the execution stack overflows because it contains too many nested method calls, such as in a deep recursion
- `String`—Provides string creation and manipulation
- `Type`—Provides basic support for `System.Reflection` functionality and is the primary way to access metadata
- `UInt16`—Contains an unsigned 16-bit number
- `UInt32`—Contains an unsigned 32-bit number
- `UInt64`—Contains an unsigned 64-bit number

System.Collections

- **Module:** mscorlib.dll
- **Description:** Contains classes and interfaces that define collection objects such as `ArrayList`, `BitArray`, `Queue`, `HashTable`, and so on
- **Key types:**
    - `ArrayList` — Provides an array implementation. The array size is managed automatically.
    - `BitArray` — Manages a compact array of bit values
    - `Hashtable` — Implements a hash table
    - `Queue` — Implements a queue
    - `Stack` — Implements a stack

System.Collections.Generic

- **Module:** mscorlib.dll
- **Description:** Contains interfaces and classes that define generic collections
- **Key types:**
    - `List` — Provides a generic equivalent of the `System.Collection.ArrayList` class
    - `Queue` — Provides a generic equivalent of the `System.Collections.Queue` class
    - `Stack` — Provides a generic equivalent of the `System.Collections.Stack` class

System.Collections.ObjectModel

- **Module:** mscorlib.dll
- **Description:** Contains classes that can be used as collections in the object model of a reusable library
- **Key types:**
    - `Collection` — Provides a base class for the generic collection
    - `KeyedCollection` — Provides the abstract base class for a collection whose keys are embedded in the values

System.Collections.Specialized

- **Module:** system.dll
- **Description:** Contains specialized and strongly typed collection objects

- **Key types:**
    - `BitVector32` — Contains a vector of Boolean values and 32-bit integers
    - `ListDictionary` — Implements an `IDictionary` interface using a singly linked list
    - `NameValueCollection` — Represents a collection of associated string keys and string values that can be accessed with either the key or the index
    - `StringCollection` — Represents a collection of strings
    - `StringDictionary` — Implements a hash table with strongly typed strings as keys and values

System.ComponentModel

- **Module:** system.dll
- **Description:** Contains classes that are used to implement the runtime and design-time behavior of components and controls
- **Key types:**
    - Component — Provides the base implementation for the `IComponent` interface
    - IContainer — Provides functionality for containers

System.Configuration.Assemblies

- **Module:** mscorlib.dll
- **Description:** Contains two classes that are used to configure an assembly — `AssemblyHashAlgorithm` and `AssemblyVersionCompatibility`.
- **Key types:**
    - `AssemblyHashAlgorithm` — Specifies all the hash algorithms used for hashing files and generating strong names
    - `AssemblyVersionCompatibility` — Defines the different types of assembly version compatibility

System.Data

- **Module:** system.data.dll
- **Description:** Contains classes for common typed data access
- **Key types:**
    - `DataColumn` — Represents a column in a `DataTable`
    - `DataColumnChangeEventHandler` — Provides an event handler when a `DataColumn` is changed

- DataColumnCollection — Represents a collection of DataColumn objects in a DataTable
- DataRelation — Represents a parent-child relation between two DataTable objects
- DataRow — Represents a row in a DataTable
- DataRowChangeEventHandler — Provides an event handler when a DataRow is changed
- DataRowCollection — Provides a collection of DataRow objects in a DataTable
- DataRowView — Represents a customized view of a data row
- DataSet — Represents an in-memory cache of data sets that may contain multiple DataTable objects.
- DataTable — Represents a data table
- DataTableCollection — Represents a collection of DataTable objects
- DataView — Represents a customized view of a DataTable for sorting, filtering, searching, editing, and navigation
- SqlDbType — Specifies SQL Server–specific data types of a field or property, for use in a SqlParameter
- StatementType — Defines types of SQL statements such as Delete, Insert, Select, Update, or Batch
- XmlReadMode — Defines enums of various XML read modes
- XmlWriteMode — Defines enums of various XML write modes

System.Data.Common

- **Module:** system.data.common.dll
- **Description:** Contains types that are shared by .NET data providers
- **Key types:**
  - DataAdapter — Represents a set of SQL commands and a databases connection for a DataSet
  - DbCommand — Represents a SQL statement or stored procedure
  - DbConnection — Represents a connection to a database
  - DbDataReader — Provides read-only, forward-only access to a data source

System.Data.SqlClient

- **Module:** system.data.sqlclient.dll
- **Description:** Contains classes for accessing SQL Server

- **Key types:**
    - `SqlCommand` — Represents a Transact-SQL statement or stored procedure to execute against a SQL Server database
    - `SqlCommandBuilder` — Used to automatically generate Transact-SQL statements for single-table updates if you set the `SelectCommand` property of the `SqlDataAdapter`
    - `SqlConnection` — Represents an open connection to a SQL Server database
    - `SqlDataAdapter` — Represents a set of data commands and a database connection, which are used to fill the `DataSet` and update a SQL Server database
    - `SqlDataReader` — Provides a way of reading a forward-only stream of rows from a SQL Server database
    - `SqlTransaction` — Represents a Transact-SQL transaction to be made in a SQL Server database

System.Data.SqlServerCe

- **Module:** system.data.sqlserverce.dll
- **Description:** Contains the .NET Data provider for SQL Server Mobile
- **Key types:**
    - `SqlCeCommand` — Represents a SQL command issued to the SQL Server Mobile database. No batch command support.
    - `SqlCeCommandBuilder` — Provides a means of automatically generating single-table commands used to reconcile changes made to a `DataSet`
    - `SqlCeConnection` — Represents an open connection to the database
    - `SqlCeDataAdapter` — Represents a set of data commands and a database connection that are used to fill the `DataSet` and update the data source
    - `SqlCeRemoteDataAccess` — Provides remote data access
    - `SqlCeTransaction` — Represents a SQL transaction

System.Data.SqlTypes

- **Module:** system.data.common.dll
- **Description:** Contains structures of native SQL Server data types
- **Key types:**
    - `SqlBytes` — Represents a reference type of a buffer or a stream.
    - `SqlInt32` — Represents a 32-bit numeric value
    - `SqlInt64` — Represents a 64-bit numeric value

- `SqlMoney`—Represents a concurrency value
- `SqlSingle`—Represents a floating-point number
- `SqlString`—Represents a variable-length stream of characters

System.Diagnostics

- **Module:** mscorlib.dll
- **Description:** Contains classes for accessing traces and debuggers
- **Key types:**
    - `Debug`—Provides a set of methods and properties that help debug your code
    - `Debugger`—Provides interaction with a debugger
    - `Trace`—Provides a set of methods and properties that help you trace the execution of your code

System.Drawing

- **Module:** system.drawing.dll
- **Description:** Contains types for basic GDI+ functionality
- **Key types:**
    - `Bitmap`—Encapsulates a GDI+ bitmap
    - `Brush`—Encapsulates a GDI+ brush
    - `Color`—Represents an ARGB color
    - `Font`—Defines a font
    - `FontFamily`—Defines a group of typefaces sharing a similar basic design and certain variations in styles
    - `Graphics`—Represents a drawing surface
    - `Pen`—Represents a GUI+ pen
    - `Point`—Defines a point on a two-dimensional plane
    - `Rectangle`—Stores a set of four integers that represent the location and size of a rectangle
    - `RectangleF`—Defines a rectangle of four floating-point numbers
    - `Region`—Describes the interior of a graphics shape composed of rectangles and paths

System.Drawing.Drawing2D

- **Module:** system.drawing.dll
- **Description:** Contains the `CombineMode` enumerations.
- **Key types:**
    - `CombineMode` — Specifies how to combine various clipping regions

System.Drawing.Imaging

- **Module:** system.drawing.dll
- **Description:** Contains classes that provide advanced GDI+ imaging functionality
- **Key types:**
    - `ImageAttributes` — Controls how image colors are rendered

System.Drawing.Text

- **Module:** system.drawing.dll
- **Description:** Provides advanced GDI+ typography functionality
- **Key types:**
    - `FontCollection` — Provides a base class for font collections
    - `InstalledFontCollection` — Represents the fonts installed on the system

System.Globalization

- **Module:** mscorlib.dll
- **Description:** Contains classes that define culture-related information, including language, country/region, calendars in use, format patterns for dates, currency, and numbers, and the sort order for strings
- **Key types:**
    - `CultureInfo` — Provides information about a specific locale
    - `DateTimeStypes` — Defines various date and time styles
    - `NumberFormatInfo` — Defines how numeric values are formatted and displayed

System.IO

- **Module:** mscorlib.dll
- **Description:** Contains types that enable file and stream I/O
- **Key types:**
  - `Directory`—Exposes static methods for creating, moving, and enumerating through directories and subdirectories
  - `DirectoryInfo`—Exposes instance methods for creating, moving, and enumerating through directories and subdirectories
  - `File`—Provides methods for creating, copying, deleting, moving, and opening files, and aids in the creation of `FileStream` objects
  - `FileStream`—Represents a stream for a file
  - `IOException`—Represents an exception that is thrown when an I/O error occurs.
  - `StreamReader`—Implements a `TextReader` that reads characters from a byte stream in a particular encoding
  - `StreamWriter`—Implements a `TextWriter` that writes characters from a byte stream in a particular encoding
  - `TextReader`—Represents a reader that can read a sequential series of characters
  - `TextWriter`—Represents a writer that can write a sequential series of characters

System.IO.Ports

- **Module:** system.dll
- **Description:** Contains classes for controlling serial ports
- **Key types:**
  - `SerialPort`—Represents a serial port resource
  - `SerialDataReceivedEventHandler`—Represents the method that handles events of data reception of a serial port

System.Messaging

- **Module:** system.messaging.dll
- **Description:** Provides classes that enable you to connect to, monitor, and administer message queues on the network, and to send, receive, or peek messages (reads a message but does not remove it from the queue)

- **Key types:**
  - `Message` — Provides access to the properties needed to define a message queuing message
  - `MessageQueue` — Provides access to a message queue
  - `ReceiveCompletedEventHandler` — Represents a method to handle events of a message queue

System.Net

- **Module:** system.dll
- **Description:** Contains a set of classes that implement generic network protocol programming interfaces
- **Key types:**
  - `Dns` — Provides simple DNS functionality
  - `HttpWebRequest` — Provides HTTP-specific implementation of the `WebRequest` class
  - `HttpWebResponse` — Provides HTTP-specific implementation of the `WebResponse` class
  - `IPAddress` — Encapsulates an IP address (both IPv4 and IPv6)
  - `IPEndPoint` — Contains a combination of an IP address and a port number
  - `SocketAddress` — Stores serialized information of an endpoint
  - `WebException` — Provides an exception that is thrown when a web access error occurs
  - `WebProxy` — Encapsulates a web proxy setting for the `WebRequest` class
  - `WebRequest` — Provides basic functionality of a generic web request
  - `WebResponse` — Provides basic functionality of a generic web response

System.Net.Sockets

- **Module:** system.dll
- **Description:** Contains a managed implementation of the Windows Sockets (Winsock) interface
- **Key types:**
  - `AddressFamily` — Specifies a standard addressing scheme
  - `NetworkStream` — Provides a stream object for network data access
  - `SelectMode` — Defines the polling modes for the `Socket.Poll` method
  - `Socket` — Implements a Berkeley socket interface
  - `SocketException` — Provides an exception that is thrown when a socket error occurs
  - `TcpClient` — Implements a TCP client

- `TcpListener`—Implements a TCP server
- `UdpClient`—Implements a UDP client

System.Reflection

- **Module:** mscorlib.dll
- **Description:** Contains types that retrieve information about assemblies, modules, members, parameters, and other entities in managed code
- **Key types:**
    - `Assembly`—Provides the functionality of an assembly
    - `Module`—Performs reflection of a module

System.Resources

- **Module:** mscorlib.dll
- **Description:** Contains types that enable developers to create, store, and manage various culture-specific resources used in an application
- **Key types:**
    - `ResourceManager`—Provides convenient access to culture-specific resources at runtime
    - `ResourceReader`—Enumerates `.resources` files and streams, reading sequential resource name and value pairs

System.Runtime.CompilerServices

- **Module:** mscorlib.dll
- **Description:** Contains types for compiler writers using managed code to control the runtime behavior of the CLR
- **Key types:**
    - `DateTimeConstantAttribute`—Persists an 8-byte `DataTime` constant
    - `DecimalConstantAttribute`—Stores a decimal constant in metadata

System.Runtime.InteropServices

- **Module:** mscorlib.dll
- **Description:** Contains types that implement support for COM interop and PInvoke services

- **Key types:**
    - `GCHandle` — Provides a handle to the managed object pool for interop applications
    - `Marshal` — Provides a collection of methods for allocating unmanaged memory, copying unmanaged memory blocks, and converting managed types to unmanaged types, as well as other miscellaneous methods used when interacting with unmanaged code

System.Security

- **Module:** mscorlib.dll
- **Description:** Contains two exception classes related to system security
- **Key types:**
    - `SecurityException` — Provides an exception for security errors
    - `VerificationException` — Provides an exception for verification errors

System.Security.Cryptography .X509Certificates

- **Module:** mscorlib.dll
- **Description:** Contains the CLR implementation of Authenticode X.509 v.3 certificates
- **Key types:**
    - `X509Certificate` — Implements Authenticode X.509 v.3 certificates

System.Security.Policy

- **Module:** mscorlib.dll
- **Description:** Contains the `Evidence` class, which defines the set of information that constitutes input to security policy decisions

System.Text

- **Module:** mscorlib.dll
- **Description:** Contains classes representing ASCII, Unicode, UTF-7, and UTF-8 character encodings

- **Key types:**
    - `ASCIIEncoding` — Represents ASCII encoding
    - `Decoder` — Implements a decoder
    - `Encoder` — Implements an encoder
    - `Encoding` — Represents a specific encoding
    - `StringBuilder` — Represents a mutable string of characters
    - `UnicodeEncoding` — Represents Unicode encoding
    - `UTF8Encoding` — Represents a UTF-8 encoding of Unicode characters
    - `UTF7Encoding` — Represents a UTF-7 encoding of Unicode characters
    - `UTF32Encoding` — Represents a UTF-32 encoding of Unicode characters

System.Text.RegularExpressions

- **Module:** system.dll
- **Description:** Contains classes that provide access to the .NET Compact Framework regular expression engine
- **Key types:**
    - `Match` — Represents the result from a single regular expression match
    - `Regex` — Represents an immutable regular expression

System.Threading

- **Module:** mscorlib.dll
- **Description:** Contains types that provide threading support, including thread synchronization and access to the system thread pool
- **Key types:**
    - `AutoResetEvent` — Provides an event used to notify a waiting thread for synchronization. The event automatically returns to the non-signaled state when a waiting thread is released
    - `Interlocked` — Provides atomic operations for variables that are shared by multiple threads
    - `ManualResetEvent` — Provides an event used to notify one or more waiting threads. Once it has been signaled, `ManualResetEvent` remains signaled until it is manually reset
    - `Monitor` — Implements a monitor for thread synchronization. The `Monitor` class controls access to objects by granting a lock for an object to a single thread
    - `Mutex` — Implements a mutually exclusive access primitive

- `Thread` — Implements threading functionality such as creating and controlling a thread, setting its priority, etc.
- `ThreadPool` — Provides access to the system thread pool that can be used to perform work items
- `ThreadStart` — Provides a delegate for the thread procedure
- `WaitCallback` — Represents a callback method to be executed by a thread pool thread
- `WaitHandle` — Contains a basic class for synchronization types

System.Web.Services

- **Module:** system.web.services.dll
- **Description:** Contains supporting classes to access XML web services
- **Key types:**
    - `WebMethodAttribute` — Provides support for creating web services methods
    - `WebServiceBindingAttribute` — Provides support for specifying binding parameters

System.Web.Services.Description

- **Module:** system.web.services.dll
- **Description:** Contains the `SoapBindingUse` enumeration, which specifies whether the message parts are encoded as abstract type definitions or concrete schema definitions

System.Web.Services.Protocols

- **Module:** system.web.services.dll
- **Description:** Contains classes that define the protocols used to transmit data across the wire during the communication between XML Web services clients and XML Web services
- **Key types:**
    - `SoapClientMessage` — Represents the data in a SOAP request sent or a SOAP response received by an XML Web services client
    - `SoapMessage` — Represents the data in a SOAP request or response

System.Windows.Forms

- **Module:** system.windows.forms.dll
- **Description:** Contains a set of Windows Forms components for GUI applications

- **Key types:**
    - `Application` — Provides static methods and properties to manage an application
    - `Button` — Provides a Windows Button control
    - `CheckBox` — Provides a Windows CheckBox control
    - `ComboBox` — Provides a Windows ComboBox control
    - `Control` — Provides a base control class
    - `DataGrid` — Provides a control for displaying ADO.NET data
    - `Form` — Represents a window or dialog box
    - `HScrollBar` — Provides a Windows HScrollBar control
    - `ImageList` — Provides methods to manage a collection of images
    - `Label` — Provides a Windows Label control
    - `ListBox` — Provides a Windows ListBox control
    - `ListControl` — Provides a common control for implementing ListBox and ComboBox controls
    - `ListView` — Provides a Windows ListView control
    - `MainMenu` — Represents the menu structure of a form
    - `MenuItem` — Represents a single menu item
    - `MessageBox` — Provides a Windows MessageBox control
    - `Panel` — Windows panel used to group controls
    - `PictureBox` — Provides a Windows PictureBox control
    - `RadioButton` — Provides a Windows RadioButton control
    - `Screen` — Represents a display device
    - `ScrollBar` — Provides a Windows ScrollBar control
    - `StatusBar` — Provides a Windows StatusBar control
    - `TabControl` — Provides a control for displaying tab pages
    - `TextBox` — Provides a Windows TextBox control
    - `Timer` — Implements a timer that raises an event at user-defined intervals
    - `ToolBar` — Provides a Windows ToolBar control
    - `TreeView` — Provides a Windows TreeView control that displays a hierarchical collection of labeled items, each represented by a `TreeNode`
    - `VScrollBar` — Provides a Windows VScrollBar control

System.Xml

- **Module:** system.xml.dll
- **Description:** Contains classes and interfaces that provide support for standards-based XML processing
- **Key types:**
  - `XmlDocument` — Encapsulates an in-memory XML document or stream
  - `XmlNode` — Represents a single XML node in an XML document or stream
  - `XmlNodeReader` — Represents a stream reader that provides fast, non-cached forward-only access to XML data
  - `XmlReader` — Provides a base class for `XmlNodeReader` and `XmlTextReader`
  - `XmlTextReader` — Represents a stream reader that provides fast, non-cached, forward-only access to XML data
  - `XmlTextWriter` — Represents a stream writer that provides a fast, non-cached, forward-only way of generating XML streams or files
  - `XmlWriter` — Provides a base class for `XmlTextWriter`

System.Xml.Schema

- **Module:** system.xml.dll
- **Description:** Contains classes that provide standards-based support for XSD (XML Schema Definition) schemas
- **Key types:**
  - `XmlSchema` — Provides an in-memory representation of an XML Schema
  - `XmlSchemaAttributes` — Represents the attribute element from the XML Schema
  - `XmlSchemaXPath` — Represents the W3C schema selector element

System.Xml.Serialization

- **Module:** system.xml.dll
- **Description:** Contains classes that are used to serialize any objects into XML format documents or streams
- **Key types:**
  - `XmlAttributes` — Represents a collection of XML attribute objects that control serialization and deserialization
  - `XmlElementAttributes` — Represents a collection of XML element objects that control serialization and deserialization

- `XmlSerializer`—Provides the basic functionality to serialize and deserialize objects into and from XML documents, along with related attribute classes, such as `XmlElementAttributes` and `XmlAttributes`

System.Xml.XPath

- **Module:** system.xml.dll
- **Description:** Contains support for the XQuery 1.0 and XPath 2.0 data models
- **Key types:**
    - `XPathItem`—Represents an item in the XQuery 1.0 and XPath 2.0 data models
    - `XPathException`—Provides the exception thrown when an error occurs while processing an XPath expression

The Smartphone Bootstrapping Process

For a newly deployed Windows Mobile–based Smartphone, one of the first tasks to perform is the bootstrap procedure. Without the bootstrap process, an unconfigured Smartphone can make only voice calls but not data calls. The bootstrap process will do the following:

- Provision the device with data connectivity and enable direct data communication to the Internet or via proxies
- Configure Service Indication (SI) and Service Loading (SL)
- Enable the device to accept over-the-air (OTA) configurations from a list of trusted IP addresses
- Configure the device security model and policy settings

The bootstrap process can be initiated by one of the following methods:

- Controls on the Smartphone's UI
- Remote API and ActiveSync when the Smartphone is cradled
- An OTA Wireless Access Protocol (WAP) push mechanism
- The ROM configuration XML file

Bootstrapping from the User Interface

Bootstrapping from the UI is always available as the last resort to provision a Smartphone device. Using this method, however, the configurable settings are limited to the available controls on the user interface.

What Can Be Set

- WAP (Wireless Application Protocol) settings
- HTTP proxy settings
- PPP (Point-to-Point Protocol) and GPRS (General Packet Radio Services) settings
- Synchronization and e-mail settings

How to Bootstrap from the User Interface

- To configure data connections, such as WAP proxy, HTTP proxy, SOCK proxy, dial-up connections, GPRS settings, and VPN connections, select Settings⇨Data Connections.
- To configure the Accessibility, Profiles, Home Screen, Power, Telephony, Sounds, and Security controls, go to Settings, and then select the specific settings.
- To configure e-mail, select Inbox/SMS⇨Menu⇨Options⇨Email Setup.
- To configure synchronization settings, select ActiveSync⇨Menu⇨Options.
- To configure Internet Explorer Mobile, open the Internet Explorer Mobile application, and then select Menu⇨Options.

Bootstrapping Using Remote API and ActiveSync

Remote API (RAPI) enables a desktop-based application to communicate with a Smartphone device via ActiveSync, a software application running on both the PC and the Smartphone device for data exchange. If a Smartphone device is cradled to a desktop PC, it can be bootstrapped from the desktop configuration tools using RAPI.

What Can Be Set

- WAP settings
- HTTP proxy settings
- PPP and GPRS settings
- Synchronization and e-mail settings
- Corporative-specific settings
- Mobile operator–specific settings

How to Bootstrap with the RapiConfig Tool

You can use the RapiConfig tool to feed the provisioning XML to the Configuration Manager on a Windows Mobile–based Smartphone device. The Configuration Manager passes the configuration request to the Configuration Service Provider (CSP) to facilitate the bootstrap process.

To provision the device with RapiConfig, perform the following steps:

1. Cradle the device to the PC and establish an ActiveSync connection.
2. Launch the command window on the PC and change the working directory to the Tools folder of the Windows Mobile Smartphone SDK. The default location is `c:\Program Files\Windows CE Tools\wce500\Windows Mobile 5.0 Smartphone SDK\Tools`.
3. Run the RapiConfig command. For example, the following command will bootstrap the device with the provisioning file `myProv.xml`:

```
RapiConfig.exe <provisioning.xml>
```

Note that bootstrapping a Windows Mobile–based device using RAPI is disabled by default. To enable bootstrapping with RAPI, MANAGER privileges must be granted to the SECROLE_USER_AUTH role.

A Sample XML File

Assume a corporation needs to bootstrap Smartphone devices and set the e-mail connection to use a predefined network connection to enhance security. The following XML file can be used and signed to bootstrap Smartphone devices:

```
<wap-provisioningdoc>
   <characteristic type="SecurityPolicy">
   </characteristic>

   <characteristic type="EMAIL2">
      <parm name="CONNECTIONID"
      value="{A1182988-0D73-439e-87AD-2A5B369F808B}"/>
      <characteristic type="SMTP">
         <parm name="PXADDR" value="MyOrg"/>
         <parm name="NAME" value="MyOrg"/>
         <parm name="REPLYADDR" value="john@myorg.com"/>
      </characteristic>

      <characteristic type="POP3">
         <parm name="PXADDR" value="MyOrg2"/>
         <parm name="AUTHNAME" value="john"/>
         <parm name="AUTHSECRET" value="john123"/>
         <parm name="DOMAIN" value="MyOrg"/>
      </characteristic>

   </characteristic>
</wap-provisioningdoc>
```

In the preceding example, the outgoing e-mail account `SMTP` and incoming e-mail account `POP3` are grouped under the `EMAIL2` node, which encapsulates the characteristics of all e-mail accounts. The `CONNECTIONID` parameter specified right beneath the `EMAIL2` characteristic dictates the network connection ID that will be used for e-mail communication. The `SMTP` characteristic defines the outgoing e-mail server as `MyOrg`; the display name is `Myorg`, and the replay e-mail address is `john@myorg.com`. The `POP3` characteristic defines the name of the outgoing e-mail server as `MyOrg2`, the username as `john`, the password as `john123`, and the name of the domain as `MyOrg`.

BootStrapping Using the Over-the-Air (OTA) WAP Push Method

Windows Mobile–based Smartphone devices can be bootstrapped with the over-the-air (OTA) Wireless Application Protocol (WAP) push mechanism. OTA bootstrapping is particularly useful for network operators because a device does not need to be preconfigured when it is manufactured. Rather, a Smartphone device can be bootstrapped at the point of sale.

Note that by default, OTA bootstrapping is disabled. To enable OTA bootstrapping, the OPERATOR security role must be added to the Wireless Access Protocol (WAP) Signed Message Policy and Grant Manager Policy.

What Can Be Set

- WAP settings
- HTTP proxy settings
- PPP and GPRS settings
- Synchronization and e-mail settings
- Corporate-specific settings
- Mobile operator–specific settings

How to Bootstrap with the OTA WAP Push

To bootstrap a Smartphone device with the OTA WAP push mechanism, perform the following steps:

1. Define the configuration data that is required for the bootstrap process. This may include Trusted Provisioning Server, PPP settings, security policy, corporate certificates, and the WAP Gateway.
2. Generate an XML provisioning file with UTF-8 encoding and compress the message into WAP Binary XML (WBXML) format.
3. Sign the message with a network PIN and a user PIN.
4. Send the XML file in a WAP push message to the Push Proxy Gateway (PPG).

When the message arrives at the device, it is intercepted by the SMS router and passed to the WAP stack. It is then directed to the authentication UI, where the user and the message are authenticated with the user's PIN. The Configuration Manager passes the message to the bootstrap Configuration Service Provider (CSP), and the bootstrap message is decompressed and executed.

Bootstrapping Using a ROM Configuration XML File

The Windows Mobile ROM is divided into a number of discrete regions to isolate different types of applications and configuration settings. The OPERATOR region is owned by the mobile operator and contains all mobile operator customization files. The XML provisioning file can be pre-burned to the OPERATOR region to provision the device during the bootstrap process. Microsoft also provides a default provisioning XML file in the MICROSOFT ROM region to dictate default settings for Microsoft applications.

What Can Be Set

- WAP settings
- HTTP proxy settings
- PPP and GPRS settings
- Synchronization and e-mail settings
- Corporate-specific settings
- Mobile operator–specific settings
- Any settings that are supported

How to Bootstrap from a ROM Configuration File

When a Smartphone device is cold booted, it is provisioned with the XML provisioning files stored in the ROM. The name of the XML provisioning file in the OPERATOR region is `.provxml`, and the XML provisioning file in the MICROSOFT region is `mxip_SMARTFON_1.provxml`. The characteristics of bootstrapping from a ROM configuration file are as follows:

- Only the manufacturer can burn the UTF-8 encoded `.provxml` XML file to the OPERATOR ROM region.
- The Microsoft Provisioning file, `mxip_SMARTFON_1.provxml`, is also burned to the Microsoft ROM region during the manufacturing process. It defines the following:
    - Default network and mapping table entries
    - Default Microsoft Internet Explorer Mobile favorites
- The preconfigured settings burned in the ROM can be reconfigured by the network operator if ROM updating is supported.

The Default XML Provisioning File Format

The following code illustrates how the favorite settings of Internet Explorer Mobile are configured in the `mxip_SMARTFON_1.provxml` provisioning file, which is burned to the MICROSOFT ROM region:

```
<wap-provisioningdoc>
   <characteristic type="BrowserFavorite">
      <characteristic type="Smartphone">
         <parm name="URL"
         value="http://go.microsoft.com/fwlink/?LinkId=5980"/>
      </characteristic>
      <characteristic type="MSN Mobile">
         <parm name="URL" value="http://mobile.msn.com/pocketpc"/>
      </characteristic>
```

```
        <characteristic type="WindowsMedia.com">
           <parm name="URL"
           value="http://windowsmedia.com/redir/smartphone.asp"/>
        </characteristic>
        <characteristic type="Smartphone Web Guide">
           <parm name="URL"
           value="http://go.microsoft.com/fwlink/?LinkId=6956"/>
        </characteristic>
        <characteristic type="Smartphone How To">
           <parm name="URL"
           value="http://go.microsoft.com/fwlink/?LinkId=6946"/>
        </characteristic>
     </characteristic>
  ...

  </wap-provisioningdoc>
```

Bootstrap Security

The bootstrap process is vital to a Window Mobile–based Smartphone because it provides configuration data to the Smartphone device. The following are security features that can be leveraged to enhance security during the bootstrap progress:

- Security policy settings can be used to define levels of security and to determine whether Smartphone devices are configurable over-the-air (OTA).
- Smartphone devices rely on a PIN-based mechanism and/or a signed `.cab` file to secure provisioning.
- For an OTA WAP push bootstrap, the message is signed with a network PIN known only by the mobile operator and the device. For example, for Global System for Mobile Communications (GSM), this PIN is the International Mobile Subscriber Identity (IMSI) number from the device's Subscriber Identity Module (SIM) card.
- Using a `.cab` file for bootstrapping a corporate device over the air, the `.cab` file is signed with a private key from the corporate certificate.
- A Trusted Provisioning Server (TPS) can be defined so that a mobile device will only accept the provisioning message from the TPS. A trusted Push Proxy Gateway can also be defined to allow TPS to provide continuous provisioning. The following XML code describes how to define a TPS. In the example, TPS is set to the URL www.mytursted.com, and the proxy ID is set to myProxy. (The parameter CONTEXT-ALLOW currently accepts a value of 0 only.)

```
<wap-provisioningdoc>
    <characteristic type="BOOTSTRAP">
        <parm name="NAME" value="my Trusted TPS" />
        <parm name="PROVURL" value="http://www.mytursted.com" />
        <parm name="PROXY-ID" value="myProxy" />
        <parm name="CONTEXT-ALLOW" value="0" />
    </characteristic>
</wap-provisioningdoc>
```

Index

SYMBOLS

A

B

C

D

E

J

K

L

M

N

O

P

Q

R

S

T

U

V

W

X

Y